Model Development and Optimization

Applied Optimization

Volume 28

Series Editors:

Panos M. Pardalos
University of Florida, U.S.A.

Donald Hearn
University of Florida, U.S.A.

Model Development and Optimization

by

Viktor V. Ivanov

Glushkov Institute of Cybernetics,
Kiev, Ukraine
University of South Florida,
Tampa, U.S.A.

SPRINGER-SCIENCE+BUSINESS MEDIA, B.V.

A C.I.P. Catalogue record for this book is available from the Library of Congress.

ISBN 978-1-4613-6819-9 ISBN 978-1-4615-4062-5 (eBook)
DOI 10.1007/978-1-4615-4062-5

Printed on acid-free paper

Originally published by Kluwer Academic Publishers in 1999
Softcover reprint of the hardcover 1st edition 1999

Dedication

For my dear children: Natalya, Vladislav, and Yuriy
and all the other people
who were generous with their help

TABLE OF CONTENTS

PREFACE

At present, concerning intensive development of computer hardware and software, computer-based methods for modeling of difficult problems have become the main technique for theoretical and applied investigations. Many unsolved tasks for evolutionary systems (ES) are an important class of such problems. ES relate to economic systems on the whole and separate branches and businesses, scientific and art centers, ecological systems, populations, separate species of animals and plants, human organisms, different subsystems of organisms, cells of animals and plants, and so on.

Available methods for modeling of complex systems have received considerable attention and led to significant results. No large-scale programs are done without methods of modeling today. Power programs, health programs, cosmos investigations, economy designs, etc. are a few examples of such programs.

Nevertheless, in connection with the permanent complication of contemporary problems, existing means are in need of subsequent renovation and perfection.

In the monograph, along with analysis of contemporary means, new classes of mathematical models (MM) which can be used for modeling in the most difficult cases are proposed and justified. The main peculiarities of these MM offer possibilities for the description of ES; creation and restoration processes; dynamics of elimination or reservation of obsolete technology in ES; dynamics of resources distribution for fulfillment of internal and external functions of ES; and so on. The complexity of the problems allows us to refer to the theory and applications of these MM as the mathematical theory of development. For simplicity, the title "Model Development and Optimization" was adopted.

The monograph contains both the theoretical and applied aspects. General theory, PART I, mainly concerns the existence and uniqueness of solutions for the systems of equations of MM and for respective optimization problems. Considerable attention is also paid to asymptotic properties of the solutions and qualitative investigation of the structure of the solutions for optimization problems.

PART II of the book focuses mainly on optimal numerical methods. The solution of Volterra integral equations with pre-assigned accuracy by the optimal methods is stated and investigated. Reduction of many other problems to the solution of Volterra-type equations is analyzed.

PART III of the book, applications, are chiefly concerned with economy, biology, immunology, and health care. For this we emphasize new possibilities of our technique based on MM of development, and peculiarities of its applications in each specific case. Thus, the applied part of the book is entitled "Introduction to applications".

The monograph is a further development of the ideas first established in certain works by academicians L.V. Kantorovich and V.M. Glushkov. Comparisons of well-known MM with given classes of MM and the history of the issue are considered in detail in a special section of this book.

The ideas and results of academicians V.I. Vernadsky, V.M. Glushkov, and G.I. Marchuk as well as the famous Italian mathematician V. Volterra, had a large influence on the subject of this book. Scientific works of Professors N.S. Bakhvalov and J.F. Traub influence most on Part II of this book.

The Author is grateful to Professors L. Ljung, Masao IRI, A.G Sukharev, L.H. Garcia-Rubio, H. Salem for support of his efforts.

The Author is also grateful to Dan Price for language assistance and to Editor John R. Martindale and his Assistant, Sharon Donovan for their help and useful remarks.

St. Petersburg, FL USA October 1998

PART I:

GENERAL THEORY

CHAPTER 1

EVOLUTIONARY SYSTEMS

With a view to the best comprehension of objects investigated, various examples of *evolutionary systems* (ES) are considered. There are three classes of ES: *Artificial ES* (AES) being created or already created by human beings and functioning with their participation, *Natural ES* (NES), and *Joined ES* (JES). The principal attention of this chapter is recognition of common properties, among ES, and differences, between AES and NES. We study mainly structure-functional properties of ES that are independent on the specific character of ES, for example, on separate countries or species of plants.

1. ARTIFICIAL ES

The main elements of AES are *work places* (WP). *A work place* is usually localized in the time and space aggregate of labor functions of the respective ware: material, energy, and information, which should be fulfilled by a respective specialist. The results of WP functioning are various goods and services or products. The main characteristics or *indices* of the WP are the indices of *efficiency* of WP functioning, e g, the quantities of products produced per unit expenditure and per unit time. Different WP and AES relate to different kinds of human activities. Apparently, diverse of WP and their indices are in continuous development.

Let us consider in more detail WP and AES in industry, education, science, art, and health services.

1.1. Industry as ES

All separate enterprises or corporations and industry as a whole can be considered as ES. Of course, the specific character of WP in each case depends on the aim and the result of their functioning. There are two important classes of WP: WP, reproducing external goods with respect to ES, and WP, reproducing or creating new, more effective, WP for the ES itself. Theoretically, the case is possible that there could be WP creating new WP, which, in turn, create new WP, and so on. However, in practice this recurrence has to be interrupted while restricting the combination of creation and self-creation functions on some WP. Not without reasons do employers of many enterprises of Japan give their employees two salaries: one for goods and services production and the other for the increment of WP efficiency [3, 9].

As one can see, *self-creation* or self-perfection *functions* become inevitable. It is also possible that on some WP copies are created, which are similar to the process of cell proliferation, and on other WP new special WP are created, which are similar to the process of cell differentiation. The necessities of a new product and greater efficiency of WP stimulate creation of new WP. But what does *new WP* mean? If at least one of labor functions and/or any element of the respective "wares" including "personnel-ware" are changed, there will be *new WP*.

The importance of the theory of *labor functions* similar to the theory of algorithms should be emphasized. It is well known that any algorithm can be constructed with the help of a special complete set of small number operations. In a similar way, any *labor function* can be constructed with the help of a special set of a few labor functions. This allows us to create new WP sometimes quickly and effectively.

The branch of industry as ES creating new WP we call a *subsystem A* of ES, and the branch of Industry creating or producing goods and services that are external with respect to ES, we call a *subsystem B* of ES. Clearly, the subsystem A creates new WP for B and for itself. So, A is the subsystem of creation or reproduction of ES and B is the subsystem of production of goods and services external to ES.

The distribution of WP between the subsystems A and B is very important. A share can be used for controlling ES. It is not clear in advance which kind of consequences arise from increasing or decreasing of the share. The problem of the best distribution has arisen.

In the course of the time, *functioning WP* becomes *obsolete*, i.e., at least one of the WP elements becomes obsolete. So, these obsolete WP should be eliminated or reserved or reconstructed. The temporal boundaries for obsolete WP and/or the *intensities* of the employment of WP are also significant. These boundaries and/or intensities can also be used for controlling ES. In general, there may be two kinds of boundaries: one, for WP in the subsystem A and another, for WP in the subsystem B.

The next important feature of ES is the existence of the external resources that enter ES. It means that ES are *open dynamic systems*. We can use these external and internal resources to create balance equations for ES.

To conclude with industry as ES we may consider two cases: 1. t_0 is the moment of ES origin and the starting point for modeling and 2. $[t_0, t^\wedge]$ is the given *prehistory* of ES, $t^\wedge > t_0$, and $t^\wedge$ is the starting point.

1.2. Education as ES

Any educational center: school, college, university, etc., and education as a whole can be considered as ES.

All the features of ES in industry: WP as the main elements of ES, existence of the subsystem A for realization of the internal functions of the system development and the subsystem B for realization of the external functions, allocation of the system's resources between its internal and external functions, the out of date or obsolete WP, the inflow of resources from the outside and existence of the moment of ES origin or ES prehistory, are also suitable to the case of education. Essential difference consists of their WP and their products.

The *educational product* is the number of specialists of determined quality, and the index of efficiency of WP is the number of new specialists of determined quality per the unit of WP in the subsystem B per unit time.

The quality of specialists can be decided by means of different rules. To avoid the natural question: "And who are the judges?" different tests are not infrequently used, and their results are decided with the help of computers.

Any field of knowledge contains basic ideas and facts, experimental and/or theoretical. The usual mode of compilation of the respective tests is elaboration of the list that contains true ideas and facts (and/or questions to them) along with false ones similar in meaning; all of them are labeled. A person tested should select the right labels (and answers). Usually, if this number is not less than ¾ of the total number of labels (and questions), a person passes the test, i.e., he or she is the right specialist.

The *subsystem B in education* is the set of WP, the product of which is the number of the specialists. The *subsystem A in education* as ES is the set of WP, the products of which are new, more effective WP in both A and B subsystems of education as ES. Creation of these WP also means training of the respective, more effective specialists in the field of education. Any science-methodical center in the field of education, the public educational office, and administration of any educational center will belong to the subsystem A if their functions are the creation of new WP. Actually, any educator belongs to A while he or she is enlarging their own qualification. Thus the primary indication of WP belonging to the subsystem A or B is not a place, time or person, but a kind of fulfilling labor function.

The model of education depends on the teaching philosophy or policy to be adopted. The quantity and value of new evidence, familiarities with this evidence, and getting the ability to use it estimate the qualities of teaching. Of course, the notion "value" requires more precise definition, but this is out of the present frame of work.

1.3. Science as ES

The *science product* is a new, more effective technology. Since new technology results in higher level values of indices of WP efficiency, we can assume that the science external products are these indices of WP efficiency for different ES and both their subsystems A and B. The internal product of the *subsystem A in science* as ES is the new technology, which is the new WP and/with their efficiency indices in A for creation of the new technology, including creation or restoration of A itself. The external product of the *subsystem B in science* is the new technology for the other branches of human activities. So, we can accept that indices of WP efficiency in the subsystem A of science and WP main products are actually the same.

Any scientific center, any branch of science and science as a whole can be considered as ES. At all events the subsystem A of these ES is the set of WP the products of which are new, more effective WP creating new technology for itself and external ES. Any scientist, while his or her labor function is perfection of his or her own WP, belongs to the subsystem A of science as ES.

Thus, it is not difficult to see that essence of development of science (and the other AES) is presence of the *subsystem self-perfection*.

1.4. Art as ES

We only would like to emphasize that writer, artist or musician also has his or her own work place and own products. Any organization in art and art as a whole can be considered as ES with separation of the subsystems A and B depending on WP functions. The modern means of technology and especially computers can considerably enlarge efficiency of labor functions in art. The problem of the best distribution of these means between the subsystems A and B in art and the problem of determination of the best temporal boundary for obsolete WP in art are also important for art.

1.5. Health care as ES

We dwell on this subject in more detail, since it is a vital issue and important parts of the applications in this book relate to health services or health care (HC).

The *HC external product* is the cure, i.e., the number of people who have improved their health to such extent that they do not need providers of HC, or who have prolonged their active life for a sufficiently long time, say not less than a year, after curing. Unfortunately, for example, for the USA, which spends more on HC than any other nation, "There are thousands of billing codes in U.S. insurance for health care procedures; there is not one single billing code for a cure." (see [5], p. 3).

To improve the situation in HC a systematic approach considered in [5], in particular, a commitment to Continuous Quality Improvement that "means that the outcome of a process and the process itself are under continuous review, with a formal approach used to continuously improve both process and results." ([5], p. 186).

In the ES language it means that we need to improve indices of efficiency of functioning for both subsystems A and B in HC as ES. In this connection, the *subsystem A in HC* is, as usual, the subsystem creating new, more effective WP for itself and for the *subsystem B in HC* that, in turn, is the aggregate or team of WP with curing as the main labor function. In addition, the subsystem A comprises not only all doctors, the so-called organizers of HC system, but also any doctor who enlarges his/her own qualification and any person while helping to perfect HC system as ES.

A special feature of HC as ES is the fact that on its WP in the subsystem B, "raw" materials that should be cured are living organisms or their subsystems and organs that in turn can be considered as NES. So, we have to deal with interaction between AES and NES. Later we will consider different types of interaction between two ES. We would like to emphasize that according to the theory of NES the most difficult cases of pathology and diseases and, what is primary, the reasons for these diseases are related to an infraction in the organism subsystem namely A-type.

According to the theory of AES, for successful development of HC (under the same resources inflow) it is important that we distribute its internal resources between its subsystems A and B in a reasonable way, and in the same way we determine the temporal boundary for obsolete WP in the subsystems A and B, and, of course, that we have a competitive market of interaction between HC and its consumers, etc.

The respective formalized, mathematical apparatus allows us to consider and investigate such kinds of problems both on the qualitative and quantitative level.

2. NATURAL ES

The main elements of NES are *reaction* or active *centers* (RC). A *reaction center* is usually localized in the time and space aggregate of chemical reactions (or appropriate physiological functions) with the respective ware: material, energetic, and information, medium of that is usually *bio-mass*. The results of RC functioning are some products and/or bio-mass. The main characteristics of RC are the indices of *efficiency* of RC functioning, e g, the quantities of products and/or bio-mass created per unit of the respective ware expenditure and per unit of time. Sometimes, it is convenient to assume that RC are cells of an organism or a plant, in other cases RC are parts of a cell, e g, organelles, macromolecules or ionic particles [2].

Let us consider in more detail NES such as a separate plant, a separate organism, a population of animals, and the biosphere.

2.1. Plant as ES

There are two important classes of RC of a plant as ES: RC in the *subsystems B in a plant* creating some product for fulfillment of a plant interaction with its environment and RC in the *subsystem A in a plant* creating new RC, among them RC that create new RC. In the cell's case as RC, proliferation and differentiation of cells are processes of creation of new RC in the subsystem A.

The share or ratio of RC distributed between the subsystems A and B is very important in the case of plants too. This ratio can be used, in particular, for controlling productivity of plants. The problem of the best distribution has also arisen.

In time, earlier created RC become obsolete. So, these obsolete RC should be eliminated or reserved or renovated. Temporal boundaries for obsolete RC seem also to be very important and useful for controlling NES.

Of course, a plant as ES cannot exist without existence of external resources which a plant consumes. We can use external and internal resources to create balance equations for NES. Using the comparatively simple balance of the process of *photosynthesis*, we can obtain the additional equation for new bio- or better *phyto-mass* of a plant. Indeed, the main phyto-mass is created during the process of *photosynthesis* from carbon dioxide (CO_2), hydrogen (H), and oxygen (O_2). During the hydrolysis of H_2O hydrogen H and oxygen O_2 goes to the energy need of the plant and oxygen O_2 from CO_2 is released and goes outside [1]. As a result we have [4, 8]

phyto-mass = 6massCO_2 + 6massH_2O - 6massO_2 = 6massC +12massH + 6massO, (*)

from which it follows that

$$6\text{mass}O_2/\text{phyto-mass} \cong 1 \qquad (^{**})$$

It should be noted that in the daytime and especially at night the process of plant's breathing takes place with the consumption of O_2 and secretion of CO_2. However, these processes practically do not change mass of plant.

It should be also noted that for plants there are essentially two different phases of plant development: 1. from the instant of germination of a seed to the instant of appearance of leaves when the process of photosynthesis does not take place and 2. after appearance of leaves. In each of these phases we can consider two cases: i) t_0 is the moment of ES origin and starting point for modeling and ii) $[t_0, t^\wedge]$ is the given prehistory of ES, $t^\wedge > t_0$, and $t^\wedge$ is the starting point for modeling.

2.2. Organism of an animal as ES

All the features of ES in a plant: RC as the main elements of ES, existence of the subsystem A for realization of the internal functions of the ES development and the subsystem B for realization of the external functions of interaction between ES and its environment, allocation of ES resources between its internal and external functions, the obsolete RC, the temporal bounds for the obsolete RC, consumption of the external resources, presence of two different phases of development (1. from the instant of conception to the instant of birth and 2. after birth), and existence of the moment of ES origin and/or ES prehistory also take place in an animal as ES.

We dwell here on *differences between an animal and a plant* and *NES and AES*.

Determination of the balance equations for an animal, which is similar to (*), (**) are essentially different and more difficult because an animal usually does much work and exudes a significant quantity of heat and the respective waste matter. Indices of efficiency for RC are usually higher by far for animals than for plants. And, of course, interpretation and composition of RC are different for different NES.

Some differences between NES and AES are the following [1, 2, 6, 10, 12]:

i) There are different compositions of RC and WP ware. "All life forms employ two categories of chemical, one that stores information and a second that acts, based upon that information, to duplicate the organism." ([6], p.1).

ii) Indices of efficiency for RC of NES are considerably higher than for WP of AES. For example, some cells recreate cells during minutes with a negligible outlay, while WP create new similar WP during years with great expenditure. It means that efficiency of RC can be approximately a billion times higher than WP. A reason is the very prolonged period of evolution. "Any living cell carries with it the experience of a billion years of experimentation by its ancestors" (Max Debrick, 1944). But the rate of growth of WP indices is usually very high. Presupposing their growth about 2% on the average during each year, it is easy to obtain the increase of efficiency indices on 8-10 orders during about 10^3 years. Therefore AES can probably achieve that level of a high efficiency during about a thousand of years for which NES required a billion years.

iii) The life spans for some NES are considerably shorter than for most AES, but for some NES are considerably longer. It appeared advantageous for nature to provide the best survival of population at the expense of the comparatively high death rate of its species. So, the life span for the population as ES is considerably longer than for most AES, and for separate species of population it is much shorter.

According to [1], living things:

i) are highly organized;

ii) are homeostatic, "staying the same";

iii) reproduce themselves;

iv) grow and develop;

v) take energy from the environment and change it;

vi) respond to stimuli;

vii) are adaptable.

2.3. Population as ES

Now the internal product is the birthrate of the species and the external product is the result of interaction between the population and its environment; the *subsystem A in population* is that of reproduction of species, and index of efficiency of RC in A is the number of new species that appeared per the unit of the resource expenditure for the unit of the time; RC itself in A can be interpreted as the population species; therefore, obsolete RC are dead species; the *subsystem B in population* is that of production of an external product; the instant of ES origin is that time when due to mutation and recombination of the genetic information we have a new population.

The important meaning for the development and control of population as ES is the distribution of its internal resources between its subsystems A and B and a temporal bound for obsolete RC. The natural hypothesis is that a temporal bound is mainly determined by the genetic information, whereas distribution of resources depends mainly on the external effects from the environment. If population has severe environmental conditions its reproduction will decrease.

The theory of development and development modeling with the help of the respective MM of ES permit us to evaluate the internal and external effects more precisely (see Ch. 11, 12).

2.4. Biosphere as ES

According to [10] the *subsystem A in the biosphere* as ES is the subsystem of recreation of the *living substance* of the planet, mainly the *phyto-mass* by way of photosynthesis; *the subsystem B in the biosphere* is the subsystem of creation of the so-called *inorganic substance*, mainly the oxygen; the obsolete product is the *dead* or bio-boned *substance* which is mainly equal by mass to the humus.

The important exception of the biosphere as ES is the composition of its external resources. There is the *bio-geo-chemical cycle* of substances in the biosphere, which results in the biosphere "consuming" a part of its own product: inorganic substance and the obsolete product. The other parts of its external resources are resources of the atmosphere and the hydrosphere, solar energy, and cosmic radiation.

It appears that the instant of the biosphere's origin is the time when the conditions on our planet were sufficient for self-creation of the living substance. This problem can be also analyzed with the help of MM of development (see Ch. 14).

3. JOINED ES

AES are the results of human activities. The size of these activities is now such that land and its depths, air, water and the whole biosphere have begun not infrequently to suffer from them. Destiny and the dignity of human beings are now inseparable from the fate of their environment. So, academician V.I. Vernadsky came to consider human activities and their environment as the unit and named this unit by the *neo-sphere* [10]. Thus, the *neo-sphere* includes as its parts the atmosphere, human, geo, hydro and biosphere. In other words, the neo-sphere is "Managing Planet Earth" [7].

We can consider the *neo-sphere* as one of NES in which the *subsystem A* recreates the living and bio-boned substance of our planet, the *subsystem B* creates the inorganic resources (parts of them are leaving the planet), and the external resources are parts of NES own products, solar energy, and cosmic radiation. We can also consider the *neo-sphere* as JES that is the combination of two ES, one of which is human activities as AES (ES1) and another is the other part of our planet as NES (ES2).

The principal meaning for interactions between ES1 and ES2 is the composition of their external resources. The external resources of ES1 are parts of the products of ES2 and parts of all the external resources of the bio-sphere. The external resources of ES2 are the remaining parts of the external resources of the bio-sphere and the portion of the obsolete products of ES1 leaving ES1.

The critical value is the size of the parts and the composition of the obsolete products leaving ES1. Starting from the determined parts of the parts, the processes of recreation in ES2 can become irreversible, and even a small quantity of special obsolete products of human activities can be the reason for environmental disaster.

Thus, it is necessary to create strategies for sustainable human development on our planet with finite resources and a fragile environment. The development modeling allows us to determine such strategies more accurately (see Ch. 9).

4. GENERALIZED STRUCTURE OF ES

According to the description above, the generalized *structure of ES* has the form of Fig.1. Here **f** is the external resource of ES; **m** is the new internal product (resource) of the subsystem **A**; **c** is the new external product of the subsystem **B**; **y** is a relative share of distribution of the internal resource between **A** and **B**, $0 \leq y \leq 1$; **G** is the obsolete product of ES; γ is the intensity of the subsystem **A** functioning along the channel **ym--m**, $0 \leq \gamma \leq 1$; α is the index of efficiency of functioning along the channel **ym--m**; δ is the intensity of the subsystem **B** functioning along the channel **(1-y)m--c**, $0 \leq \delta \leq 1$; β is the index of efficiency of functioning along the channel **(1-y)m--c**; new technology or the respective functions α and β can be created outside ES (the exogenous case) or inside ES (the endogenous case).

It should be emphasized that the subsystems **A** and/or **B** can be, in turn, ES with the same structure as initial ES, and so on.

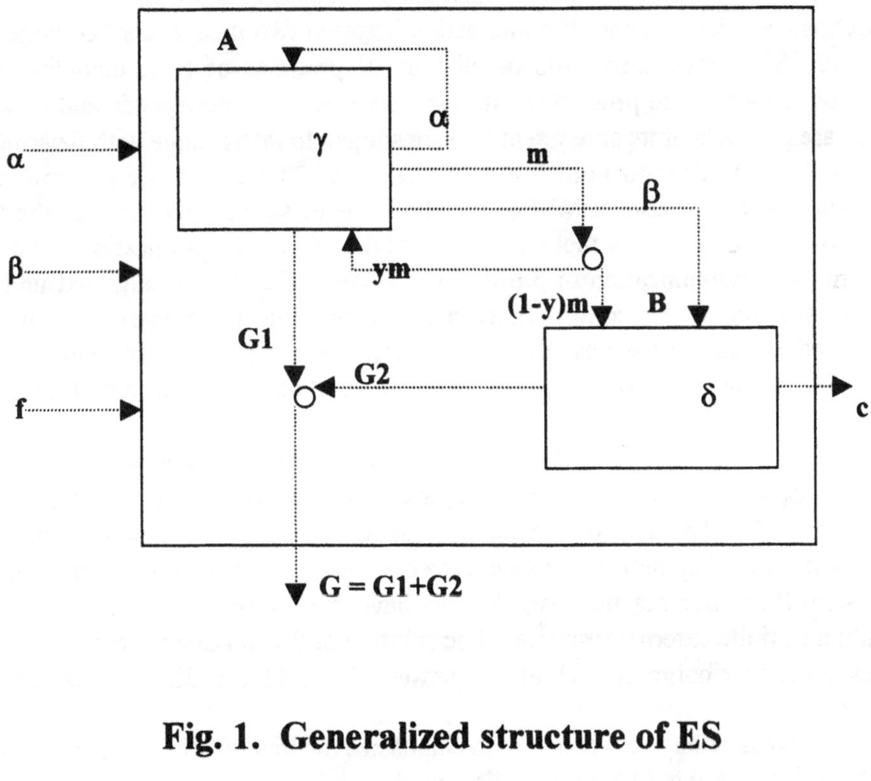

Fig. 1. Generalized structure of ES

5. GENERALIZED STRUCTURE OF TWO ES INTERACTION

We already considered two ES interaction into the neo-sphere.

Another example of interaction in JES is the interaction between health care as AES (ES1) and patients as NES (ES2). The main functions of ES1 are perfection of their own WP and curing of patients. As the response to that curing, ES2 gives to ES1 a certain part of its external products. This part is also an important additional control function that can be determined by means of an estimate of the maximal number of cured patients.

One more example of two ES interaction is the interaction between directing ES (ES1) and executing ES (ES2) in economy as ES. Of course, we can consider both kinds of ES as one ES, but it is not so for consideration in more detail.

The main functions of ES1 are the creation of new WP for itself and for ES2. So, external products of ES1 are the new additional WP for ES2 including a new "personnel-ware" for ES2 as the results of training for ES2 persons with new professions. In response, ES2 gives to ES1 a certain part of its external products. This part is an important additional control function the best value of which can be found, for example, from analysis of joined maximal profit as ES1 and ES2.

Let us consider, in short, the interaction between two independent countries as ES1 and ES2. Determined parts of all kind of products of ES1, including new technologies (which are products of its subsystem A1) and new goods and services (which are products of its subsystem B1) are subject to interchange with determined parts of all kind of products of ES2 and vice versa. These parts are the important additional control functions. MM of ES allows us to select reasonable or the best values, in a certain sense, of these parts. One of the criteria of optimization is the idea of coordinated maximization of profits for ES1 and ES2. This means that under a certain variation of the control functions it is possible to increase both profits simultaneously until we achieve the so-called *Pareto set* on which increasing of both profits is impossible. So, we should try to find the best values of the control functions outside the Pareto set.

Let us consider, at last, the classical example of interactions between two populations, beasts or birds of prey (ES1) and their victims (ES2) [11, 12]. Here the subsystems A1 and A2 are correspondingly subsystems of reproduction of plunderers and victims, the subsystem B1 consumes the maximal possible part of victims, and the subsystem B2 minimizes that part. So, we have the mini-max problem under the condition of finite external resources. The solution of this problem gives determined values to the distribution in each of ES: between A1 and B1 in ES1 and between A2 and B2 in ES2.

In this connection, we would like to emphasize the high quality of the books [11, 12] that are by far ahead of their publication dates.

According to the description in this section and s. 3., the generalized *structure of two ES interactions* has the form shown in Fig. 2. and 3. Values **fi, mi, ci, yi,** and **Gi, i = 1** for ES1 and **i = 2** for ES2, have the same sense as values **f, m, c, y,** and **G** in Fig. 1. Fig. 2 is related best to the above-mentioned example of peace interaction between two independent countries. Unfortunately, interaction between human activities and the rest of our planet (the example of the neo-sphere) is related more likely to Fig. 3, which is certainly about interaction between plunderers and victims.

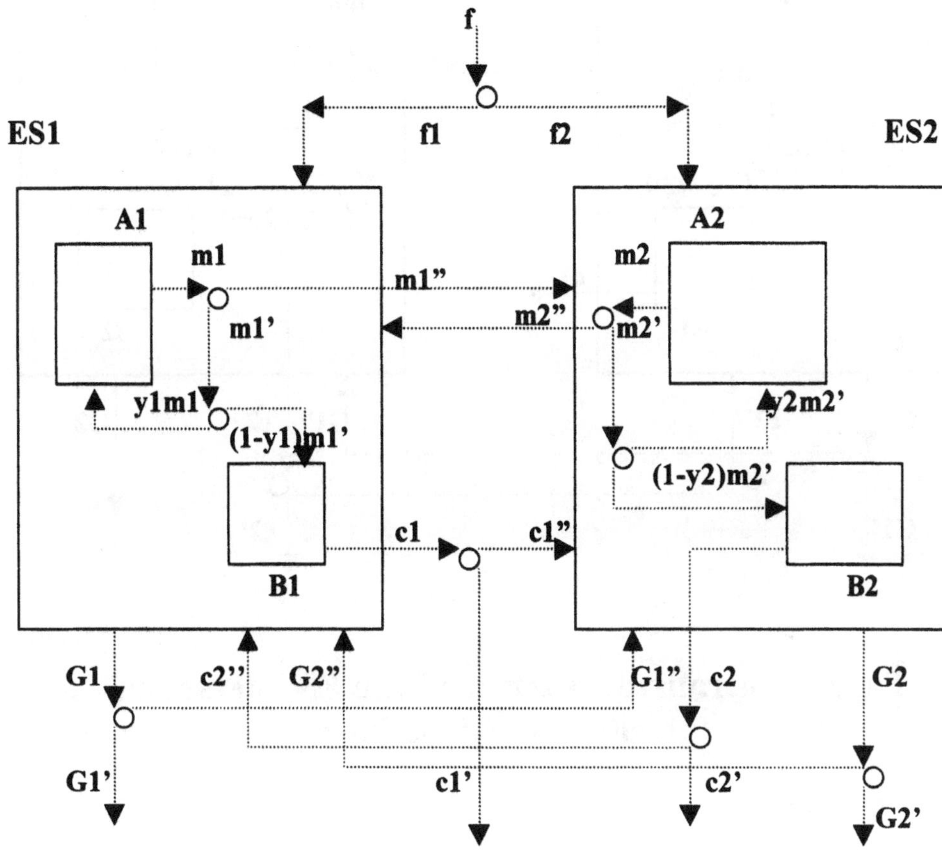

Fig 2. Generalized structure of two ES interaction (peace coexistence)

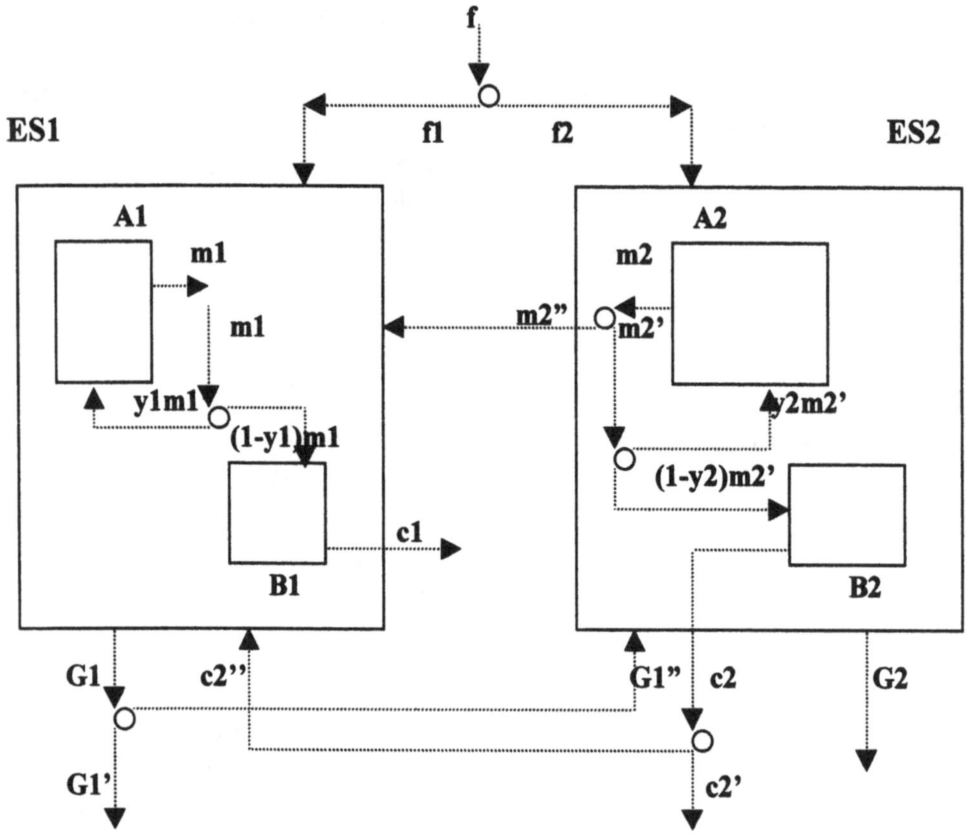

Fig 3. Generalized structure of two ES interaction (plunderers and victims)

REFERENCES

1. Curtis H., Biology, Worth Publishers, Inc, 1984, 1159 pp.
2. Davydov A.S., Biology and Quantum Mechanics, Kiev: Naukova dumka, 296 pp. (in Russian)
3. Economic Policy and Development: New Perspective, Auburn House Publishers Co., 1985, 320 pp.
4. Element, The Encyclopedia Americana, Intern. Edition, Grolier Inc., 10, 1990, 203-209.
5. Halvorson G.G., Strong Medicine, Random House, 1993, 253 pp.
6. Levine A.J., Viruses, Science Amer. Library, 1992, 240 pp.
7. Managing Planet Earth, Reading from Science Amer. Magazine, 1990, 146 pp.
8. Photosynthesis, The Encyclopedia Americana, Intern. Edition, Grolier Inc., 22, 1990, 38-41.
9. Tsvetkov V.Ya., Fifteen Stone of Reandzi's Garden, Politizdat, 1991, 414 pp. (in Russian).
10. Vernadsky V.I., Living Substance, M.: Nauka, 1978, 240 pp. (in Russian).
11. Volterra V., Theory of Functional and of Integral and Integro-Differential Equations, Dover Publications, 1959, 340 pp.
12. Volterra V., Mathematical Theory of Fighting for Existence, M.: Nauka, 1976, 222 pp. (in Russian).

CHAPTER 2

MATHEMATICAL MODELS OF DEVELOPMENT

According to the examples of ES and ES structure (see Ch. 1), mathematical models (MM) of ES or development are introduced. These MM are exact by definition and different from each other depending on interpretation and the extent of detailing or collection. Some additional MM are introduced concerning the fiscal side of human activities for AES and the natural balance relations for NES. A comparison with the well-known MM is considered. The problem of estimating input data and the problem of MM completeness are analyzed. Statements of other problems are given.

1. BASE MM

The *base MM* of ES have the form

$$m(t) = \int_0^t \alpha(t,s)\lambda(t,s)y(s)m(s)ds + \Delta m(t),\ 0 \le y(s),\ \lambda(t,s) \le 1,$$

$$c(t) = \int_{t^\wedge}^t \beta(t,s)\mu(t,s)z(s)m(s)ds + \Delta c(t),\ 0 \le z(s),\ \mu(t,s) \le 1,$$

$$P(t) = \int_0^t [\lambda(t,s)y(s)+\mu(t,s)z(s)]m(s)ds,\ y(s) + z(s) = 1,\ t \ge t^\wedge \ge 0,$$

$$M(t) = \int_0^t m(s)ds,\ G(t) = M(t) - P(t),\ f^\wedge(t) \ge f(t) = m(t) + c(t), \qquad (1)$$

where $m(t)$ is the rate of creation of the first kind new *generalized product* (resource) quantity at the time instant t, which provides the fulfillment of the internal functions of ES, that is, restoration of itself and creation of the second kind product; $y(t)m(t)$ is a share of $m(t)$ for fulfillment of internal functions in the subsystem A of restoration and perfection of the system as a whole; $\lambda(t,s)$ is a relative share of the *intensity* of $y(s)m(s)$ product use at the instant t; $\alpha(t,s)$ is the *efficiency* index for functioning of the subsystem A along the channel $\lambda(t,s)y(s)m(s)$--$m(t)$, i.e., the number of units of $m(t)$ created in the unit of time starting from the instant t per one unit of $\lambda(t,s)y(s)m(s)$; $\Delta m(t)$ is an increment of $m(t)$ (may be negative); $c(t)$ is the rate of creation of the second kind new *generalized product* quantity at the instant t, which provides the realization of the external functions of ES; $z(s)m(s)$, $\mu(t,s)$, $\beta(t,s)$, and $\Delta c(t)$ are similar to ym, λ, α, and Δm respectively but for the subsystem B of creation of the second kind product; $P(t)$ is the total quantity of the first kind product functioning at the instant t;

$M(t)$ is the total quantity of the first kind product to be created during the time t - 0; $G(t)$ is the total quantity of the out-of-date or *obsolete product* at the instant t; $f^\wedge(t)$ is the rate of the resource inflow from the outside ($m(t)$ and $c(t)$ are measured in the units of $f^\wedge(t)$); $t^\wedge$ is the starting point for modeling; $[0,t^\wedge]$ is the *prehistory* of ES for which all the functions are given (their value will be noted by the same symbols but with the index "^," e g, $m(t) = m^\wedge(t)$, $t \in [0,t^\wedge]$).

It is obvious that all the relations' (1) are faithful representations by definition. The *indices* α and β depend on m, Δm, c, Δc, λ, μ, y, P, $f^\wedge$ in the general case.

Thus, (1) is the system of nonlinear integro-functional relations, which consists of 7 equalities and 11 inequalities connecting 18 values, namely: m, Δm, c, Δc, α, β, λ, μ, y, z, P, M, G, t, $t^\wedge$, f, $f^\wedge$, 0, all of them are nonnegative (except Δm and Δc).

The typical suitable examples to interpret all the values can be as follows:

1. The economy as a whole. Then (see [5 - 10] and Ch. 1) A is a subsystem of the capital goods industry and B is a subsystem of the consumer goods industry; $m(t)$ is the rate of production of new work places (WP) number in A and B; $\alpha(t,s)$ is a *labor productivity* in the subsystem A, i.e., the number of WP created for the unit of the time starting from the time instant t by one worker from A at WP created at the time instant s; $\beta(t,s)$ is a *labor productivity* at the instant t by one worker at WP created for the subsystem B at the instant s; $P(t)$ is the total number of functioning WP at the instant t, which can be equal to the quantity of labor resources; $G(t)$ is the total number of the obsolete (and lying in reserve) WP at the time t; $f^\wedge(t)$ is the rate of the inflow of resources from the outside, e g, from the biosphere and the cosmos.

2. The biosphere. Then (see [10] and Ch. 1) A is a subsystem of re-creation of the *living substance* of the planet, mainly the *phyto-mass* by way of photosynthesis; B is a subsystem of creation of the so-called *inorganic substance*, mainly the oxygen; $m(t)$ is the rate of re-creation of the new living substance quantity; $P(t)$ is the total quantity of the functioning living substance at the time t; $G(t)$ is the total quantity of bio-boned or *dead substance* at the time t, which is mainly equal by mass to the humus; $\alpha(t,s)$ is the specific rate of reproduction of $m(t)$ at the instant s; β is the specific rate of production of $c(t)$, and so on.

We note three special cases of MM (1) when λ and μ have the form

$$\lambda = \lambda(t-s),\ \mu = \mu(t-s); \tag{2}$$

$$\lambda, \mu = \begin{cases} 0, & 0 \le s < a(t), \\ 1, & a(t) \le s \le t; \end{cases} \tag{3}$$

$$\lambda = \begin{cases} 0, & 0 \le s < a(t), \\ \lambda(t-s), & a(t) \le s \le t; \end{cases} \qquad \mu = \begin{cases} 0, & 0 \le s < b(t), \\ \mu(t-s), & b(t) \le s \le t. \end{cases} \tag{4}$$

The case (2) corresponds to the *stationary process* of the intensity; the case (3) means that the product created before a threshold $a(t)$, $a(t) \le t$, is never used at the instant t, but that created after $a(t)$ is used entirely; the case (4) extends the previous ones.

Instead of the relations (1), in particular, we have ($\Delta m = \Delta c = 0$)

$$m(t) = \int_{a(t)}^{t} \alpha(t,s)y(s)m(s)ds,\ P(t) = \int_{a(t)}^{t} m(s)ds,\ G(t) = M(t) - P(t); \tag{5}$$

$$c(t) = \int_{a(t)}^{t} \beta(t,s)[1-y(s)]m(s)ds,\ f^{\wedge}(t) \ge f(t) = m(t) + c(t),\ \ t \ge t^{\wedge} \ge 0, \tag{6}$$

and

$$a'(t) \ge 0,\ t \ge 0. \tag{7}$$

Denoting $P'(t)$ by $p(t)$, we convert (5) to the form

$$m(t) = \int_{a(t)}^{t} \alpha(t,s)y(s)m(s)ds,\ a'(t) = [m(t)- p(t)]/m(a(t)). \tag{8}$$

As one can see, even in the simplified formulation, MM (1) is reduced to the relations (8) in which along with the nonlinear integral equation of the unusual form (where a variable lower bound $a(t)$ can be the desired unknown function) there appears the *nonlinear functional equation.*

The important particular case of MM (1), which extends (5)-(8), is

$$m(t) = \int_{a(t)}^{t} \alpha(t,s)y(s)m(s)ds,\ M(t) = \int_{0}^{t} m(s)ds,\ c(t) = \int_{b(t)}^{t} \beta(t,s)[1-y(s)]m(s)ds,$$

$$0 \le y(s) \le 1, P_1(t) = \int_{a(t)}^{t} y(s)m(s)ds,\ P_2(t) = \int_{b(t)}^{t} [1-y(s)]m(s)ds,\ G(t) = M(t) - P(t),$$

$$P(t) = P_1(t) + P_2(t), f(t) = m(t) + c(t),\ 0 \le a(t), b(t) \le t,$$

$$t \ge t^{\wedge} \ge \min\,[a(t^{\wedge}),\ b(t^{\wedge})] = 0. \tag{9}$$

For MM (5)-(8) the functions α, β, y, and P or α, β, y, and f are usually to be given and the other elements to be determined. For MM (9) the functions α, β, y, P_1, and P_2 or α, β, y, f, and c are usually to be given and the other elements to be found.

2. MORE GENERAL MM

It is not hard to introduce different additions and generalizations of MM (1) [4-6, 8, and 9]. The *n-product MM* of ES, $n > 2$, can be formally written almost in the same form (1) under the condition that m and c (and Δm, Δc) are the vector functions: $m = (m_1,..., m_r)$, $c = (c_1,..., c_s)$, $r+s = n$; y and z are the column matrices and $\alpha\lambda$ and $\beta\mu$ are appropriate *matrix functions* (where the inequalities for the vectors and matrices signify the inequalities of the same name for all of their appropriate components).

However, the relations for *f*, *P*, and *M* should be replaced by

$$f(t) = \sum_{i=1}^{r} m_i(t) + \sum_{i=1}^{s} c_i(t),\ r+s = n,$$

$$P(t) = \sum_{i=1}^{r} P_i(t),\ M(t) = \sum_{i=1}^{r} M_i(t),\ G_i(t) = M_i(t) - P_i(t),\ M_i(t) = \int_0^t m_i(s)ds,$$

$$P_i(t) = \int_0^t [\sum_{j=1}^{r} \lambda_{ij}(t,u)y_{ij}(u) + \sum_{j=1}^{s} \mu_{ij}(t,u)z_{ij}(u)]m_i(u)du. \quad (10)$$

We can consider also the more general case when *y* and *z* are nonnegative matrices with the restriction

$$\sum_{j=1}^{r} y_{ij}(t) + \sum_{j=1}^{s} z_{ij}(t) = 1,\ i = 1,\ldots, r. \quad (11)$$

It is also not hard to describe a *continuous MM* of ES in a similar form considering t and s as many-dimensional independent variables and examining the appropriate integrals as multivariate ones. For example, if there are different products in each point $r = (x_1, x_2, x_3)$ of 3-dimensional space we will have

$$m(r,t) = \int_0^t \int_0^r \alpha(t,r,s,v)\lambda(t,r,s,v)y(s,v)m(s,v)dvds,\ v = (v_1, v_2, v_3), \quad (12)$$

instead of the first equation in (1). The other equations in (1) can be rewritten in a similar way.

Stochastic MM similar to (1) can be obtained by considering α, β, λ, μ, and y as the functions of a random factor ω. *Discrete MM* can be represented in the same form if the integrals in (1) are understood in the sense of Stieltjes.

3. ADDITIONAL MM

For AES, regarding the fiscal side of human activity, we consider, in addition, *MM of cost, prices*, market trends, and quality control.

Let $q(t)$ be cost of the unit of the product (goods and services) $c(t)$ at the instant t and let $c(t)$, $c_n(t)$ and $c_r(t)$ be the quantity of the whole, *social-necessary*, and realized external *product* in the time unit starting from t respectively. Then

$$q^\wedge(t) = q(t)k^\wedge_n(t),\ k^\wedge_n(t) = c_n(t)/c(t), \quad (13)$$

is the *social-necessary cost* of the product unit at the instant t;

$$d(t) = q^\wedge(t) - q(t) = q(t)[k^\wedge_n(t) - 1] \quad (14)$$

is the so-called differential return;

$$r(t) = q^\wedge(t)l^\wedge{}_r(t) = q(t)l_r(t),\ l^\wedge{}_r(t) = c(t)/c_r(t),\ l_r(t) = c_n(t)/c_r(t), \tag{15}$$

is the market price of the product unit at the instant t.

In turn,

$$q(t) = q^{(1)}(t)q^{(2)}(t)q^{(3)}(t), \tag{16}$$

where $q^{(1)}$ is the cost of the new labor, $q^{(2)}$ is the factor cost growth at the expense of a labor enclosed previously, $q^{(3)}$ is the factor cost growth at the expense of amortization of all the means of a labor under creation of the product unit at the instant t.

The product quality is equivalent to the product growth with the factor $qu(t)$, where $qu(t) \geq 1$. This means that the real amount of the realized products

$$c_{qu}(t) = qu(t)c_r(t). \tag{17}$$

Let dT be the segment of the time during which the number of the social-necessary goods and services $c_n(t)$ becomes invariable in practice. Then the market income from selling the product obtained during $[t,t+dT]$ will be

$$dI = \int_t^{t+dT} r(t)c_{qu}(t)dt = \int_t^{t+dT} q(t)qu(t)c_n(t)dt \tag{18}$$

and the *profit* will be

$$dPr = dI - dE = \int_t^{t+dT} q(t)[qu(t)c_n(t)/c(t) - 1]c(t)dt,\ dE = \int_t^{t+dT} q(t)c(t)dt, \tag{19}$$

where dE is the total cost of the product during $[t,t+dT]$.

It is clear that the *accumulated profit* $Pr(T)$ and *cost* $E(T)$ can be received by integration of dPr and dE.

The variability of the market conjuncture results in the structure and composition of the variable $c(t)$, which results in necessity of new WP creation or introduction.

For NES we consider, in addition, the *balance equation* in the form

$$f(t) = k(f^*,t)c(t),\ 1 \leq k(f^*,t), \tag{20}$$

where the coefficient $k = k(t) = k(f^*) = k(f^*,t)$ depends on the qualitative structure $f^* = f^*(t)$ of the external resources $f^\wedge(t) = f(t)$. It is shown in Ch.1 that for the static of a plant as ES we have $k(t) \approx 2$. For an animal as ES the equation (20) can be obtained from the laws of conservation and the conversion of matter and energy [15, 23].

The equation (20) means that for NES usually c, f, α, and β are given and y, z, m, P, and λ, μ (or $a(t)$) to be found. If the equation in (20) is indeterminate, then for NES usually f, α, and β are given and y, z, m, c, P, and λ, μ (or $a(t)$) to be found.

We can also introduce the *equation* of *homeostasis*

$$P'(t) = 0 \tag{21}$$

For MM (1), it has the form

$$[\lambda(t,t)y(t)+\mu(t,t)z(t)]m(t) + \int_0^t [\lambda'_t(t,s)y(s)+\mu'_t(t,s)z(s)]m(s)ds = 0, \tag{21'}$$

and for MM (5), the form

$$m(t) - m((a(t))a'(t) = 0. \tag{21''}$$

We will apply this equation in Ch. 11 later.

In the case of two ES interactions according to Fig. 2 (see Ch. 1), one can obtain the respective *MM of interactions* if besides the equations of the (1)-(12)-type for each ESi, $i = 1, 2$, the following bonds are added:

$$mi = mi' + mi'', \; ci = ci' + ci'', \; Gi = Gi' + Gi'', \; f = f1 + f2,$$

$$\Delta mi = mi{+}1'' - mi'', \; \Delta ci = ci{+}1'' - ci'', \; fi = fi + Gi{+}1'', \; i = 1, 2, \tag{22}$$

where $i+1$ should be comprehended as addition by module 2. Besides, all the equations should be considered with the substitution of $yi'mi$ and $zi'mi$ for $yimi$ and $zimi$ respectively.

In the case of the interaction of two ES according to Fig. 3 (Ch. 1), we have the same MM as just above, regarding $m1'' = c1'' = 0$.

The more general MM of two ES interactions see [4].

4. COMPARISON WITH WELL-KNOWN MM

We dwell on the connection of MM in question with the *classical MM* and some well-known new ones.

Everybody is familiar with the approach of the so-called "black box" when only the input $X = (x_1,..., x_n)$ and the output $Y = (y_1,..., y_m)$ of a dynamic system are given. We have in the linear approximation

$$Y(t) = \int_{t-T}^t K(t,s)X(s)ds. \tag{23}$$

Here T is the upper bound for all the transient end times, $K(t,u)$ is the matrix of the pulse transition functions $k_{ij}(t,u)$ that are the response functions of the system when $x_j(s) = \delta(u-s)$, $t > u > t-T$, where δ is Dirac δ-function, and all the other $x_k(s) = 0$. A nonlinear dynamic system can also be represented as (23), but K will depend on X.

The MM (1) deals with the so-called "gray box" when the structure of a dynamic system is partly revealed. Indeed, it is possible to say that in (1) the matrix K has been factored into three parts α, λ, and y or β, μ, and z such that each of them has its own applied sense. In addition, several outputs of the system in (1) have served as its inputs. At last, due to the functions $P(t)$, $G(t)$, and $f^{\wedge}(t)$ we deal with the so-called open dynamic system in the case of (1). The essential differences are more in that all the values in (1) are nonnegative by definition and a diminution of the output values is regulated not by a sign, but, for example, by the rate of $a(t)$ growth.

We can find certain analogies of MM (1)-(22) among MM in economics (so-called *macroeconomics MM*) [5, 9, 12, 16, 22] and in population modeling [18, 20, 24].

Detailed comparison of MM (1) with some integral dynamic *macroeconomics MM* can be seen in [9, 11]. In particular, MM of L.V. Kantorovich and V.I. Zhiyanov [13] have the form

$$P(t) = \int_{m(t)}^{t} e^{\delta s} U[\lambda(s),1]\,\varphi(s)ds, \quad \varphi(t) = T'(t) + \varphi[m(t)]m'(t), \tag{24}$$

where P is the product output of an economic system, m is the temporal bound for creation of WP, φ is the number of WP in the unit of the time, $e^{\delta s}U$ is the productive function (δ is a parameter), λ is an investment, T is the number of workers.

Let $\beta(t) = e^{\delta t}U[\lambda(t),1]$. Then after integration in (24) we obtain MM that are equivalent to the particular case of MM (1), which are the one-product models of the (5)-(6)-type.

MM of A.A. Petrov and I.G. Pospelov [17] have the form

$$Y(t) = \int_{\nu(t)}^{\xi(t)} m(t,\lambda)d\lambda, \quad R(t) = \int_{\nu(t)}^{\xi(t)} \lambda m(t,\lambda)d\lambda, \tag{25}$$

where $Y(t)$ is the product output, $m(t,\lambda)$ is the output distribution per unit of the labor-consuming character λ, $R(t)$ is the number of workers, $\nu(t)$ is the smallest labor-consuming character and $\xi(t)$ is the current one. One can see that $\beta = \lambda^{-1}$. After the substitution $\lambda = \lambda(s)$ we obtain MM that are also equivalent to the one-product models of (5)-(6)-type.

The first similar models for ecological systems were proposed in [20] (see also [18]). They have the form

$$B(t) = \int_{t}^{T} m(u)l(u)\varphi(u-t)du + \int_{0}^{t} m(u)l(u)B(t-u)du, \; T \geq t \geq 0, \tag{26}$$

where $B(t)$ is the birthrate: intensity of the number of individuals being born in the unit of the time at the instant t, $m(u)$ is the specific age birthrate: the average number of descendants being born by an individual of age u in the unit of the time, $l(u)$ is the fatigue function: a share of individuals attaining the age u, $\varphi(u)$ is the initial (under

$t = 0$) distribution of individuals by the age u, T is the upper bound of the age.

The equation (26) allows us to find the dynamics of the birthrate B under given m, l, and φ. After substitution $s \rightarrow t\text{-}u$ we obtain the particular case of MM (1), namely

$$B(t) = \int_{t-T}^{t} m(t-s)l(t-s)B(s)ds + \Delta B. \tag{27}$$

Differential MM with partial derivatives are alternatives to *integral MM*. For the linear integral-type model (27) the respective analogies [18] have the form

the evolutionary equation

$$\partial x/\partial t + \partial x/\partial u = - d(u)x(t,u) \tag{28}$$

and the birthrate equation

$$x(t,0) = \int_{0}^{T} m(u)x(t,u)du, \tag{29}$$

where $x(t,u)$ is the number of individuals of the age u at the instant t, $x(0,u) = \varphi(u)$; $m(u)$ is the same as in (27) and $d(u)$ is the specific age death rate. It is easy to see that the equation (28) is equivalent to

$$x(t,u) = x(t-u,0)\exp[- \int_{0}^{t-u} d(v)dv]. \tag{30}$$

Let us substitute (30) into the equation (29) and introduce the desired function $B(t) = x(t,0)$. Then

$$B(t) = \int_{t-T}^{t} m(t-s)\exp[- \int_{0}^{t-s} d(v)dv]B(s)ds + \Delta B. \tag{31}$$

From (31) it follows that the fatigue function is connected with the death rate by

$$l(u) = \exp[- \int_{0}^{-u} d(v)dv]. \tag{32}$$

The transition from *differential MM* to *integral ones* in the case of a non-stationary external environment, where the given functions $m(t,u)$, $d(t,u)$, and $l(t,u)$ depend else on t, is quite similar. In the more general cases when the scantiness of resources, intro-specific competitions, etc. are also considered, the transition from nonlinear differential MM to integral ones can be found in [7, 9, 18, 20].

One of the advantages of *integral MM* compared to *differential ones* is the reduction of dimensionality of the desired functions. If one needs to find, for example, the density $x(t,u)$ one can do it by the formula (30), where $x(t-u,0) = B(t-u)$.

Note that ES are not monotone dynamical systems, in general, and MM considered are different from MM in [21]. The compositions of mappings corresponding to the equations of the models are not forming the so-called semi-group [24], in general.

From the position of functional analysis, MM considered are rather general and strongly nonlinear. The only common restriction on the corresponding mappings or operators is their non-negativity. Of course, it would be useful if we considered our models not only in the general case, but also with the additional restriction, for example, in the form of semi-group. However, it would exceed the limits of the book.

MM of (1)-(12)-types were introduced in the paper [5] by academician V.M. Glushkov. Shortly after, the similar models were also introduced in [8]. Since then, successors of Glushkov continue his work and investigation. Hundreds of scientific works have been published some of which are summarized in [7, 9, 10]. The author of this book reported the general theory of this technique to the World Congress of Nonlinear Analysts in Tampa, FL, August 1992, and to the Meeting of the AMS, Atlanta, GA, October 1997. The participants agreed that the respective direction of the research was new for the USA and the western countries. The applications to health care were reported to the Conference on Mathematical Models in Medical and Health Sciences, Nashville, TN, May 1997. It was emphasized by M. Novak, Shanks lecturer (Oxford University) that one of the important problems should be the problem of creation of the evolutionary theory of a disease progression. But the evolutionary theory and its applications, in particular, to health care are the cornerstones of this book.

5. METHODS FOR ESTIMATION OF MM INPUT DATA

The estimates of efficiency *indices* of the α, β-types are the most difficult. Let us consider several MM of α and some methods of the estimate of parameters of these MM in both AES and NES. The construction will be similar for β.

We divide the given ES prehistory $[0,t\hat{}]$ in two segments: $[0,t1]$ and $[t1,t2]$, $0 < t1 < t2 \le t\hat{}$. If α has the form

$$\alpha = \alpha(t,s) = \alpha(s)\exp[-c(t-s)],\ c \ge 0, \tag{33}$$

where c is a parameter considered as given for the prehistory $[0,t1]$, then $\alpha(s)$ can be found on $[0,t1]$ from the relations

$$m\hat{}(t) = m(0) + \int_{a(t)}^{t} \alpha(s)y(s)m\hat{}(s)ds = m(t)\exp(ct),\ P(t) = \int_{a(t)}^{t} m(s)ds \tag{34}$$

under the condition that m, y, and P are given. Indeed, one can find $a(t)$ from the second equation in (34) which is equivalent to $P'(t) = m(t) - m(a(t))a'(t)$, $a'(0) = 0$. Let M be the anti-derivative for m. Then $P(t) = M(t) - M(a)$ and

$$a(t) = M^{-1}[M(t) - P(t)],\ a'(0) = [m(0) - P'(0)]/m(0), \tag{35}$$

where M^{-1} is the reciprocal function of M (which always exists since m is nonnegative). After this one can find α from the first equation in (34):

$$\alpha(t) = [m^{\wedge\prime}(t)+\alpha(a(t))y(a(t))m^{\wedge}(a(t))a'(t)]/[y(t)m((t)],\ \alpha(0) = m^{\wedge\prime}(0)/[y(0)P'(0)]. \quad (36)$$

We prolong $\alpha(t)$ on the segment $[t1,t2]$ with the help of one of the methods of prediction and find $m(t)$ (and $a(t)$) under the given $y(t)$ from the system of equations (34). This system with respect to $m(t)$ and $a(t)$ has the unique solution under very general conditions (it will be shown in the next Ch. 3). Let us denote this solution $m^*(t)$. The parameter c we can select with the help of the method of simple scanning from the condition

$$I1 = \int_{t1}^{t2} [m(t) - m^*(t)]^2 dt = \min \quad (37)$$

or the condition $I1' = \|m - m^*\| = \min$, where m is given on $[t1,t2]$ and $\|.\|$ is any norm or measure of disagreement between the functions m and m^*. After this we prolong $\alpha(t)$ on the given future segment $[t^{\wedge},T]$.

In a similar way we can select, for example, three parameters, $c1$, $c2$ and $c3$ if α has the form

$$\alpha = \alpha(s)\exp[-c1(t-s)-c2(t-s)^2-c3(t-s)^3],\ c1,\ c2,\ c3 \geq 0. \quad (38)$$

If α has the form of the so-called *degenerate kernel*:

$$\alpha = \sum_{k=1}^{n} \phi_k(t)\psi_k(s) \quad (39)$$

we can find all the functions ϕ_k and ψ_k from the condition [2]

$$I2 = \int_0^{t2} [m(t) - \sum_{k=1}^{n} \phi_k(t)\int_{a(t)}^{t} \psi_k(s)y(s)ds]^2 dt = \min, \quad (40)$$

where $a(t)$ is given by the formula (35). In the particular case of (39) when $\phi_k(t) = c_k\chi_k(t)$, c_k are parameters, and the functions χ_k and ψ_k are given, we can find all the parameters c_k from (40) by the well-known *method of least squares* [1, 3].

Let $\alpha(t,s) = \alpha(t-s)$ and let P, m, and y be given in the prehistory. To find $\alpha(t-s)$ we can solve the Volterra-type equations of the first kind:

$$m(t) = \int_0^{t-a(t)} \alpha(s)y(t-s)m(t-s)ds. \quad (41)$$

In the general case it is required to find the function α from the condition:

$$\min \|m - \int_{a(t)}^{t} \alpha y m ds\|\ (\alpha \in V) \quad (42)$$

where V is a reasonable set given, $\|\bullet\|$ is a sign of the measure of $\bullet$.

After the estimate of the functions α and β-types on the prehistory we can obtain their estimates on the given future period $[t^{\wedge}, T]$ by the well-known *prediction methods* (see, for example, [1, 3, and 7]). Dependence of α, β on the various internal and external factors can be found by *data processing methods* based on [1, 3, and 7].

Until now we have a deal with the *problem of identification* for *ES operating*. For *ES design*, all elements of the ES MM have to be given and the ES itself has to be design. This can be done by analogy with the general theory of synthesis or design of dynamic systems. That is, the ES MM have to be approximated by MM with sufficient accuracy such that they can be realized physically [1, 7] and/or biologically [19].

6. ON THE PROBLEM OF MM COMPLETENESS

We still need to improve our approach. In the theoretical plan the most important problem is that of completion of the given class of MM. The *problem of MM completeness* means the constriction of an additional reasonable set of relations such that it would be possible to find all the elements of ES in the future by knowing the prehistory of ES and predicting only separate elements or parameters of the exogenous nature. In other words, one of our objectives is to have to do with so-called "light box." This completion can be based at least on the following approaches:

i) system *de- and aggregation methods*;

ii) detailing of the structure of ES;

iii) using some extreme concepts.

6.1. Methods of ES de- and aggregation

The function of efficiency $\alpha(t,s)$ is usually increasing of s and decreasing of t. Anyone can accept many probable reasons for increasing and decreasing of efficiency for any particular ES. The simpler the system the more reliable the reasons.

Let

$$m_i(t) = \int^t_{a(t)} \sum^n_{j=1} \alpha_{ij}(t,s) y_{ij}(s) m_j(s) ds, \; i = 1,\dots, n;$$

$$m(t) = \sum^n_{i=1} m_i(t) = \int^t_{a(t)} \alpha(t,s) y(s) m(s) ds. \tag{43}$$

Then

$$\alpha(t,s) = [\sum^n_{ij=1} \alpha_{ij}(t,s) y_{ij}(s) m_j(s)] / [y(s) \sum^n_{i=1} m_i(s)]. \tag{44}$$

An example of *elementary indices* of efficiency is

$$\alpha_{ij}(t,s) = \alpha_{ij}\hat{}\exp[-d_i(t-s)],\ \alpha_{ij}\hat{},\ d_i > 0,\ i,j = 1,\dots, n, \tag{45}$$

where $\alpha_{ij}\hat{}$, and d_i are constants. Then

$$m_i'(t) + d_i m_i(t) = \sum{}^{n}_{j=1} \alpha_{ij}\hat{}\{y_{ij}(t)m_j(t) - y_{ij}(a(t))\exp[-d_i(t-a(t))]a'(t)\},\ i = 1,\dots, n. \tag{46}$$

Thus, the matter reduces to the solution of the linear system with the prehistory. In particular, if $y_{ij}(t) = y_{ij}$, where y_{ij} are constants, then the matter reduces to the solution of the linear stationary system with its prehistory.

Another example of *elementary indices* of efficiency is

$$\alpha_{ij}(t,s) = \alpha_{ij}\hat{}/y_{ij}(s)s/t,\ \alpha_{ij}\hat{} > 0,\ i,j = 1,\dots, n,\ t > 0. \tag{47}$$

Then the matter reduces to the solution of the linear stationary system with the prehistory

$$z_i'(t) = \sum{}^{n}_{j=1} \alpha_{ij}\hat{}\{z_j(t) - z_j(a(t))a'(t)\},\ z_j(t) = tm_j(t),\ i = 1,\dots, n. \tag{48}$$

6.2. Detailing of the structure of ES

Another way for the construction of MM of the functions α and β is detailing of the structure of ES. Decreasing of α and β of t depends on many factors in the theory of deterioration (aging). As the reasonable simple hypotheses we accept

$$\alpha_1(t,s) = \alpha_1(s)\exp[-c_1(t-s)],\ \beta_1(t,s) = \beta_1(s)\exp[-d_1(t-s)]; \tag{49}$$

or

$$\alpha_2(t,s) = \alpha_2(t-s)\exp(c_2 t),\ \beta_2(t,s) = \beta_2(t-s)\exp(d_2 t), \tag{50}$$

where c_i and d_i are given constants, $c_i \geq 0$.

Reasoning about the structure of ES as above results in

$$\alpha_\mu{}^{\nu}(t) = \int{}^{t}_{a(t)} \alpha_\mu(t,s)x(s)m(s)ds,\ \nu = 0, 1(\equiv');\ m(t) = \int{}^{t}_{a(t)} \alpha_\mu(t,s)y(s)m(s)ds,$$

$$M(t) = \int{}^{t}_{0} m(s)ds;\ \beta_\kappa(t) = \int{}^{t}_{a(t)} \alpha_\mu(t,s)z(s)m(s)ds,\ \mu, \kappa = 1, 2;$$

$$c(t) = \int{}^{t}_{a(t)} \beta_\kappa(t,s)[1-x(s)-y(s)-z(s)]m(s)ds;\ P(t) = \int{}^{t}_{a(t)} m(s)ds,$$

$$G(t) = \int{}^{a(t)}_{0} m(s)ds = M(t) - P(t);\ f(t) = \alpha_\mu{}^{\nu}(t) + m(t) + \beta_\kappa(t) + c(t);$$

$$0 \leq x, y, z, 1-x-y-z \leq 1;\ 0 \leq a(t) \leq t,\ T \geq t \geq t\hat{} \geq 0, \tag{51}$$

where $\alpha_\mu{}^\nu(t)$ is νth derivative of $\alpha_\mu(t)$, $\alpha_\mu{}^0 = \alpha_\mu$, $\alpha_\mu(t^\wedge) = \alpha_\mu{}^\wedge$, $x(s)m(s)$--$\alpha_\mu{}^\nu(t)$ and $z(s)m(s)$--$\beta_\kappa(t)$ are the channels of creation of the new technology quantities in the respective subsystem (for AES it is *Science* - *S* and for NES it is likely *Genome* - *G*), and the remainder notations previously noted.

Clearly $\alpha_\mu(t)$ is always increasing in the case $\nu = 1$, since $\alpha_\mu'(t) > 0$. For NES x, y, z, and G (or m) are usually given. For AES x, y, z, and P are usually given. The distributions x, y, and z usually are found from the solution of some extreme problems. For NES we can probably count on the fact that those extreme problems have been usually solved by nature during its evolution.

Note that for the simplicity here and below the *dimensionless equations* have been considered. In order that *dimensionless equations* have been introduced it is sufficient that any element $x(t)$ of the equations has been rewritten in the form

$$x(t) = x(t[t]/[t])[x]/[x] = x_1(t_1)[x],\ t_1 = t/[t], \tag{52}$$

where $[x]$ and $[t]$ are dimensions of x and t and x_1 is the respective dimensionless element. After this all dimensions and the subscript $_1$ can be omitted.

The natural generalization of MM (51) is

$$\alpha_\mu{}^\nu(t) = \int^t_{a1(t)} \alpha_\mu(t,s)x(s)m(s)ds,\ \nu = 0,\ 1;\ m(t) = \int^t_{a2(t)} \alpha_\mu(t,s)y(s)m(s)ds,$$

$$M(t) = \int^t_0 m(s)ds;\ \beta_\kappa(t) = \int^t_{a3(t)} \alpha_\mu(t,s)z(s)m(s)ds,\ \mu,\ \kappa = 1, 2;$$

$$c(t) = \int^t_{a4(t)} \beta_\kappa(t,s)[1\text{-}x(s)\text{-}y(s)\text{-}z(s)]m(s)ds;\ P_i(t) = \int^t_{ai(t)} m_i(s)ds,\ G_i = \int^{ai(t)}_0 m_i(s)ds,$$

$$i = 1, 2, 3, 4;\ m_1 = xm,\ m_2 = ym,\ m_3 = zm,\ m_4 = (1\text{-}x\text{-}y\text{-}z)m;$$

$$P(t) = P_1(t)+P_2(t)+P_3(t)+P_4(t),\ G(t) = M(t)\text{-}P(t);\ f(t) = \alpha_\mu{}^\nu(t)+m(t)+\beta_\kappa(t)+c(t);$$

$$0 \le x, y, z, 1\text{-}x\text{-}y\text{-}z \le 1;\ 0 \le ai(t) \le t,\ i = 1, 2, 3, 4;\ T \ge t \ge t^\wedge \ge 0, \tag{53}$$

For MM (53) x, y, z, and P_i (or m_i) are given.

We can generalize the models (51), (53) in the way similar to (1) considering all the elements in (51), (53) as appropriate functional matrices (1 is the unit matrix), interpreting integrals in the sense of Stieltjes and so on (see s. 2 for details).

Let us note two particular cases. The first is the case when the *extractive subsystem* (*E*) is introduced besides (53). Then

$$\alpha_\mu{}^\nu(t) = \int^t_{a1(t)} \alpha_\mu(t,s)x(s)m(s)ds,\ \nu = 0,\ 1;\ m(t) = \int^t_{a2(t)} \alpha_\mu(t,s)y(s)m(s)ds,$$

$$M(t) = \int^{t}_{0} m(s)ds;\ \beta_\kappa(t) = \int^{t}_{a3(t)} \alpha_\mu(t,s)z(s)m(s)ds,\ \mu,\ \kappa = 1, 2;$$

$$\gamma_\kappa(t) = \int^{t}_{a4(t)} \alpha_\mu(t,s)e(s)m(s)ds,\ \gamma_1(t,s) = \gamma_1(s)\exp(e_1(t-s)),\ \gamma_2(t,s) = \gamma_2(t-s)\exp(e_2 t);$$

$$f(t) = \alpha_\mu{}^{\nu}(t) + m(t) + \beta_\kappa(t) + \gamma_\kappa(t) + \mathrm{cost}(f) + c(t) = \int^{t}_{a5(t)} \gamma_\kappa(t,s)g(s)m(s)ds;$$

$$c(t) = \int^{t}_{a6(t)} \beta_\kappa(t,s)[1-x(s) - y(s) - z(s) - e(s) - g(s)]m(s)ds;\ P_i(t) = \int^{t}_{ai(t)} m_i(s)ds,$$

$$G_i = \int^{ai(t)}_{0} m_i(s)ds,\ i=1, \ldots, 6;\ m_1 = xm,\ m_2 = ym,\ m_3 = zm,\ m_4 = em,\ m_5 = gm,$$

$$m_6 = (1-x-y-z-e-g)m;\ P(t) = P_1(t)+P_2(t)+P_3(t)+P_4(t)+P_5(t)+P_6(t),\ G(t) = M(t)-P(t);$$

$$0 \le x, y, z, e, g, 1-x-y-z-e-g \le 1;\ 0 \le ai(t) \le t,\ i = 1, \ldots, 6;\ T \ge t \ge t^\wedge \ge 0, \tag{54}$$

where cost(f) is the cost or the necessary resource for f extraction, which should be naturally less than f itself; $\gamma_\kappa(t,s)$ is the new technology for the subsystem E; the remaining notations have the same sense.

The second is the case when the *recreation subsystem* (R) is introduced:

$$\alpha_\mu{}^{\nu}(t) = \int^{t}_{a1(t)} \alpha_\mu(t,s)x(s)m(s)ds,\ \nu = 0, 1;\ m(t) = \int^{t}_{a2(t)} \alpha_\mu(t,s)y(s)m(s)ds,$$

$$M(t) = \int^{t}_{0} m(s)ds;\ \beta_\kappa(t) = \int^{t}_{a3(t)} \alpha_\mu(t,s)z(s)m(s)ds;\ \gamma_\kappa(t) = \int^{t}_{a4(t)} \alpha_\mu(t,s)e(s)m(s)ds;$$

$$\delta_\kappa(t) = \int^{t}_{a5(t)} \alpha_\mu(t,s)r(s)m(s)ds,\ \delta_1(t,s) = \delta_1(s)\exp(r_1(t-s)),\ \delta_2(t,s) = \delta_2(t-s)\exp(r_2 t);$$

$$N(t) = \int^{t}_{a6(t)} \delta_\kappa(t,s)w(s)m(s)ds;\ f(t) = \alpha_\mu{}^{\nu}(t) + m(t) + \beta_\kappa(t) + \gamma_\kappa(t) + \delta_\kappa(t) + P(t) +$$

$$\mathrm{cost}(f) + c(t) = \int^{t}_{a7(t)} \gamma_\kappa(t,s)g(s)m(s)ds;\ c(t) = \int^{t}_{a8(t)} \beta_\kappa(t,s)[1-x(s) - y(s) - z(s) - e(s)$$

$$- g(s) - r(s) - w(s)]m(s)ds,\ \mu,\ \kappa = 1, 2;\ P_i(t) = \int^{t}_{ai(t)} m_i(s)ds,\ G_i = \int^{ai(t)}_{0} m_i(s)ds,$$

$$i = 1, \ldots, 8;\ m_1 = xm,\ m_2 = ym,\ m_3 = zm,\ m_4 = em,\ m_5 = gm,\ m_6 = rm,\ m_7 = wm,\ m_8 =$$

$$(1-x-y-z-e-g-r-w)m;\ P(t) = P_1(t) + P_2(t) + P_3(t) + P_4(t) + P_5(t) + P_6(t) + P_7(t) + P_8(t),$$

$$G(t) = M(t) - P(t);\ 0 \le x, y, z, e, g, r, w, 1-x-y-z-e-g-r-w \le 1;\ 0 \le ai(t) \le t,$$

$$i = 1, \ldots, 8;\ T \ge t \ge t^\wedge \ge 0, \tag{55}$$

where $N(t)$ is the number of specialists created in the subsystem R, $\delta_\kappa(t,s)$ is the new technology for the subsystem R, the remaining notations have the usual sense. We can consider different cases here: $N(t) < P(t)$, $N(t) > P(t)$, and $N(t) = P(t)$.

The idea of *factorization* of the matrix $K(t,s)$ of the pulse transition functions (compare (1) and (23)) can be also strengthened. First, we should distinguish the stationary factor $\alpha^*(t-s)$ of the function $\alpha(t,s)$-type and the non-stationary factor $\alpha(t,s)/\alpha^*(t-s)$. The last one in many important cases can be represented in the form

$$\alpha(t,s)/\alpha^*(t-s) = u(s)v(t)v^{-1}(s), \tag{56}$$

where $u(s)$ is a control factor corresponding to the non-stationary amplified or relaxed unit, $v(t)v^{-1}(s)$ is another non-stationary control factor signifying the easiest way for changing of values of ES internal element α and m without changing of ES structure. The last statement follows from the fact that after substitution of (56) for (1) or (51)-(55) and replacement of m by vm we will have actually the same models.

On the whole we have the following *factorization*

$$K(t,s) = \alpha(t,s)\lambda(t,s)\alpha^*(t-s)u(s)v(t)v^{-1}(s)y(s), \tag{57}$$

where $\alpha(t,s)$ means remained non-stationary factors including the delay unit, non-stationary sluggishness, etc. We hope that with a certain permissible degree of accuracy typically we can put $\alpha(t,s) = 1$ in (57).

It should be noted that for NES, in particular for an organism, the factor $u(t)$ is probably connected with a reaction of the endocrine subsystem of the organism and realized with the help of hormones, the factor $v(t)v^{-1}(s)$ is connected with the nervous subsystem and realized by different kind of nerves, and the factor $\alpha^*(t-s)$ with the *genome* of the organism and realized with the help of the code in DNA of cells.

6.3. Using some extreme concepts

Let us return to the models (5)-(7) and consider some *extreme concepts*. For AES, we suppose that $\alpha(t,s)$ has to maximize $m(t)$ under given $P(t)$, constants a, c, and

$$P''(t) > 0,\ P''(t) \geq 0;\ 0 \leq a(t) \leq at,\ a < 1;\ \alpha(t,s) \leq \alpha(s)\exp[-c(t-s)],\ 0 \leq y(t) \leq 1, \tag{58}$$

Theorem 1. *The desired* $y(t)$, $a(t)$, $m(t)$, *and* $\alpha(t,s)$ *are*

$$y(t) = 1,\ a(t) = at,\ m(t) = \sum{}_{i=0}^{\infty} P'(a^i t)a^i,$$

$$\alpha(t,s) = \exp(-ct)\sum{}_{i,j=0}^{\infty} [P''(a^{i+j}s)a^{2i+j} + cP'(a^{i+j}s)a^{i+j}]\exp(cas)/m(s). \tag{59}$$

Proof. From the conditions of the theorem it follows that the desired $m(t)$ and $\alpha(s)$

are the solutions of the equations in (5), where $y(s) = 1$, $a(t) = at$, and $\alpha(t,s) = \alpha(s)\exp[-c(t-s)]$. Solving these equations, we get (61). □

For NES, we suppose that $\alpha(t,s)$ has to maximize $P(t)$ under given $m(t)$, $a(t)$, constants b, d, and the conditions

$$bt \leq a(t) \leq t,\ 0 \leq b < 1;\ \alpha(t,s) \leq \alpha(s)\exp[-d(t-s)],\ d \geq 0,\ 0 \leq y(t) \leq 1, \tag{60}$$

Theorem 2. *The desired $y(t)$, $a(t)$, $\alpha(t,s)$, and $P(t)$ are $y(t) = 1$, $a(t) = bt$,*

$$\alpha(t,s) = \exp(-dt)\textstyle\sum_{i=0}^{\infty} m'(b^i s)b^i\exp(db^i s)/m(s),\ P(t) = \int_{bt}^{t} m(s)ds. \tag{61}$$

The proof of (61) is similar to the proof of (59).

Suppose yet, for NES, that in the MM (5)-(7) α, β have to maximize $P(t)$ under given $f(t)$, $c^\wedge(t)$, and the restrictions

$$bt \leq a(t) \leq t,\ 0 \leq b < 1;\ \beta = \alpha \leq \alpha(s)\exp[-d(t-s)], d \geq 0,\ 0 \leq y(t) \leq 1,\ c(t) \geq c^\wedge(t), \tag{62}$$

Theorem 3. *The desired $a(t)$, $y(s)$, and $\alpha(t,s)$ are*

$$a(t) = bt,\ y(s) = A/(A+B),\ \alpha(t,s) = (A+B)\exp[-d(t-s)],\ A = q'(s) + \textstyle\sum_{i=0}^{\infty} b^i q'(b^i s)/q(s),$$

$$B = r'(s) + \textstyle\sum_{i=0}^{\infty} b^i r'(b^i s)/q(s),\ q(s) = \exp(ds)[f(s) - c^\wedge(s)],\ r(s) = \exp(ds)c^\wedge(s). \tag{63}$$

Proof. It is obvious that maximal $P(t) = \int_{bt}^{t} m(s)ds$, $m(t) = f(t) - c^\wedge(t)$. So, we have

$$q(t) = \int_{bt}^{t} A(s)q(s)ds,\ r(t) = \int_{bt}^{t} B(s)q(s)ds,\ \alpha(s)y(s) = A,\ \alpha(s)[1-y(s)] = B.$$

Solving this system, we obtain (63). □

7. STATEMENT OF THE MODELING PROBLEMS

We can solve at least the following *modeling problems*:

A. The *prediction* of ES functioning [1, 3, 7].

B. The solution of problems of the type "*If*. . ., *then* . . . " and/or "In order to . . . , it is sufficient . . . ."

C. The solution of *optimization problems*.

The examples of modeling problems B. can be

B.1. If efficiency indices, α and β, increase by 2 percent per year, then how fast does $m(t)$ increase for certain meanings of the distribution function $y(s)$?

B.2. In order for the NES functioning bio-mass $P(t)$ to decrease, that is $P'(t) < 0$, it is sufficient that $y(s)$ be assigned a certain value.

Examples of *optimization problems* are

C.1. The external ES function maximization problem:

$$I = I1 = \int_{t^\wedge}^{T} c(t)dt = \int_{t^\wedge}^{T} \{\int_{a(t)}^{t} \beta(t,s)[1-y(s)]ds\}dt = \max \quad (64)$$

using x, y, z provided that the relations (1) or (51), (53) are observed and the accumulated cost $E \le E^+$, where E^+ is given.

C2. The accumulated profit maximization problem over x, y, and z:

$$I2 = Pr(T) = \max \quad (65)$$

C.3. The average functioning ES recourse minimization problem over x, y, z:

$$I3 = \int_{t^\wedge}^{T} P(t)dt = \int_{t^\wedge}^{T} \{\int_{a(t)}^{t} m(s)ds\}dt = \min \quad (66)$$

provided that the restrictions on $c(t)$: $c(t) \ge \hat{c}(t)$, where $\hat{c}(t)$ is given, are observed.

C.4. The problem of ES high speed of operation over x, y, z:

$$I4 = T - t^\wedge = \min \quad (67)$$

provided that the conditions $\int_{t^\wedge}^{T} c(t)dt \in C^*$ and $P(t) \in P^*$, where the sets C^* and P^* are given, are observed.

C.5. ES reliability maximization problem over x, y, z:

$$I5 = \min[m^+(t) - m(t), m(t) - m^-(t), c^+(t) - c(t), c(t) - c^-(t)] = \max \quad (68)$$

provided that the inequalities $m^-(t) \le m(t) \le m^+(t)$ and $c^-(t) \le c(t) \le c^+(t)$ with the assigned functions $m^{\pm}(t)$ and $c^{\pm}(t)$, are observed.

C.6. The problem of out-of-date ES products minimization over x, y, z:

$$I6 = \int_{t^\wedge}^{T} G(t)dt = \int_{t^\wedge}^{T} [\int_{0}^{a(t)} m(s)ds]dt = \min \quad (69)$$

provided that (1) or (51), (53) are observed.

C.7. The external ES function minimization problem over x, y, and z:

$$I7 = \int_{t^\wedge}^{T} c(t)dt = \min \quad (70)$$

C.8. The active ES "life" maximization problem over x, y, z:

$$I8 = T - t^\wedge = \max \quad (71)$$

provided that the restriction $c(t) \ge c^0(t)$, where $c^0(t)$ is given, are observed.

In the case when ES is the economy as a whole, (64) is the maximization problem for the number of consumer goods and services during the design period $T - t^{\wedge}$; (66) is the minimization problem for the average labor inputs. For NES, the problems (70) and (71) can make more sense than (64) and (66). If ES are populations of viruses in human organism, e.g. the population of HIV, $c(t)$ is the aggressive factor, then just the problem (70) makes sense. If ES are human beings, attractiveness of (71) is obvious.

By analogy, we can also consider the problems (64)-(71) when optimization can be obtained at the expense of not only the distributions x, y, z, but also the distributions e, g, r, w, and the control factors u and v (see (51)-(57)).

Considering the problems of design or synthesis of AES, we can obtain AES optimization at the expense of the functions α and β-types themselves.

Bearing in mind the opportunity of the biotech revolution [19], we can also obtain NES modeling and optimization at the expense of the functions α and β-types.

REFERENCES

1. Astrom K.J., Eykhoff P., System Identification - A Survey, Automatica, 7, 1971, 123-167.
2. Baglay R.D., Smirnov K.K., Handling Two Dimensional Signals on Computers, ZVMiMF, 15, N 1, 1975, 241-247 (in Russian).
3. Bethea R.M., Duran B.S., Boullion T.L., Statistical Methods for Engineers and Scientists, Decker, 1975, 583 pp.
4. Girlin S.K., Ivanov V.V., Modeling of Developing Systems Interaction, DAN Ukr.SSR, N 1, 1986, 58-60 (in Ukrainian).
5. Glushkov V.M., On a Class of Dynamic Macroeconomic Models, Upravlyayushchie. Sistemy and Mashiny, 2, 1977, 3-6 (in Russian).
6. Glushkov V.M., Ivanov V.V., JanenkoV.M., One Class of Nonlinear Dynamic Models and Its Applications, Physica, 2D, 1981, 61-72.
7. Glushkov V.M., Ivanov V.V., Janenko V.M., Developing Systems Modeling, M.: Nauka, 1983, 352 pp. (in Russian).
8. Glushkov V.M., Pshenichnyj B.N., On Mathematical Models of Economic Growth, Cybernetics, 13, N 4, 1977, 1-6.
9. Hritonenko N., Yatsenko Yu., Modeling and Optimization of the Lifetime of Technologies, Kluwer Academic Publishers, 1996, xxxvi + 250 pp.
10. Ivanov V.V., Systems Development Simulation Problems and C. Caratheodory's Concepts, in the book: Constantin Caratheodory: An International Tribute, World Science., 1, 1991, 501-526.
11. Ivanov V.V., Yatsenko Yu.P., Galiev U.E., Comparison of Some Integral Macroeconomic Models, Avtomatika, 4, 1986, 47-53 (in Russian).
12. Kantorovich L.V., Gor'kov L.I., On Functional Equations under Analysis of One-Product Economic Model, DAN SSSR, 129, N 4, 1959, 732-736 (in Russian).
13. Kantorovich L.V., Zhiyanov V.I., One-Product Dynamic Model of Economy Considering Variation of Fund Structure under Technical Progress, DAN SSSR, 211, N 6, 1973, 1280-1283 (in Russian).
14. Krasnoshchokov P.S., Petrov A.A., Principles of Model Construction, Moscow Univ., 1983, 264 pp. (in Russian).
15. Law, Dornald's Illustrated Medical Dictionary, W. B. Saunders Co., 1988, 899- 902.
16. Malcomson J.M., Replacement and the Rental Value of Capital Equipment Subject to Obsolescence, Journal of Economic Theory, 10, 1975, 24-41.
17. Petrov A.A., Pospelov I.G., Systematic Analysis of Developing Economy, Izv. AN SSSR, Techn. kibernetika, 1979, N 2, 18-27; N 3, 28-38; N 4, 11-23; N 5, 13-24 (in Russian).
18. Poluectov R.A., Pykh Yu.A., Shvytov I.A., Dynamic Model of Ecological Systems, L.: Gidrometioizdat, 1980, 288 pp. (in Russian).
19. Rifkin J., The Biotech Century, Penguin Putnam Inc., 1998, 272 pp.
20. Sharpe F.R., Lotka A.J., A Problem in Age-Distribution, Philos. Mag., 21, 1911, 435-438.
21. Smith H.L., Monotone Dynamical Systems: An Introduction to the Theory of Competitive and Cooperative Systems, Mathematical Surveys and Monographs, vol. 41, AMS, 1995, x+174 pp.
22. Solow R., Investment and Technical Progress, Stanford Univ. Press, 1960, 89-104.
23. Thermodynamic Principles, McGraw-Hill Encyclopedia of Science & Technology, Vol. 13, 1982, 645-650.
24. Webb G., Theory of Nonlinear Age Dependant Population, Decker, 1985, 294 pp.

CHAPTER 3

INVESTIGATION OF EQUATIONS

Here are special simple cases of MM of ES and examples of analytic solutions. The conditions for existence and uniqueness of solutions for the systems of equations and also stability of solutions are established. Some cases of non-uniqueness of solutions are considered. Cases of oscillating and wave solutions are found. Asymptotic properties of solutions are investigated.

1. EXAMPLES OF ANALYTIC SOLUTIONS

Let in (1, Ch. 2) $\alpha\lambda = A(s) \geq 0$. Then the first equation in (1, Ch. 2) is a linear differential equation (or a system of equations) of the form

$$m'(t) = A(t)y(t)m(t) + \Delta m'(t),\ m(t^\wedge) = m^\wedge. \tag{1}$$

If $\alpha\lambda = A(s)m(s)$, $A(s) \geq 0$, then we have a *bilinear differential equation* of the form

$$m'(t) = A(t)y(t)m^2(t) + \Delta m'(t),\ m(t^\wedge) = m^\wedge, \tag{2}$$

or a system of such equations

$$m'(t) = A(t)y(t)m(t)m^T(t) + \Delta m'(t),\ m(t^\wedge) = m^\wedge, \tag{2'}$$

where $m(t)$ is a column vector and $m^T(t)$ is a row vector. The solutions of the equations (1)-(2') can be obtained by standard methods [1, 5, 9].

Let us consider now the solutions of the system of equations (5, 6, Ch2) under the conditions: for AES α, β, y, and P are given and also α, $y^\wedge$, and P are constants, and for NES f is given and α, β, and y are given constants, $\alpha y + \beta(1-y) \neq 0$.

It is easy to see that for $t \in [t^\wedge, a^{-1}(t^\wedge)]$, $a(a^{-1}(t^\wedge)) = t^\wedge$, for AES

$$m(t) = \alpha y^\wedge \int_{a(t)}^{t^\wedge} m^\wedge(s)ds + \alpha \int_{t^\wedge}^{t} y m(s)ds,\ P(t) = \int_{a(t)}^{t^\wedge} m^\wedge(s)ds + \int_{t^\wedge}^{t} m(s)ds, \tag{3}$$

from which

$$m(t) = m^\wedge \exp[\alpha(y-y^\wedge)(t-t^\wedge)],\ M(t) = M(t^\wedge) + m^\wedge\{\exp[\alpha(y-y^\wedge)(t-t^\wedge)] - 1\}/[\alpha(y-y^\wedge)],$$

$$a(t) = M^{\wedge-1}[M(t) - M(t^\wedge)],\ M^\wedge(t) = \int_0^t m^\wedge(s)ds,\ M(t^\wedge) = M^\wedge(t^\wedge),$$

$$c(t) = \beta\^(1-y\^)P + \int_{t\^}^{t} [\beta(1-y) - \beta\^(1-y\^)]m(s)ds, \quad (4)$$

where $M\^^{-1}$ is the *reciprocal* function or *operator* for $M\^$.

For NES $f = m + c = [\alpha y + \beta(1-y)]P(t)$ and therefore

$$P(t) = f(t)/[\alpha y + \beta(1-y)],\ m(t) = \alpha y P(t),\ c(t) = \beta(1-y)P(t). \quad (5)$$

For the interaction of two ES according to Fig. 3, Ch.1, let $\alpha 2, y2, y2\^, P2$ be given constants, and let $m2'' = qm2$ (see (22, Ch.2)), q be also constants. Then $m2'(t) = [\alpha 2(y2 - y2\^) - q]m2$ and therefore

$$m2(t) = m2\^\exp\{[\alpha 2(y2 - y2\^) - q](t-t\^)\}. \quad (6)$$

From (6) it follows that $m2(t)$ is decreased under

$$q > \alpha 2(y2 - y2\^). \quad (7)$$

The inequality (7) can be called the *condition of irreproducible loss* for the 2nd ES.

For the system (51, Ch. 2) we assume that x, y, and P are positive constants. Then in the case $\nu = \mu = 1$, $c_1 = 0$ we have

$$\alpha(t) = \alpha(t\^) + x/yM,\ M'' = xPM',\ M(t) - M(a(t)) = P,\ M' = m, \quad (8)$$

from which

$$\alpha(t) = \alpha(t\^) + x/y\{\exp[xP(t-t\^)] - 1\}/xP,\ m = m\^\exp[xP(t-t\^)],$$

$$a(t) = M^{-1}[M(t) - P] = t + 1/xP\ln\{1-xP\exp[-xP(t-t\^)]\}. \quad (9)$$

For arbitrary $P(t)$, $P(t) > 0$, and positive constants x, y using the relations $M - M(a) = P$, $m - m(a)a' = P'$, we reduce the case $\nu = \mu = 1$, $c_1 = 0$ to the second order linear differential equation of the form

$$M'' - xPM' - xP'M = (y\alpha(t\^) - xP)P' \quad (10)$$

the solutions and properties of which have been investigated in details by many mathematicians (see, for example, [5, 8, and 9]).

2. EXISTENCE AND UNIQUENESS OF SOLUTIONS

Let us consider MM (1, Ch. 2).

Theorem 1. *If $\alpha\lambda$, $\beta\mu$, y, and Δm are given nonnegative integrable bound functions, then the system of equations in* (1, Ch. 2) *concerning m, c, P, G, and f has the unique nonnegative bound solution, and also m(t) has the form*

$$m(t) = u(t) + \int_{t^\wedge}^{t} g(t,s)u(s)ds,\ t \geq t^\wedge,\ g(t,s) = \sum_{k=0}^{\infty} g_{k+1}(t,s),\ g_1(t,s) = \alpha(t,s)\lambda(t,s)y(s),$$

$$g_{k+1}(t,s) = \int_{s}^{t} g_1(t,z)g_k(z,s)dz,\ k = 1,2,...,\ u(t) = \int_{0}^{t^\wedge} g_1^\wedge(t,s)m^\wedge(s)ds + \Delta m(t), \tag{11}$$

and c, P, G, and f are obtained by the respective replacement of m(t).

This result is a corollary of the well-known theory of Volterra linear integral equations [1]. The similar theorem and the same relations (11) take place when $\alpha\lambda, \beta\mu$, and y are given matrixes and m and c are desired vector functions.

Let us say

$$m(f,[a,b]) = \min |f(t)|\ (a \leq t \leq b),\ M(f,\ [a,b]) = \max |f(t)|\ (a \leq t \leq b) \tag{12}$$

for any function f and any segment $[a,b]$. And also

$$m(f,[0,t]) = m(f,t),\ M(f,[0,t]) = M(f,t);\ m(f,[0,T]) = m(f),\ M(f,[0,T]) = M(f). \tag{13}$$

Theorem 2. *Let in the systems* (5, 6, Ch. 2) *the elements αy, $\beta(1-y)$, P, and $m^\wedge(t)$, $t \in [0,t^\wedge]$, be given positive functions, and here*

$$m^\wedge \in C([0,t^\wedge]),\ \alpha y,\ \beta(1-y) \in C^1([0,T] \times [0,T]),\ P \in C^1([t^\wedge,T]), \tag{14}$$

where C and C^1 are continuous and continuously differentiable spaces of functions respectively. Then in order that these systems have the unique nonnegative solutions $m, a, c, f \in C^1([t^\wedge,T])$, $0 \leq t^\wedge < T < \infty$, it is sufficient that the following condition holds:

$$0 \leq P(t) \leq m^\wedge(t^\wedge) + \int_{t^\wedge}^{t} m(\alpha y,s)P(s)ds,\ 0 \leq t^\wedge \leq t \leq T < \infty. \tag{15}$$

In addition, the desired m(t) and a(t) can be found by the iterative process

$$m_{i+1}(t) = \int_{a_i(t)}^{t} \alpha(t,s)y(s)m_i(s)ds,\ P(t) = \int_{a_i(t)}^{t} m_i(s)ds,\ m_i(s) = m^\wedge(s),\ s \in [0,t^\wedge],$$

$$i = 1, 2, \dots,\ m_1(t) = m(\alpha y,t)P(t),\ t^\wedge \leq t \leq T, \tag{16}$$

which converges at the rate exceeding that of geometrical progressions.

Proof. On the strength of (15), (16) the equation

$$\int_{ai(t)}^{t} m_i(s)ds = P(t) \tag{17}$$

has a solution. Indeed, if $\int_{ai-1(t)}^{t} m_{i-1}(s)ds = P(t)$, then $m_i(t) \geq m(\alpha y,t)P(t)$ and therefore

$$\int_0^t m_i(s)ds = m^\wedge(t^\wedge) + \int_{t^\wedge}^t m_i(s)ds \geq P(t),\ \int_t^t m_i(s)ds = 0 \leq P(t). \tag{18}$$

But $\int_{a1(t)}^t m_1(s)ds = P(t)$ since $\int_0^t m_1(s)ds = m^\wedge(t^\wedge) + \int_{t^\wedge}^t m(\alpha y,s)P(s)ds \geq P(t)$. Further,

$$m_2(t) - m_1(t) = \int_{a1^\wedge}^t [\alpha y - m(\alpha y,t)]m_1 ds \leq M(\alpha y)P(t)(t-t^\wedge),\ a1^\wedge = \max(t^\wedge, a1), \tag{19}$$

And also for $i \geq 2$

$$|m_{i+1} - m_i| \leq M(\alpha y)[\int_{ai-1^\wedge}^t |m_i - m_{i-1}|ds + |\int_{ai}^{ai-1} m_i ds|],\ \int_{ai-1}^t m_{i-1}ds = \int_{ai}^t m_i ds, \tag{20}$$

from which $|\int_{ai-1}^{ai} m_i ds| \leq \int_{ai-1^\wedge}^t |m_i - m_{i-1}|ds,\ ai-1^\wedge = \max(t^\wedge, ai-1)$. So,

$$|m_i - m_{i+1}| \leq 2M(\alpha y)\int_{ai-1^\wedge}^t |m_{i-1} - m_i|ds \leq P(t)[2M(\alpha y)(t-t^\wedge)]^i/i!,$$

$$2M(\alpha y)m(\alpha y)m(P)|ai - ai-1| \leq P(t)[2M(\alpha y)(t-t^\wedge)]^i/i!. \tag{21}$$

From (15), (20), and (21) it follows rather easily the validity of Theorem 2. □

Note that the condition (7, Ch. 2), which is $a'(t) \geq 0$, $t \geq 0$, holds iff $m(t) \geq P'(t)$. It follows from $P'(t) = m(t) - m(a)a'$ and $m(a) > 0$. For $a'(t) \geq 0$, $t \geq 0$, to hold it is sufficient that $m(\alpha y) \geq P'(t)$. If $m(\alpha) \geq P'(t)$, then there exists y, $0 < y \leq 1$, for which $m(\alpha y) \geq P'(t)$. If $P'(t) \leq 0$, then for any y, $0 \leq y \leq 1$, $m(t) > P'(t)$.

The initial results about the existence and uniqueness of the solutions of the equations under consideration have been related to the case $t^\wedge > 0$. The present, more general formulation of Theorem 2, is based on the paper [3], considering originating ES. If $t^\wedge = 0$, we will have obviously $m(t) = a(t) = 0$. It will not be the case if we consider the system similar to the (5, 6, Ch. 2) when Δm, $\Delta c \neq 0$ or we assume $\alpha y = \delta(t-s)$, t, $s \sim 0$, where δ- Dirac delta-function. The proof itself has been different in the case $t^\wedge > 0$. Here, briefly, is an outline of the main idea of that proof.

Let us introduce the relations

$$m(t) = \Psi(m)(t) = \Phi\{M^{\wedge -1}[M(t)-P(t)],t\} + \int_{t^\wedge}^t \alpha(t,s)y(s)m(s)ds,\ M(t) = \int_0^t m(s)ds,$$

$$a(t) = M^{\wedge -1}[M(t)-P(t)],\ \Phi(a,t) = \int_a^{t^\wedge} \alpha(t,s)y^\wedge(s)m^\wedge(s)ds,\ M^\wedge(t) = \int_0^t m^\wedge(s)ds, \tag{22}$$

Since Φ and $M^\wedge$ and hence $M^{\wedge -1}$ are continuously differentiable monotone increasing functions, the *Volterra-type operator* Ψ effects contracted mapping of $C([t^\wedge, t^*])$ into itself, where $t^* > t^\wedge$, and has the property: $a(t) \leq t^\wedge$, $t \in [t^\wedge, t^*]$. By virtue of (22) and $M(t^\wedge) - P(t^\wedge) = 0$ such t^* will always be found. It follows that the unique positive solution of the equation $m = \Psi(m)$ on the segment $[t^\wedge, t^*]$ can be found by the *method of simple iteration*

$$m_{i+1} = \Psi(m_i),\ i = 0, 1, \ldots;\ m_0 = 0;\ m = \Sigma^{\infty}_{i=0} (m_{i+1} - m_i). \tag{23}$$

On the strength of the Weierstrass' theorem, we have from (23) that m is the continuous function. But then it follows from (22) that $m, a, c, f \in C^1([t^\wedge, t^*])$. The first instant t^* for which $a(t^*) = t^\wedge$, is until now unknown and found in the process of solving. If $a(t) \leq t^\wedge$ *for* $t^\wedge \leq t \leq T$, then it follows that the constructed solution is valid over the whole given segment and the theorem is proved. In the similar way, it can be proved that for $T > t^*$ the problem has the unique positive solution over $[t^*, t_1]$, $a(t_1) = t^*$, then over $[t_1, t_2]$, $a(t_2) = t_1$, and so on until the solution over the whole segment $[t^\wedge, T]$ has been obtained. The latter are possible since by virtue of the conditions of the theorem it is easy to show that

$$t - a(t) \geq d > 0,\ t \in [t^\wedge, T], \tag{24}$$

where d is a certain constant.

<u>Theorem 3</u>. *Under the assumption of Theorem* 2 *and under the condition*

$$0 < f(t) \leq \int_0^{t^\wedge} [\alpha y + \beta(1-y)] m^\wedge ds + \int_{t^\wedge}^{t} f(s)[\alpha y + \beta(1-y)][1 + M(\beta(1-y), s)/m(\alpha y, s)]^{-1} ds \tag{25}$$

instead of (15), *where f is a given positive continuously differentiable function, the system of equations* (5, 6, Ch. 2) *has the unique nonnegative differentiable solution m, a, c, P for* $0 \leq t^\wedge \leq T < \infty$. *In addition, m(t) and a(t) can be found by the iterative process*

$$f(t) = m_{i+1}(t) + c_{i+1}(t),\ c_{i+1}(t) = \int_{ai(t)}^{t} \beta(t,s)[1 - y(s)] m_i(s) ds,\ m_i(s) = m^\wedge(s),\ s \in [0, t^\wedge],$$

$$i = 1, 2, \ldots,\ m_1(t) = [1 + M(\beta(1-y), t)/m(\alpha y, t)]^{-1},\ t \in [t^\wedge, T], \tag{26}$$

m_{i+1} *is expressed in the terms of* m_i *the same way as in* (16), *which converges at the rate exceeding that of geometrical progressions.*

<u>Proof</u>. The equation (26) with respect to $ai(t)$ has the unique solution. Indeed,

$$\int_0^t [\alpha y+\beta(1-y)]m_i ds = \int_0^{t^\wedge} [\alpha y+\beta(1-y)]m^\wedge ds + \int_{t^\wedge}^t [\alpha y+\beta(1-y)]m_i(s)ds$$

$$\geq \int_0^{t^\wedge} [\alpha y+\beta(1-y)]m^\wedge ds + \int_{t^\wedge}^t f(s)[\alpha y+\beta(1-y)][1+M(\beta(1-y,s))/m(\alpha y,s)]^{-1}ds \geq f, \quad (27)$$

since $c_i \leq m_i M(\beta(1-y),t)/m(\alpha y,t)$, $m_i + c_i = f$, and $\int_0^0 [\alpha y+\beta(1-y)]m_i ds = 0 \leq f$. Then on the strength of (25) and (26) we can receive the estimates similar to (21) by the similar way to the proof of Theorem 2. □

In a similar way, we can prove the theorem when instead P or f the function c is given.

If f is given, then the condition $a'(t) \geq 0$, $t \geq 0$, takes place iff

$$\{\alpha(t,t)y(t)+\beta(t,t)[1-y(t)]\}m(t) + \int_{a(t)}^t \{\alpha'_t(t,s)y(s)+\beta'_t(t,s)[1-y(s)]\}m(s)ds \geq f'(t). \quad (28)$$

It follows from

$$f(t) = \int_{a(t)}^t \{\alpha(t,s)y(s) + \beta(t,s)[1-y(s)]\}m(s)ds. \quad (29)$$

For $a'(t) \geq 0$, $t \geq 0$, to hold it is sufficient that

$$\alpha(t,t)y(t)+\beta(t,t)[1-y(t)] \geq \{f'(t)/f(t)\text{-min}\,[m(\alpha'_t/\alpha),m(\beta'_t/\beta)]\}[1+M(\beta(1-y)/m(\alpha y)]. \quad (30)$$

If c is given, then the condition $a'(t) \geq 0$, $t \geq 0$, takes place iff

$$\beta(t,t)[1-y(t)]m(t) + \int_{a(t)}^t \beta'_t(t,s)[1-y(s)]m(s)ds \geq c'. \quad (31)$$

And we can receive the sufficient condition that is similar to the above.

Theorem 4. *Under the assumptions of Theorem 2 and under the condition*

$$0 < P_1(t) \leq \int_0^{t^\wedge} ym^\wedge ds + m(\alpha)\int_{t^\wedge}^t yP_1(s)ds,$$

$$0 < P_2(t) \leq \int_0^{t^\wedge} (1-y)m^\wedge ds + m(\alpha)\int_{t^\wedge}^t (1-y)P_1(s)ds, \quad (32)$$

where P_1 and P_2 are given differentiable functions, the system of equations (9, Ch.2) *has the unique nonnegative differentiable solution m, a and b, which can be found by the iterative process*

$$P_1(t) = \int_{ai(t)}^t y(s)m_i(s)ds,\ P_2(t) = \int_{bi(t)}^t [1-y(s)]m_i(s)ds,$$

$$m_i(s) = m^\wedge(s),\ s \in [0,t^\wedge],\ i = 1, 2, \ldots,\ m_1(t) = m(\alpha)P_1(t),\ t \in [t^\wedge,T], \quad (33)$$

$m_{i+1}(t)$ are expressed in terms of $m_i(t)$ the same way as in (16), *converging faster than*

geometrical series.

Proof. It is similar to the proof of Theorem 2. □

Now, the condition $a'(t) \geq 0$, $b'(t) \geq 0$, $t \geq 0$, takes place iff $m(t)y(t) \geq P_1'(t)$, $m(t)[1-y(t)] \geq P_2'(t)$. For this to hold it is sufficient that $m(\alpha)y(t)P_1(t) \geq P_1'(t)$, $m(\alpha)[1-y(t)P_1(t) \geq P_2'(t)$.

Theorem 5. *Under the assumption of Theorem 2 and under the condition*

$$P(t) < \min (\int^{t\wedge}_0 ym\wedge ds, \int^{t\wedge}_0 (1-y)m\wedge ds), f(t) < \min (\int^{t\wedge}_0 \alpha ym\wedge ds, \int^{t\wedge}_0 \beta(1-y)m\wedge ds),$$

$$[m(\alpha,t) > f(t)/P(t) > M(\beta,t)] \vee [m(\beta,t) > f(t)/P(t) > M(\alpha,t)] \quad (34)$$

the system of equations (9, Ch.2) *has the unique nonnegative differentiable solution m, a, b, c, which can be found by the iterative process*

$$\Phi(ai,bi) = m_{i+1}(t)+c_{i+1}(t) = f(t), \Psi(ai,bi) = \int^t_{ai(t)} y(s)m_i(s)ds + \int^t_{bi(t)} [1-y(s)]m_i(s)ds =$$

$$P(t), c_{i+1}(t) = \int^t_{bi(t)} \beta(t,s)[1-y(s)]m_i(s)ds, \; i = 1, 2, ..., m_1(t) = 0, t \in [t\wedge,T], \quad (35)$$

m_{i+1} *are expressed in the terms of* m_i *in the same way as above, converging faster than geometrical progressions.*

Proof. It is not difficult to determine that $\Phi(t,t) = \Psi(t,t) = 0 < f, P$; $\Phi(0,0) > f$, $\Psi(0,0) > P$. So, there exist values ai and bi for which $\Phi(ai,bi) = f$ and $\Psi(ai,bi) = P$. These values will meet the conditions: $0 < ai, bi < t$. Indeed, we can rewrite the system (35) in the form $x\wedge = \int^t_{ai} ym_i ds = (Pc_{i+1}/z\wedge - f)/d$, $z\wedge = \int^t_{bi} (1-y)m_i ds = (f - Pm_{i+1}/x\wedge)/d$, $d = c_{i+1}/z\wedge - m_{i+1}/x\wedge$, from which on the strength of (34), it follows rather easily that $d \neq 0$ and if $\beta > \alpha$ then $x\wedge > P[m(\beta,t) - f/P]/[M(\beta) - m(\alpha)] > 0$, $z\wedge > P[f/P - M(\alpha,t)]/[M(\beta) - m(\alpha)] > 0$, and if $\alpha > \beta$ then

$$x\wedge > P[f/P - M(\beta,t)]/[M(\alpha) - m(\beta)] > 0, z\wedge > P[m(\alpha) - f/P]/[M(\alpha) - m(\beta)] > 0.$$

These inequalities mean that $ai, bi < t$. If ai or bi is equal to 0, then respectively

$$\Phi(0,bi)>\Phi(0,t)>f, \Psi(0,bi)>\Psi(0,t)>P \text{ or } \Phi(ai,0)>\Phi(t,0)>f, \Psi(ai,0)>\Psi(t,0)>P.$$

It means that $ai, bi > 0$. This solution is unique since Jacobian of the system (35)

$$J = (ym)|_a((1-y)m)|_b(\beta(t,b) - \alpha(t,a)) \quad (36)$$

has the property of having fixed sign. The further proof is similar to the above. □

The condition $a'(t), b'(t) \geq 0, t \geq 0$, takes place iff

$$z_1\beta(t,b) \geq z_2 \geq z_1\alpha(t,a)\ (J>0)\ or\ z_1\beta(t,b) \leq z_2 \leq z_1\alpha(t,a)\ (J<0), \quad (37)$$

where

$$z_1 = m(t)-P'(t),\ z_2 = \{\alpha(t,t)y(t)+\beta(t,t)[1-y(t)]\}m(t)-f'(t)+\int^t_{a(t)} \alpha'_t(t,s)y(s)m(s)ds$$

$$+\int^t_{b(t)} \beta'_t(t,s)[1-y(s)]m(s)ds. \quad (38)$$

We can receive the respective sufficient condition by the similar way to the above.

Note that in the case of NES $G(t)$,

$$G(t) = \int^{a(t)}_0 y(s)m(s)ds + \int^{b(t)}_0 [1-y(s)]m(s)ds, \quad (39)$$

can be given instead of $P(t)$. We can receive the result similar to Theorem 5 with the replacement of $P(t)$ by $G(t)$.

<u>Theorem 6</u>. *In the case of* (51, Ch. 2), *except when* $\mu=1,\ \nu=0$, *we have the system, which has the unique differentiable and positive solution* α_μ, m, a, β_ν, c, M, G, f *on any segment* $[t^\wedge,T]$, $0<t^\wedge<T<\infty$, *under the assumptions which are similar to those of Theorem* 2 *and under the condition that*

$$0 < P(t) \leq \int^{t^\wedge}_0 m^\wedge(s)ds\ ,\ y(t) > 0,\ t \in [t^\wedge, T]. \quad (40)$$

In addition, the desired functions α_μ, m, *and* a *can be found by the iterative process*

$$\int^t_{a_i(t)} m_i(s)ds = P(t),\ \alpha_\mu{}^\nu{}_{i+1}(t) = \int^t_{a_i(t)} \alpha_{\mu i}(t,s)x(s)m_i(s)ds,$$

$$m_{i+1}(t) = \int^t_{a_i(t)} \alpha_{\mu i}(t,s)y(s)m_i(s)ds,\ m_i(s) = m^\wedge(s),\ \alpha_{\mu i}(s) = \alpha_\mu{}^\wedge(s),\ s \in [0,t^\wedge],$$

$$i = 1, 2, \ldots,\ m_1(s) = \alpha_{\mu 1}(s) = 0,\ s \in [t^\wedge,t], \quad (41)$$

which converges faster than geometrical progressions.

<u>Proof.</u> Let $c_i = d_i = 0$, $i = 1, 2$, in (51, Ch. 2). Otherwise, the proof is similar. We eliminate the case $\mu=1,\ \nu=0$ since it seems senseless. Indeed, we have in this case under x *and* y are positive constants and $m > q > 0$, $P = \int^t_{a(t)} m(s)ds > xd > 1$

$$y/x\alpha(t) = m(t) = \int^t_{a(t)} xm^2(s)ds > qxd,\ q(xd)^2, \ldots,\ q(xd)^n, \ldots \quad (42)$$

and under $m < q\hat{}, P < xd < 1$

$$m(t) = \int_{\alpha(t)}^{t} xm^2(s)ds < q\hat{}xd,\ q\hat{}(xd)^2, ..., q\hat{}(xd)^n, ... \tag{43}$$

It means that our equations are almost inconsistent.

On the strength of (40), (41) the equation $\int_{ai(t)}^{t} m_i(s)ds = P(t)$ always has the unique solution $ai(t)$. The respective equations in (51, Ch. 2) have the unique positive differentiable solution on the segment $[t\hat{}, T]$. Indeed, consider in detail the case $\nu = 1$, $\mu = 1$. Here we have

$$\alpha_1'(t) \le M(x)\alpha_1(t)P(t),\ m(t) \le M(y)\alpha_1(t)P(t), \tag{44}$$

from which

$$\alpha_1(t) \le \alpha_1(t\hat{})\exp[M(x)\int_{t\hat{}}^{t} P(s)ds],\ m(t) \le M(y)P(t)\alpha_1(t\hat{})\exp[M(x)\int_{t\hat{}}^{t} P(s)ds]. \tag{45}$$

It means that the functions m and α_1 are bounded on any segment $[t\hat{}, T]$. The desired result follows from the general theory of Volterra-type equations. And the iterative process (42) converges faster than geometrical series on the strength of the estimates which is similar to (21).

In the case of $\nu = 0$, $\mu = 2$ we have

$$\alpha_2(t) \le M(x/y)m(t),\ m(t) \le m\hat{} + M(x/y)\int_{t\hat{}}^{t} m(t-s)m(s)ds, \tag{46}$$

from which consecutively

$$m(t) \le m\hat{} + M(x/y)\int_{t\hat{}}^{t} m\hat{}(t-s)m^+(s)ds,\ t \in [t\hat{}, 2t\hat{}],$$

$$m(t) \le m^{+}\hat{} + M(x/y)\int_{2t\hat{}}^{t} m^{+}\hat{}(t-s)m^+(s)ds,\ t \in [2t\hat{}, 4t\hat{}], \tag{47}$$

and so on. It means that $m(t)$ and hence $\alpha_2(t)$ are bound on any segment $[t\hat{}, T]$. The further proof is similar to the above. The case $\nu = 1$, $\mu = 2$ can be considered by analogy with that of $\nu = 1$, $\mu = 1$. □

After determination of the functions α_{μ}, m, and a, we can find β_{κ}, c, M, G, and f on the basis of their respective equations in (51, Ch. 2).

It should be noted that instead of (40) we can accept the more general sufficient conditions, which are similar to (15). For example, in the case $\nu = 1$, $\mu = 1$ we can accept the condition

$$0 < P(t) \le m\hat{} + \alpha_1(0)m(y)\int_{t\hat{}}^{t} \exp(-c_1 s)P(s)ds \tag{48}$$

and in the case $\nu = 1$, $\mu = 2$ the condition

$$0 < P(t) \leq m^{\wedge} + \alpha_2(0)m(y)\int_{t^{\wedge}}^{t} \exp(c^{\wedge}_2 s)P(s)ds,\ c^{\wedge}_2 = \min(c_2, 0). \tag{49}$$

The condition $a'(t) \geq 0$ takes place iff $m(t) \geq P'(t)$. For $\nu = 1$ it is sufficient that

$$\alpha_\mu(0)m(y)P(t) \geq P'(t). \tag{50}$$

It follows from

$$m(t) \geq \alpha_\mu(0)m(y)P(t). \tag{51}$$

The remaining cases can be considered in a similar way.

Theorem 7. *Under the conditions of Theorems* 2 *and* 6 *and under the condition*

$$0 < f(t) \leq \int_0^{t^{\wedge}} \{\alpha_\mu^{\wedge}(t,s)[x(s)+y(s)+z(s)] + \beta_\kappa^{\wedge}(t,s)[1-x(s)-y(s)-z(s)]\}m^{\wedge}(s)ds, \tag{52}$$

where f is given, the system of equations (51, Ch. 2) *has the unique differentiable solution with respect to m, a,* α_μ*, c,* β_κ*, P. In addition, the desired functions m, a,* α_μ*, c, and* β_κ *can be found by the iterative process*

$$f(t) = m_{i+1} + c_{i+1} + \alpha_\mu{}^{\nu}{}_{i+1} + \beta_{\kappa i+1},\ \beta_{\kappa i+1} = \int_{ai(t)}^{t} \alpha_{\mu i}(t,s)z(s)m_i(s)ds,$$

$$c_{i+1} = \int_{ai(t)}^{t} \beta_{\kappa i}(t,s)[1-x(s)-y(s)-z(s)]m_i(s)ds,\ i = 1, 2, \ldots, \tag{53}$$

m_{i+1}, $\alpha_\mu{}^{\nu}{}_{i+1}$ *are the same as in* (41), *converging faster than geometrical progressions.*

Proof. It is similar to the proof of Theorem 6. □

Theorem 8. *In the case of MM* (53, Ch. 2), *except when* $\mu = 1$, $\nu = 0$, *under the condition*

$$0 < P_j(t) \leq \int_0^{t^{\wedge}} m_j^{\wedge}(s)ds,\ j = 1, 2, 3, 4, \tag{54}$$

where $P_j(t)$ *and* $m_j^{\wedge}(s)$, $0 \leq s \leq t^{\wedge}$, $t^{\wedge} \leq t \leq T$, *are given positive differentiable functions, the system of respective equations has the unique positive differentiable solution. In addition, the desired functions aj, m, and* α_μ *can be found by the iterative process*

$$P_j(t) = \int_{aji}^{t} m_{ji}(s)ds,\ \alpha_\mu{}^{\nu}{}_{i+1}(t) = \int_{a1i(t)}^{t} \alpha_{\mu i}(t,s)m_{1i}(s)ds,$$

$$m_{i+1}(t) = \int^{t}_{a2i(t)} \alpha_{\mu i}(t,s)m_{2i}(s)ds,\ i = 1, 2,...;\ \alpha_{\mu i}(t,s) = \alpha_{\mu}\hat{}(t,s),\ m_{ij}(s) = m_j\hat{}(s),$$

$$0 \le s \le t\hat{},\ m_{ji} = x_j m_i,\ x_1 = x,\ x_2 = y,\ x_3 = z,\ x_4 = (1\text{-}x\text{-}y\text{-}z), \tag{55}$$

where x, y, and z are given, converging faster than geometrical progressions.

Proof. It is similar to the proof of Theorems above. □

The condition $aj'(t) \ge 0$, $t \ge 0$, takes place iff $m_j(t) \ge P_j'(t)$, $j = 1, 2, 3, 4$. For $\mu = 1$, $\nu = 1$ it is sufficient that

$$x_j \alpha_1(0) P_2(t) \ge P_j'(t),\ j = 1, 2, 3, 4. \tag{56}$$

It follows from

$$m_j(t) \ge x_j \alpha_1(0) P_2(t),\ j = 1, 2, 3, 4. \tag{57}$$

Theorem 9. *In the case of MM* (54, Ch. 2), *unless* $\mu = 1$, $\nu = 0$, *if*

$$0 < P_j(t) \le \int^{t\hat{}}_{0} m_j\hat{}(s)ds,\ j = 1, ..., 6, \tag{58}$$

where $P_j(t)$ and $m_j\hat{}(s)$, $0 \le s \le t\hat{}$, $t\hat{} \le t \le T$, are given positive differentiable functions, then the system of respective equations has the unique positive differentiable solution. In addition, the desired functions aj, m, and α_μ can be found by the iterative process

$$P_j(t) = \int^{t}_{aji} m_{ji}(s)ds,\ \alpha_{\mu}{}^{\nu}{}_{i+1}(t) = \int^{t}_{a1i(t)} \alpha_{\mu i}(t,s)m_{1i}(s)ds,\ m_{i+1}(t) = \int^{t}_{a2i(t)} \alpha_{\mu i}(t,s)m_{2i}(s)ds,$$

$$i = 1,2,... ;\ \alpha_{\mu i}(t,s) = \alpha_{\mu}\hat{}(t,s),\ m_{ji}(s) = m_j\hat{}(s),\ 0 \le s \le t\hat{},$$

$$m_{ji} = x_j m_i,\ x_1 = x,\ x_2 = y,\ x_3 = z,\ x_4 = e,\ x_5 = g,\ x_6 = 1\text{-}x\text{-}y\text{-}z\text{-}e\text{-}g, \tag{59}$$

where *x, y, z, e, and g are given, converging faster than geometrical progressions. After determination of the functions aj, m, and α_μ, we can find the functions β_κ, c, γ_κ, f, and* cost(*f*) *from their respective equations in* (54, Ch. 2).

Proof. It is similar to the above. □

Note that we can consider some other variants in the case MM (54, Ch. 2). For example, we can suppose cost(*f*) to be given and P_6 not to be given. Then we can find *a6* from the equation for *f* (under the condition that *f* appears less than $\int^{t}_{0} \gamma_\kappa m ds$), after this we can find P_6.

Theorem 10. *For MM* (55, Ch. 2), *unless* $\mu = 1$, $v = 0$, *under the condition*

$$0 < P_j(t) \leq \int^{t^\wedge}_0 m_j^\wedge(s)ds, j = 1, ..., 8, \tag{60}$$

where $P_j(t)$ *and* $m_j^\wedge(s)$, $0 \leq s \leq t^\wedge$, $t^\wedge \leq t \leq T$, *are given positive differentiable functions, the system of respective equations has the unique positive differentiable solution. In addition, the desired functions aj, m,* and α_μ *can be found by the iterative process*

$$P_j(t) = \int^t_{aji} m_{ji}(s)ds, \ \alpha_\mu{}^v{}_{i+1}(t) = \int^t_{a1i(t)} \alpha_{\mu i}(t,s)m_{1i}(s)ds, \ m_{i+1}(t) = \int^t_{a2i(t)} \alpha_{\mu i}(t,s)m_{2i}(s)ds,$$

$$i = 1,2,... ; \ \alpha_{\mu i}(t,s) = \alpha_\mu{}^\wedge(t,s), \ m_{ji}(s) = m_j^\wedge(s), \ 0 \leq s \leq t^\wedge, \ m_{ji} = x_j m_i,$$

$$x_1 = x, x_2 = y, x_3 = z, x_4 = e, \ x_5 = g, x_6 = r, x_7 = w, x_8 = (1\text{-}x\text{-}y\text{-}z\text{-}e\text{-}g\text{-}r\text{-}w), \tag{61}$$

where x, y, z, e, g, r, and w are given, converging faster than geometrical progressions. After determination of the functions aj, m, and α_μ *we can find the functions* β_κ, *c*, γ_κ, *f*, δ_κ, *N, and* cost(*f*) *from their respective equations in* (57, Ch. 2).

Proof. It is similar to the above. □

For MM (55, Ch. 2) we can also consider various variants. If, for example, we need to put $N(t) = P(t)$ then P_8 cannot be given, in general. We can find $a8$ from the equation

$$P(t) - P_8(t) = \int^t_{a8(t)} [\delta_\kappa(t,s)w(s)m(s) - m_8(s)]ds = P_1 + ... + P_7, \tag{62}$$

after this we can find P_8 and P. The sufficient condition for solvability of (62) is

$$P_1 + ... + P_7 \leq \int^{t^\wedge}_0 [\delta_\kappa(t,s)w(s)m(s) - m_8(s)]ds. \tag{63}$$

We can interpret the relation (63) as the restriction under determination of the distributions x_j, $j = 1, ..., 8$, which are usually found from the solution of certain optimization problems (see s. 7, Ch. 2).

3. CASES OF NON-UNIQUENESS SOLUTIONS

Let us consider the system of the equations

$$f(t) = m(t) + c(t) = \int^t_{a(t)} \alpha(t,s)y(s)m(s)ds + \int^t_{b(t)} \beta(t,s)[1\text{-}y(s)]m(s)ds,$$

$$P(t) = P_1(t) + P_2(t) = \int^t_{a(t)} y(s)m(s)ds + \int^t_{b(t)} [1\text{-}y(s)]m(s)ds, \ t^\wedge < t, \tag{64}$$

with respect to a and b, $0 < a, b < t$.

Let $\alpha = \beta = \alpha(t) > 0$. Then the system (64) has no solution under $f/\alpha \neq P$ and infinite number of solutions under $f/\alpha = P$, $P(t) < \int_0^t m(s)ds$.

We find a case when the system (64) has two coincident solutions or a *two-multiple solution*, which is probably related to a case of NES reproduction.

Let

$$\alpha = \alpha(t),\ (\beta-\alpha)(1-y)m = 2u(t)s + 2v(t),\ u(t) > 0,\ s \in [0, t^\wedge],\ t^\wedge > 0;\ \alpha = \beta = \alpha(t),$$

$$t \geq s > t^\wedge;\ 0 < (f - \alpha P)/u < t^{\wedge 2},\ t^\wedge < t < t^*,\ f(t^*) - \alpha(t^*)P(t^*) = 0;$$

$$0 \leq (f - \alpha P)/u \leq (t-t^\wedge)^2,\ t^* \leq t;\ \int_0^t [1-y(s)]m(s)ds < P(t) < \int_0^t y(s)m(s)ds. \tag{65}$$

Then

$$f(t) = \int_{b(t)}^t (\beta - \alpha)(1-y)mds + \alpha P = ut^{\wedge 2} + 2vt^\wedge - ub^2 - 2vb + \alpha P \tag{66}$$

from which $b(t) = \{- v \pm [v^2 - u(f - \alpha P - ut^{\wedge 2} - 2vt^\wedge)]^{1/2}\}/u$. Therefore under $v^2 - u(f - \alpha P - ut^{\wedge 2} - 2vt^\wedge) = 0$, from which $v = - ut^\wedge \pm [(f - \alpha P)u]^{1/2}$, we have a *two-multiple solution*

$$b = b(t) = t^\wedge - [(f - \alpha P)/u]^{1/2},\ t^\wedge < t < t^*,\ b(t) = t^\wedge,\ t^* \leq t,\ b(t^*) = t^\wedge. \tag{67}$$

We should still verify that the equation

$$P(t) - \int_{a(t)}^t y(s)m(s)ds = \int_{b(t)}^t [1-y(s)]m(s)ds,\ t^\wedge < t, \tag{68}$$

where $b(t)$ is given by the relation's (67) has a solution $a(t)$, $0 < a(t) < t$. Let

$$h(a) = \int_a^t y(s)m(s)ds + \int_{b(t)}^t [1-y(s)]m(s)ds,\ t^\wedge < t. \tag{69}$$

Then on the strength of (65) $h(0) > P(t)$ and $h(t) < P(t)$. So, the equation (68) always has a solution $a(t)$, $0 < a(t) < t$. Thus, the system (64) under the condition (65) has a two-multiple solution.

Let us consider a case of *two-multiple repeated* solutions for the model

$$m(t) = \int_{a(t)}^t \alpha(t,s)y(s)m(s)ds,\ c(t) = \int_{b(t)}^t \beta(t,s)[1-y(s)]m(s)ds,$$

$$P(t) = \int_{a(t)}^t y(s)m(s)ds + \int_{b(t)}^t [1-y(s)]m(s)ds,\ 0 < t^\wedge \leq t. \tag{70}$$

It is not difficult to verify that under the condition

$$m(t) = m\hat{} = \text{constant}, y(t) = y\hat{} = 0, t \in [0,t\hat{}]; \int^{t}_{(2k-1)t\hat{}} \alpha(t,s)ds = 1,$$

$$\beta(t,s) = 2[s-(t-t\hat{})]/m\hat{}, s \in [t-t\hat{}, (2k-1)t\hat{}], y(t) = 1, c(t) = (2kt\hat{}-t)^2,$$

$$t \in [(2k-1)t\hat{},2kt\hat{}]; \int^{2kt\hat{}}_{t-t\hat{}} \alpha(t,s)ds = 1, \beta(t,s) = 2(t-s)]/m\hat{}, s \in [2kt\hat{},t], y(t) = 0,$$

$$c(t) = (2kt\hat{}-t)^2, t \in [2kt\hat{},(2k+1)t\hat{}]; k = 1, 2, ..., \tag{71}$$

we have *two-multiple repeated* solutions

$$b(t) = t-t\hat{}, m(t) = m\hat{}, P(t) = m\hat{}t\hat{}, t \in [(2k-1)t\hat{},2kt\hat{}],$$

$$a(t) = t-t\hat{}, m(t) = m\hat{}, P(t) = m\hat{}t\hat{}, t \in [2kt\hat{},(2k+1)t\hat{}], k = 1, 2, ... \tag{72}$$

Note that the condition (71) is valid for

$$\alpha(t,s) = 2\delta[s-(2k-1)t\hat{}], s \in [(2k-1)t\hat{},2kt\hat{}]; \alpha(t,s) = 2\delta(2kt\hat{}-s), s \in [2kt\hat{},(2k+1)t\hat{}], \tag{73}$$

where δ is Dirac generalized function.

The relations (70)-(73) can be interpreted as imitation of unlimited reproduction of NES, in particular, unlimited reproduction of cancer cells. From (71) it follows that for preventing the process (72) NES should be busy enough with fulfilling their external functions in order for the cases of the type $y = 1$, $\beta = 0$, and $c = 0$ to be impossible.

We can summarize the results of this section as the following

Theorem 11. *In the class of MM considered under the solution of the determined system of equations may be encountering none solutions, a unique solution, two-multiple solutions, and unlimited number solutions.*

4. ON STABILITY OF SOLUTIONS

We show that solutions of equations of MM under consideration have the *technical stability*, that is, any infinitesimal variations of the input data result in infinitesimal variations of the solutions.

Let δx be a variation for any x. For the equations (5, 6, Ch. 2) and under the condition of Theorem 2 we have, regarding the main terms,

$$\delta m(t) = \int^{t}_{a(t)} [\alpha(t,s)y(s)-\alpha(t,a(t))y(a(t))]\delta m(s)ds + \int^{t}_{a(t)} \delta[\alpha(t,s)y(s)]m(s)ds +$$

$$\alpha(t,a(t))y(a(t))\delta P(t), \delta c(t) = \int^{t}_{a(t)} \{\beta(t,s)[1-y(s)]-\beta(t,a(t))[1-y(a(t))\}\delta m(s)ds +$$

$$\int_{a(t)}^{t} \delta\{\beta(t,s)[1-y(s)]\}m(s)ds + \beta(t,a(t))[1-y(a(t))]\delta P(t),$$

$$\delta a(t) = [\int_{a(t)}^{t} \delta m(s)ds - \delta P(t)]/m(a(t)), \qquad (74)$$

where m, a is the solution of the system in question, and $\delta m(t)$ is given for $0 \le t \le t^\wedge$.

In a similar way, we can receive the estimations of $\delta m(t)$, $\delta c(t)$ and $\delta a(t)$ under the conditions of Theorems 3 and 4.

For the equations (9, Ch. 2) and under the condition of Theorem 5 we have

$$\delta f(t) = -\alpha ym|_{a(t)}\delta a(t) - \beta(1-y)m|_{b(t)}\delta b(t) + \int_{a(t)}^{t} \delta(\alpha ym)ds + \int_{b(t)}^{t} \delta[\beta(1-y)m]ds,$$

$$\delta P(t) = -ym|_{a(t)}\delta a(t) - (1-y)m|_{b(t)}\delta b(t) + \int_{a(t)}^{t} \delta(ym)ds + \int_{b(t)}^{t} \delta[(1-y)m]ds, \delta m(t) =$$

$$-\alpha ym|_{a(t)}\delta a(t) + \int_{a(t)}^{t} \delta(\alpha ym)ds,\ \delta c(t) = -\beta(1-y)m|_{b(t)}\delta b(t) + \int_{b(t)}^{t} \delta[\beta(1-y)m]ds. \qquad (75)$$

The determinant of the system (75)

$$\Delta = ym|_{a(t)}(1-y)m|_{b(t)}\delta b(t)[\alpha(t,a(t)) - \beta(t, b(t))] \neq 0. \qquad (76)$$

So, the system (75) has the unique solution with respect to $\delta a(t)$, $\delta b(t)$. After solving this system it is not difficult to determine the desired relations for $\delta m(t)$ and $\delta c(t)$.

For the equations (51, Ch. 2, $v = 0$) and under the condition of Theorem 6 we have, regarding the main terms,

$$\delta\alpha_\mu(t) = \int_{a(t)}^{t} [\alpha_\mu(t,s)x(s) - \alpha_\mu(t,a(t))x(a(t))]\delta m(s)ds + \int_{a(t)}^{t} \delta[\alpha_\mu(t,s)x(s)]m(s)ds +$$

$$\alpha_\mu x|_{t,a}\delta P(t), \delta m(t) = \int_{a(t)}^{t} [\alpha_\mu(t,s)y(s) - \alpha_\mu(t,a(t))y(a(t))]\delta m(s)ds +$$

$$\int_{a(t)}^{t} \delta[\alpha_\mu(t,s)y(s)]m(s)ds + \alpha_\mu y|_{t,a}\delta P(t), \delta\beta_\kappa(t) = \int_{a(t)}^{t} [\alpha_\mu(t,s)z(s) -$$

$$\alpha_\mu(t,a(t))z(a(t))]\delta m(s)ds + \int_{a(t)}^{t} \delta[\alpha_\mu(t,s)z(s)]m(s)ds + \alpha_\mu z|_{t,a}\delta P(t), \delta c(t) =$$

$$\int_{a(t)}^{t} \{\beta_\kappa(t,s)[1-x(s)-y(s)-z(s)] - \beta_\kappa(t,a(t))[1-x(a(t))-y(a(t))-z(a(t))\}\delta m(s)ds +$$

$$\int_{a(t)}^{t} \delta\{\beta_\kappa(t,s)[1-x(s)-y(s)-z(s)]\}m(s)ds + \beta_\kappa(t,a(t))[1-x(a(t))-y(a(t))-z(a(t))]\delta P(t),$$

$$\delta a(t) = [\int_{a(t)}^{t} \delta m(s)ds - \delta P(t)]/m(a(t)). \qquad (77)$$

So, we have the linear system of two equations for determination of $\delta\alpha_\mu(t)$, $\delta m(t)$ after which we can determine $\delta\beta_\kappa(t)$, $\delta c(t)$, and $\delta a(t)$.

In a similar way, we can receive the estimates of $\delta\alpha_\mu(t)$, $\delta m(t)$, $\delta\beta_\kappa(t)$, $\delta c(t)$, $\delta\gamma_\kappa(t)$, $\delta f(t)$, $\delta\delta\kappa(t)$, $\delta N(t)$, and $\delta aj(t)$, $j = 1, ..., 10$, under the conditions of Theorems 7-10.

We can summarize the results of this section as the following

Theorem 12. *For the class of MM considered we have the technical stability that can be estimated by the formulae* (74) - (77)-*type.*

5. CASES OF OSCILLATING AND WAVE SOLUTIONS

Let α, β, and y be constants. Then for (5, 6, Ch. 2) we have

$$m(t) = \alpha y P(t),\ c(t) = \beta(1-y)P(t),\ m(t) + c(t) = f(t),\ \int^{t}_{a(t)} m(s)ds = P(t), \tag{78}$$

from which

$$P(t) = f(t)[\alpha y + \beta(1-y)],\ m(t) = \alpha y f(t)[\alpha y + \beta(1-y)],$$

$$c(t) = \beta(1-y)f(t)[\alpha y + \beta(1-y)],\ \int^{t}_{a(t)} f(s)ds = f(t)/[\alpha y]. \tag{79}$$

It means if f is a periodic function, then the functions P, m, c, and maybe a are also periodic.

Let

$$\alpha(t,s) = \alpha[1-\cos \omega(t-s)],\ a(t) = t-a,\ y = \text{constant.}, \tag{80}$$

where α, ω, and a are also constants. Then we have

$$m(t) = \alpha y \int^{t}_{t-a} [1-\cos \omega(t-s)]m(s)ds,\ P(t) = \int^{t}_{t-a} m(s)ds, \tag{81}$$

from which

$$m''(t) + \omega^2 m(t) = \alpha y \omega^2 P(t) - \alpha y(1-\cos \omega a)m^{\wedge\prime}(t-a) - \alpha y \omega \sin \omega a m^{\wedge}(t-a),\ t \in [a,2a],$$

$$m(a) = \alpha y \int^{a}_{0} [1-\cos \omega(a-s)]m^{\wedge}(s)ds,\ m'(a) = -\ \alpha y(1-\cos \omega a)m^{\wedge}(0) +$$

$$\alpha y \omega \int^{a}_{0} \sin \omega(a-s)m^{\wedge}(s)ds, \tag{82}$$

The *characteristic equation* for (82) and its roots are

$$k^2 + \omega^2 k = 0,\quad k_{1,2} = \pm i\omega,\ i = (-1)^{1/2}. \tag{83}$$

It means that the *solution* $m(t)$ of the system (81) is rather general *oscillating function* depending on the right side of the equation and initial data in (82).

Let us consider the system

$$m = \int^{a}_{t-a} \alpha y m\hat{}ds + \int^{t}_{a} \alpha ymds,\ c = \int^{a}_{t-a} \beta(1-y)mds + \int^{t}_{a} \beta(1-y)mds,\ t \in [a,2a], \quad (84)$$

For $\alpha = \alpha(s)$, $\beta = \beta(s)$ it follows from (84) that

$$m' = (\alpha ym)(t) - (\alpha ym)(t-a),\ c' = (\beta(1-y))(t) - (\beta(1-y))(t-a). \quad (85)$$

It means that, in particular, when αym and $\beta(1-y)$ are arbitrary constants for $t \in [0,a]$ and αym and $\beta(1-y)m$ are arbitrary positive functions of m, c for $t \in [a,2a]$, the system (86) is an arbitrary autonomous system of differential equations of the second order. One can apply to this system the well-known qualitative methods of investigation based on construction of the phase picture of the solutions (see, for example, [5]).

Let, in particular,

$$\alpha m = A - A'm + A''c,\ \beta m = B - B'm + B''c,$$

$$Ay - \alpha ym(t-a) = d',\ B(1-y) - (\beta(1-y))(t-a) = d'', \quad (86)$$

where $A, A', A'', B, B', B'', d', d''$ are positive constants, $\Delta = A''B' - A'B'' \neq 0$. Then

$$m' = -d' + y(-A'm + a''c),\ c' = -d'' + (1-y)(-B'm + B''c). \quad (87)$$

The *singular point* of the system (87) is

$$m^* = [B''d'/y - A''d''/(1-y)]/\Delta,\ c^* = [B'd'/y - A'd''/(1-y)]/\Delta. \quad (88)$$

The sufficient condition for positiveness of m^*, c^* is

$$B'/A' > B''/A'' > d''y/[d'(1-y)] > 0,\ \Delta > 0. \quad (89)$$

We can rewrite the system (87) in the form

$$d(m-m^*)/dt = y[-A'(m-m^*)+A''(c-c^*)],\ d(c-c^*)/dt = (1-y)[-B'(m-m^*)+B''(c-c^*)]. \quad (90)$$

The *characteristic equation* for (90) is

$$k^2 - k[B''(1-y) - A'y] + \Delta y(1-y) = 0, \tag{91}$$

the roots of which

$$k_{1,2} = [B''(1-y) - A'y]/2 \pm \{[B''(1-y) - A'y]^2 - 4\Delta y(1-y)\}^{1/2}/2. \tag{92}$$

We consider the phase picture's variation while y is changed from 0 to 1. We need to find y_1, y_2 for which discriminant of the equation (91)

$$D(y) = [B''(1-y)-A'y]^2 - 4\Delta y(1-y) =$$

$$y^2[(A'-B'')+4A''B'] - 2[2A''B'-B''(A'-B'')]y + B''^2 = 0. \tag{93}$$

We have

$$y_{1,2} = [A''B' + B''^2 + \Delta \mp 2(A''B'\Delta)^{1/2}]/(2A''B' + 2\Delta + A'^2 + B'^2). \tag{94}$$

We have also y critical,

$$y_{cr} = B''/(A' + B''), \tag{95}$$

for which $B''(1-y) - A'y = 0$.

It is not difficult to verify that

$$0 < y_1 < y_{cr} < y_2 < 1. \tag{96}$$

Thus, we have the following result.

Theorem 13. *The singular point* (m^*, c^*) (88) *of the system* (87) *is*
i) *the unsteady node for* $0 < y \leq y_1$;
ii) *the unsteady focus (increasing by amplitude oscillations) for* $y_1 < y < y_{cr}$;
iii) *the center (undamped oscillations) for* $y = y_{cr}$;
iv) *the steady focus (decreasing by amplitudes oscillations) for* $y_{cr} < y < y_2$;
v) *the steady node for* $y_2 \leq y < 1$.

If we put

$$m' = -d' + \alpha y m = c,\ c' = -d'' + \beta(1-y)m = -m + \delta(1-m)c, \tag{97}$$

where δ is a parameter, then we have the classic system of *Van-der- Pol's equations* which describes steady auto-oscillations of variable m or c [5]. On the plane (m,c) this system has the phase picture in the form of the steady *limiting cycle*.

Setting

$$m' = (m\text{-}m^\wedge)^2 + c - c^\wedge,\ c' = \beta(1\text{-}y)m\ (\alpha ym = (m\text{-}m^\wedge)^2 + c - c^\wedge), \tag{98}$$

we obtain the case of the so-called *catastrophe of folding* in the vicinity of the point $(m^\wedge, c^\wedge)$. Setting in (1, 10, Ch.2) $n = 3, r = 1, s = 2$,

$$m' = (m\text{-}m^\wedge)^3 + (c_1 - c_1^\wedge)(m\text{-}m^\wedge) + (c_2 - c_2^\wedge),\ c_i' = \beta z_i m,\ i = 1, 2, \tag{99}$$

we obtain the case of the *catastrophe of assembling* in the vicinity of the point $(m^\wedge, c_1^\wedge, c_2^\wedge)$ and so on [7].

If we put

$$\alpha m = A + 4m'(t\text{-}d)[1\text{-}m'(t\text{-}d)],\ Ay - (ym\alpha)(t\text{-}a) = 0,\ d < a,\ t \in [a,2a], \tag{100}$$

then we have

$$m'(t) = 4ym'(t\text{-}d)[1\text{-}m'(t\text{-}d)],\ t \in [a,2a]. \tag{101}$$

It is well known [6, 7], that under increasing y from 0 to 1 the solution of the equation (101) has the limiting cycles (on the plane $(m'(t\text{-}d), m'(t))$ and the so-called *strange attractor* (with the random oscillations).

Note that in spite of rather artificial reduction (86) and (97)-(101) they make sense in the theory of MM under consideration. Indeed, input data of the models are usually the functions of α, β-types that can be arbitrary, but nonnegative. So, it is rather important to consider some special cases of these functions and by the same token to understand better that their (and m and c) non-negativity does not mean some kind of restriction on the possibility of applications of the models.

In the case of a continuous MM of ES, for example, MM (12, Ch. 2), we can obtain, in a similar way, the equations that have different *wave solutions*.

The usual method is the supplement of the so-called diffusive terms to our models:

$$\partial M(r,t)/\partial t = m(r,t) = \int_0^t\int_0^r \alpha(t,r,s,v)\lambda(t,r,s,v)y(s,v)m(s,v)dvds +$$

$$\partial/\partial l[D_1(M,C)\partial M/\partial l] + A,\ \partial C(r,t)/\partial t = c(r,t) =$$

$$\int_0^t\int_0^r \beta(t,r,s,v)\mu(t,r,s,v)z(s,v)m(s,v)dvds + \partial/\partial l[D_2(M,C)\partial C/\partial l] + B,$$

$$r = (x_1, x_2, x_3),\ v = (v_1, v_2, v_3),\ l = (l_1, l_2, l_3), \tag{102}$$

where D_1 and D_2 are coefficients of diffusion along the direction l from the point r; A

and B are constants guaranteeing non-negativity of m and c.

Note that various cases of the MM (102) were investigated in [10]. In particular, the cases corresponding to elliptic, hyperbolic and parabolic-type equations were found and examined.

6. ASYMPTOTIC PROPERTIES OF SOLUTIONS

In the case of AES and MM (5, 6, Ch. 2) we assume that the input data fulfil the condition of Theorem 2 and have the following *asymptotic properties*

$$\alpha(t,s)y(s) \sim \alpha y,\ \beta(t,s)[1-y(s)] \sim \beta(1-y),\ P(t) \sim P, \tag{103}$$

where αy, $\beta(1-y)$, and P are positive constants. Here for any x and y the expression $x \sim y$ is identical to $|x - y| = O(|y|/t)$ or, what is the same, there exist [8] positive constants C and K such that $|x - y| < C|y|/t$ whenever $t > K$.

From the condition of Theorem 2 and (103) we have

$$0 < m(\alpha y)P(t) < m(t) < M(\alpha y)P(t),$$

$$m(\alpha y)m(P)(t-a(t)) < P(t) < M(\alpha y)M(P)(t-a(t)) \tag{104}$$

and hence

$$0 < m(P)/[M(\alpha y)M(P)] < t-a(t) < M(P)/[m(\alpha y)m(P)]. \tag{105}$$

From (103) - (105) it follows that

$$m(t) \sim \alpha yP,\ c(t) \sim \beta(1-y)P,\ t-a(t) \sim 1/\alpha y. \tag{106}$$

For NES and the same MM we assume that the input data fulfil the condition of Theorem 3 and have the *asymptotic properties*

$$\alpha(t,s)y(s) \sim C1/t,\ \beta(t,s)[1-y(s)] \sim C2/t,\ f(t) \sim C3/t, \tag{107}$$

where Ci, $i = 1,2,3$, are positive constants. From Theorem 3 and (107) we have

$$0 < C4/tP(t) < m(t) < C5/tP(t),\ 0 < C6/tP(t) < f(t) < C7/P(t)t,\ C8/t < f(t) < C9/t, \tag{108}$$

where Ci, $i = 4, \ldots, 9$, are certain positive constants, and hence

$$0 < C8/C7 < P(t) < C9/C6,\ C4C8/(C7t) < m(t) < C5C9/(C6t),\ t-a(t) = t[1-a(t)/t],$$

$$D1 < \ln[t/a(t)] < D2,\ D1 = C6C8/(C5C7C9),\ D2 = C7C9/(C4C6C8). \tag{109}$$

From (107) - (109) it follows that

$$m \sim C1C3/[(C1+C2)t],\ c \sim C2C3/[(C1+C2)t], P \sim C3/(C1+C2),\ t\text{-}a \sim t(1\text{-}e^{-1/C1}). \tag{110}$$

In a similar way, we can obtain the asymptotic properties of the solutions in the case of the system (9, Ch. 2) under the conditions of Theorems 4 and 5.

For the system (51, Ch. 2) we assume that the input data fulfil the condition of Theorem 6 and have the following *asymptotic properties* $x(s) \sim x$, $y(s) \sim y$, $P(t) \sim P$, where x, y, and P are positive constants. Then in the case $\nu = 1$, $\mu = 1$, $c_1 = 0$ we have

$$\alpha \sim \alpha'(0) + x/yM,\ M'' \sim x[M - M(a)]M',\ M - M(a) \sim P, \tag{111}$$

and

$$M \sim \exp(xPt)/xP,\ \alpha \sim x/y\exp(xPt)/xP,\ a \sim t,\ m \sim \exp(xPt). \tag{112}$$

For arbitrary $P(t)$, $P(t) > 0$, and x, y are positive constants using the relations $M - M(a) = P$, $m - m(a)a' = P'$, we reduce the case $\nu = 1$, $\mu = 1$, $c_1 = 0$ to the equation of the form (10) asymptotic properties of which have been investigated in detail in [8].

In the other cases as well as for the system (53-55, Ch. 2) under the conditions of Theorems 7-10 we can obtain the similar relations.

We can summarize the results of this section as the following

Theorem 14. *In the class of MM considered we have the asymptotic properties of solutions expressed by the relations of* (103) - (112)-*types.*

The result (110) seems to be unexpected because we can hold almost a constant for the functioning bio-mass P in spite of the fact that $\alpha, \beta, m, c, f \to 0$ for $t \to \infty$.

7. A FEW ADDENDA AND REMARKS

It is natural that with the new class of MM under consideration, many problems of investigation remain open. Among them are the problems of existence and uniqueness solutions for the system of the equations (10-12, Ch. 2) and for the equations of a continuous MM of (102)-type. Some results are to be found in [10].

The respective existence and uniqueness theorems under more general assumptions in the sense of Caratheodory [2] one can see in [10, Ch. 2].

It seems interesting to consider the models of (51, 53-55, Ch. 2)-type in the cases $\nu > 1$ and, for example, in the case $\nu = -1$, which means $\alpha^{-1} = \int^{t}_{0} \alpha(s)ds$.

It is important to investigate MM for different cases of interactions among ES. Certain results in this connection one can find in Ch. 8-10, 12, and 13.

All Theorems 1-10 can be generalized in the cases when the functions of α, β-types depend on the other elements of the models. Let us dwell on the respective generalization of Theorem 2 when instead of MM (5, 6, Ch.2) we have MM

$$m(t) = \int_{a(t)}^{t} \alpha[t,s,m(s), c(s)]y(s)m(s)ds, \; P(t) = \int_{a(t)}^{t} m(s)ds,$$

$$c(t) = \int_{a(t)}^{t} \beta[t,s,m(s), c(s)][1 - y(s)]m(s)ds, \; t \in [t^{\wedge},T], \; t^{\wedge} > 0, \qquad (113)$$

where α, y, β, P, $t^{\wedge}$, T are given and m, a, and c are sought for.

<u>Theorem 2'.</u> *Let in addition to the assumptions of Theorem* 2 *α and β in* (76) *comply with the conditions of Lipschitz over m, c for which it is sufficient that* $|\partial\alpha/\partial m|$, $|\partial\alpha/\partial c|$, $|\partial\beta/\partial m|$, $|\partial\beta/\partial c|$ *be bounded. Then the system* (76) *has the unique nonnegative solution m, a, c $\in C^1([t^{\wedge},T])$, which can be found by the iterative process*

$$m_{i+1}(t) = \int_{a_i(t)}^{t} \alpha[t,s,m_i(s), c_i(s)]y(s)m_i(s)ds, \; P(t) = \int_{a_i(t)}^{t} m_i(s)ds, \; c_{i+1}(t) =$$

$$\int_{a_i(t)}^{t} \beta[t,s,m_i(s),c_i(s)][1-y(s)]m_i(s)ds, \; i = 1, 2, \ldots, \; m_1(t) = m(\alpha y,t)P(t), \; t^{\wedge} \le t \le T, \qquad (114)$$

where $c_1(t)$ is an arbitrary integrable function, $t \in [t^{\wedge},T]$, which converges at the rate exceeding that of geometrical progressions.

The vast diversity of solutions' properties of the equations is outlined in s. 5. In Part III devoted to applications of MM we will continue to consider this problem.

The problems of determination of asymptotic properties of solutions will be considered in more detail in the next Ch. (4) (see also [4]).

REFERENCES

1. Baker C. T. H., The Numerical Treatment of Integral Equations, Clarendon, 1977, 1034 pp.
2. Caratheodory C., Gesammelte Mathematische Schriften, Munhen, 1954-1957, Rd. 1-5.
3. Girlin S.K., Modeling of Originating Evolutionary Systems, Dokl. AN Ukr. SSR, 1987, 65-67 (in Russian).
4. Gluskov V.M., Ivanov V.V., Yatsenko Yu.P., Analytical Investigation of One Class of Dynamic Models, I, Cybernetics, Plenum Publish. Corp., 16, N 2, 1980, 164-177.
5. Korn G.A., Korn T.M., Mathematical Handbook for Scientists and Engineers, McGraw-Hill Book, 1968, 1130 pp.
6. Marzec C.J., Spiegel E.A., Ordinary Differential Equations with Strange Attractors, SIAM J. Appl. Math., 38, N 3, 1980, 403-421.
7. May R.M., Necessity and Chance: Deterministic Chaos in Ecology and Evolution, Bulletin (New Series) of the AMS, Vol. 32, N. 3, 1995, 291-308.
8. Murray J.D., Asymptotic Analysis, Springer, 1992, 180 pp.
9. Standard Mathematical Tables and Formulae, CRC Press, 1995, 612 pp.
10. Vuginshtein A.E., Ivanov V.V., Analytic Investigation of Continuous Models of Developing Systems, Soviet Math. Dokl., 273(2), 1984, 600-604.

CHAPTER 4

INVESTIGATION OF OPTIMIZATION PROBLEMS

Preliminary intelligence into the general theory of control and the examples of analytic solutions for some optimization problems are given. The conditions of existence of solutions are investigated. Stability of solutions and sequence of formulae for determination of gradients of specific functionals are established. The structure of solutions for different optimization problems is considered in detail. The conditions of uniqueness of solutions and the example of non-uniqueness of solutions are adduced. Asymptotic properties of some solutions are found out.

1. PRELIMINARY INTELLIGENCE

In sufficiently general case MM (5-10-types, Ch. 2) are described by the following system of equations and inequalities:

$$x_i(t) = \sum_{j=1}^{n} \int_{zj(t)}^{t} k_{ij}(t,s) y_{ij}(s) x_j(s) ds,\ i = 1,\dots, n; \tag{1}$$

$$p_i(t) = \sum_{j=1}^{n} \int_{zj(t)}^{t} k_{ij}(t,s) y_{ij}(s) x_j(s) ds,\ i = n+1,\dots, 2n; \tag{2}$$

$$0 \le y_{ij} \le 1,\ 0 \le zj(t) \le t,\ 0 \le s \le t,\ 0 < t^{\wedge} \le t \le T; \tag{3}$$

with the initial conditions

$$x_j(s) = x_j^{\wedge}(s),\ y_{ij}(s) = y_{ij}^{\wedge}(s),\ s \in [0,t^{\wedge}],\ zj(t^{\wedge}) = z_j^{\wedge},\ j = 1, \dots, n,\ i = 1, \dots, 2n. \tag{4}$$

We consider the functions k_{ij} and p_i to be given, y_{ij} to be control ones, and x_j and zj to be phase variables. Note that some of y_{ij} can be linearly dependent. Then the control y is a collection $\{y_{ij}\}$ after elimination of linearly dependant elements.

Let us consider the *problem of optimal control*: to determine the control $y(t)$, $t \in [t^{\wedge},T]$, with the restrictions (1) - (4) and the respective variables $x = (x_1, \dots, x_n)$ and $z = (z1, \dots, zn)$ for which the functional

$$I = \int_{t^{\wedge}}^{T} \Phi(x,y,z,t) dt = \min, \tag{5}$$

where Φ is a differentiable function of x, y, z.

We will assume that the given functions k_{ij} and p_i are continuous and the control y is sought in the class of piecewise continuous functions with the restrictions (3, 4): $y \in U$. If the system (1, 2) is solvable, then its solution x is continuous. If the class U of controls is turned out to be too restricted, we can assume that the vector function y is measurable, $y_{ij} \in L_\infty([0,T])$, or relaxed (generalized) [2, 6-10].

Let us introduce the *function of Lagrange*

$$L(x,y,z,\varphi) = \int^T_{t^\wedge} \{\Phi(x,y,z,t) + \Sigma^n_{i=1} \varphi_i(t)[x_i(t) - \Sigma^n_{j=1} \int^t_{zj(t)} k_{ij}(t,s) y_{ij}(s) x_j(s) ds] +$$

$$\Sigma^{2n}_{i=n+1} \varphi_i(t)[p_i(t) - \Sigma^n_{j=1} \int^t_{zj(t)} k_{ij}(t,s) y_{ij}(s) x_j(s) ds]\} dt, \tag{6}$$

where $\varphi(t) = \{\varphi_1(t), \ldots, \varphi_{2n}(t)\}$ is the so-called Lagrange's coefficients. Under the assumptions accepted, there exists the *derivative L' by Fresher* of x, y, z determined by the formula

$$\delta L = (L',(\delta x, \delta y, \delta z)) = \int^T_{t^\wedge} [\sum^n_{j=1} \partial L/\partial x_j \delta x_j(t) +$$

$$\sum^{2n}_{i=1} \sum^n_{j=1} \partial L/\partial y_{ij} \delta y_{ij}(t) + \sum^n_{j=1} \partial L/\partial zj \, \delta zj(t)] dt, \tag{7}$$

where δu means a variation for any u and

$$\partial L/\partial x_j = \partial \Phi/\partial x_j(t) + \varphi_j(t) - \sum^n_{i=1} y_{ij}(t) \int_{Jj(t)} k_{ij}(s,t) \varphi_i(s) ds, \ \partial L/\partial y_{ij} = \partial \Phi/\partial y_{ij}(t) -$$

$$x_j(t) \int_{Jj(t)} k_{ij}(s,t) \varphi_i(s) ds, \partial L/\partial zj = \partial \Phi/\partial zj(t) + \sum^{2n}_{i=1} k_{ij}(t,zj(t)) y_{ij}(zj(t)) x_j(zj(t)) \varphi_i(t). \tag{8}$$

The formulae (7), (8) are obtained by means of L increment evaluation after change of integration order, and also the $Jj(t) \subset [t,T]$ is determined in the following way. We divide $[t^\wedge, T]$ on the intervals of monotonousness of the function $zj(t)$ and denote $zj^{(1)}(t)$, $zj^{(3)}(t)$, ... as the increasing branches of $zj(t)$ and $zj^{(2)}(t)$, $zj^{(4)}(t)$, ... as its decreasing branches. After this we evaluate the reciprocal functions

$$zj^{(k)\,-1}(t) = zj^{(k)\,-1}(d_j^{k-}),\ t^\wedge \le t \le d_j^k;\ zj^{(k)\,-1}(t) = zj^{(k)\,-1}(t),\ d_j^{k-} = \min zj^{(k)} \le t \le d_j^{k+} = \max zj^{(k)},$$

$$zj^{(k)\,-1}(t) = zj^{(k)\,-1}(d_j^{k+}),\ d_j^{k+} \le t \le T. \tag{9}$$

Then

$$Jj(t) = [t, zj^{(1)\,-1}(t)] \cup [zj^{(2)\,-1}(t), zj^{(3)\,-1}(t)] \cup \ldots. \tag{10}$$

It should be noted that if $zj'(t) \ge 0$, then the formulae (9, 10) can be considerably simplified:

$$Jj(t) = [t, Zj^{-1}(t)],\ Zj^{-1}(t) = zj^{-1}(t),\ t\hat{} \le t \le zj(T),\ Zj^{-1}(t) = T,\ zj(T) \le t \le T. \tag{11}$$

We assume that the phase variables x and z pertain to the open domain. Then we obtain the *conjugate equations*

$$\partial L/\partial x_j = 0,\ \partial L/\partial zj = 0, j = 1, \ldots, n, \tag{12}$$

which will have the unique solution with respect to $\varphi_1, \ldots, \varphi_{2n}$, $t \in [t\hat{}, T]$, if

$$\det |k_{ij}(t, zj) y_{ij}(zj) x_j(zj)|^{i=n+1, \ldots, 2n}_{j=1, \ldots, n} \neq 0 \tag{13}$$

under any y_{ij} and zj from (3), (4). Then there exists the *gradient of the functional* (5) over y:

$$I'(y) = \{\partial L/\partial y_{ij}\}^{i=1, \ldots, 2n}_{j=1, \ldots, n}. \tag{14}$$

The necessary condition of optimum (minimum) has the form

$$(I'(y^*), (y-y^*)) \ge 0, y \in U, \tag{15}$$

where (u, v) means the scalar product in Hilbert space of functions u and v, y^* is any solution of the problem (1)-(5). For the intrinsic solutions, (15) is identical to

$$I'(y^*) = 0. \tag{16}$$

It should be noted that MM (51, 53-55, Ch.2) do not belong to the class of MM (1)- (4), but the above technique of the construction of the relations' (6)-(16) remains quite similar there.

Let us introduce now the main additional concepts and results following the monograph [9].

A *general problem of optimization* is one in which some general sets W, X_1, and $C_1 \subset X_1$ and functions g_0: $W \to \mathbb{R}$ and g_1: $W \to X_1$ are given, and it is desired to minimize g_0 on the set $\{w \in W | g_1(w) \in C_1\}$. W is usually a subset of the Cartesian product of three sets: the *sets Y of states*, whose elements are usually functions with values in some vector space, the *set U of control functions* on some measure space T into some space R, and the set *B of control parameters* (endowed with some structure).

Let us restrict attention to W such that points $(y, u, b) \in Y \times U \times B$ satisfy the *equation of motion* $y = F(y,u,b)$ and, among such points, consider those that "come close" to satisfying the restriction $g_1(y,u,b) \in C_1$ and, in so doing, tend to minimize $g_0(y,u,b)$. Later we reduce the problem to the case when the equation of motion is a *functional-integral equation* and $C_1 = \{0\}$.

Now the problem is to find

$$\inf g_0(y, u, b) \tag{17}$$

on the set

$$H(U) = \{(y, u, b) \in Y \times U \times B \,|\, y = F(y, u, b)\}, \tag{18}$$

where the equation of motion or state equation

$$y(t) = F(y, u, b) = \int f[t, s, \xi(y)(s), u(s), b(t)]ds, \; t \in T = [0, T], \tag{19}$$

under the restriction

$$g_1(y, u, b) = 0, \tag{20}$$

y is a state variable, u is a control function, and b is a control parameter (y, u, and b may be vectors). Let us put the set

$$A(U) = \{(y, u, b) \in Y \times U \times B \,|\, y = F(y, u, b), g_1(y, u, b) = 0\}. \tag{21}$$

Let $Y = C(T, \mathbb{R}^n)$. As a *control space* U we consider the space U of ordinary (measurable by Lebesgue) functions and the space S of *generalized functions* representing functions of the time with the values on the *set of measures*. Let $R^*(t)$ be the *admissible set* of control values and

$$U^* = \{u \in U \,|\, u(t) \in R^*(t)\}, \; S^* = \{\sigma(t, r) \in S \,|\, \int_{R^*(t)} \sigma(t, r)dr = 1\}. \tag{22}$$

If the measure σ is concentrated on the finite number of points r_i, $i = 1, ..., n$, (what is possible in our case), then

$$\sigma(t, r) = \sum{}^n_{i=1} p_i(t)\delta(r-r_i), \; \sum{}^n_{i=1} p_i(t) = 1, \tag{23}$$

where δ is delta-function of Dirac, $p_i(t)$ are probabilities that a control u accepts the value r_i at the instant t.

Following [9] we call:

i) a point $(y, u, b) \in A(U^*)$ is a *minimizing U-solution* of the problem if it minimizes the function g_0 on the set $A(U^*)$;

ii) a sequence (y_j, u_j, b_j) in $H(U^*)$ is an *approximate U-solution* if, for every neighborhood G of 0 in $\mathbb{R}$, $g_1(y_j, u_j, b_j) \in G$ for all sufficiently large $j \in \mathbb{I}$. An approximate U-solution $(y_j\hat{}, u_j\hat{}, b_j\hat{})$ is a *minimizing approximate U-solution* if $\lim_j g_0(y_j\hat{}, u_j\hat{}, b_j\hat{})$ exists in $\mathbb{R}$ and $\lim_j g_0(y_j\hat{}, u_j\hat{}, b_j\hat{}) \leq \lim_j \inf g_0(y_j, u_j, b_j)$ for every approximate U-solution;

iii) a point $(y, \sigma, b) \in A(S^*)$ is a *minimizing relaxed* (generalized) *solution* if it minimizes the function g_0 on the set $A(S^*)$.

In the last case the state equation (19) is substituted by

$$y(t) = \int\int_{R^*(s)} f[t, s, \xi(y)(s), r, b(t)]\sigma(s, r)drds = \int f[t, s, \xi(y)(s), \sigma(s), b(t)]\mu(ds). \quad (24)$$

The physical sense of the substitution consists in that for every t, s, y, and b we do not calculate a function f itself but its mean value by $r \in R^*(t)$ and here $\sigma(s, r)$ decides what kinds of value r and weight are involved in averaging.

Let us introduce some additional notions from [9] which are necessary for understanding the subsequent theorems. Let $\mathbb{R}$ be the *metric space* of real numbers or real measures, $C(T,X)$ the *metric space* of all bounded continuous functions on T to X, T a topological space, X a Banach space; (T, Σ, μ) a positive finite *measure space* (μ is finite if its values are in $\mathbb{R}$), $\mathscr{F}(T, \Sigma, \mu, X)$ the *metric space*, the set of all μ-measurable functions on T to X; $B(T, \Sigma, \mu, S; X)$ [or $B(T, S; X)$ or B] the vector space of functions $\varphi: T \times S \to X$ such that, for all $(t, s) \in T \times S$, $\varphi(t, .) \in C(S, X)$, $\varphi(., s)$ is μ-measurable and there exists a μ-integrable $\psi_\varphi : T \to \mathbb{R}$ with $|\varphi(t,.)|_{\sup} \le \psi_\varphi(t)$. The elements of B are *Caratheodory functions*, the function $|\varphi|_B = \int |\varphi(t,.)|_{\sup}\mu(dt): B \to R$ is a norm on B. The function $t \to \varphi(t, \xi(t)): T \to X$ is μ-integrable for every choice of $\varphi \in B$ and μ-measurable $\xi: T \to S$. Let more $frm(T)$ be the vector space of all Radon measures in T (Radon measure is the finite regular Borel measure, in particular, Dirac measure), $frm^+(S)$ the set of all positive measure in S, $rpm(S)$ the set of all Radon probability measure in S. A compact metric space R (or T) can be embedded in $frm(R)$ by identifying each $r \in R$ with the Dirac measure δ_r at r (or identifying r with the element l_r of functionals $C(R)^*$) such that $l_r(c) = c(r)$ for all $c \in C(R)$, $r \in R$. Let finally the sequentially closure of a set A (*seq cl*(A)) be the collection of all limits of convergent sequences in A, *seq cl*$(A) \subset A$^, which is the intersection of all the closed sets that contain A.

<u>Theorem 1</u>. *Under the relations* (17)-(24), *given* $k, n \in \mathbb{I}$, $V \in \mathbb{R}^k$, $W \in \mathbb{R}^n$, *and functions* $f: T \times T \times V \times R \times B \to \mathbb{R}^n$ *and* $\xi: C(T, W) \to L_\infty(T, \Sigma, \mu, V)$, *if* i) V *and* W *are closed*; ii) B *has a compact topology*; iii) *the function* $g = (g_0, g_1)|$ *seq cl*$(A(S^*))$ *is continuous*; iv) $A(S^*) \neq \varnothing$; v) $f \in C(T, B(T, V \times R \times B; \mathbb{R}^n))$ *and* ξ *is continuous, then the problem has the minimizing relaxed solution* $(y, \sigma, b) \in V \times S^* \times B$.

<u>Theorem 2</u>. *Assume that the conditions of Theorem* 1 *are satisfied with* $A(S^*)$ *replaced by* $H(S^*)$. *Let* (y^, σ^, b^) *be a minimizing relaxed solution,* U *an abundant subset of* $R^\#$, *and assume, furthermore, that* y^ *is the unique solution of the equation* $y = F(y$, σ^, b^), *and the equation* $y = F(y, u$, b^) *has a solution* y *for each* $u \in U^*$ *and sufficiently close to* σ^ *(in the metric space* S*). Then there exists a sequence* (y_j, u_j) *in* $Y \times U^*$ *such that* $(y_j, u_j$, b^) *is a minimizing approximate* U*-solution and* $\lim_j g(y_j, u_j$, b^) = g(y^, ω^, b^).

Theorem 3. *Let the conditions of Theorem* 1 *be satisfied and* $(y^\wedge, \sigma^\wedge, b^\wedge)$ *be any minimizing relaxed solution. If* i) $g(y, \sigma, b) = g^\wedge(y, b)$ *for all* $(y, \sigma, b) \in Y \times S^* \times B$; ii) *the set* $R^*(s)$ *is closed for all* $s \in T$; iii) $f(., ., v, r, b)$ *is* $\Sigma \otimes \Sigma$*-measurable for all* $(v, r, b) \in V \times R \times B$; *and that there exists* $v \in frm^+(T)$ *such that* iv) $v(A) > 0$ *for any open* $A \subset T$; *and* v) $\{f(., s, v, r', b\} | r' \in R^*(s)\}$ *is a convex subset of* $\mathscr{F}(T, \Sigma, v, \mathbb{R}^n)$ *for all* $(s, v, b) \in T \times V \times B$, *then the problem has the minimizing U-solution.*

Theorem 4. *Let the conditions of Theorem* 1 *be satisfied, and assume furthermore that, for all* $(t, s, v, r, b, y, \sigma) \in T \times T \times V \times R \times B \times Y \times S^*$, *we have* $g(y, \sigma, b) = g^\wedge(y, b)$, $R^*(s)$ *is closed, and*

$$f(t, s, v, r, b) = \sum_{j=1}^{l} \alpha^j(t, s, v, b)\beta^j(s, v, r, b), \tag{25}$$

where $l \in \mathbb{I}$, α^j: $T \times T \times V \times B \rightarrow \mathbb{R}$, β^j: $T \times V \times R \times B \rightarrow \mathbb{R}^n$, $\beta^j(s, ., ., .)$ *are continuous, and* $\beta^j(., v, r, b)$ μ*-measurable. Then there exists a minimizing relaxed solution* $(y^\wedge, \sigma^\wedge, b^\wedge)$ *such that, for all* $s \in T$, $\sigma^\wedge(s)$ *is a measure supported on a set of at most* $ln + 1$ *points of* $R^*(s)$.

These Theorems 1- 4 are proved in [9, pp. 408-413] (under some more general conditions and objects).

2. EXAMPLES OF ANALYTIC SOLUTIONS

Let

$$a^{-1}(t^\wedge) - t^\wedge \geq T - t^\wedge = 1/(\alpha y^\wedge),\ y \leq y^\wedge.$$

Then for AES

$$I(y) = \int_{t^\wedge}^{T} c(t)dt = \beta^\wedge(1-y^\wedge)P(T-t^\wedge) + \int_{t^\wedge}^{T} [\beta(1-y) - \beta^\wedge(1-y^\wedge)]m(t)dt \tag{26}$$

and

$$I'(y) = \int_{t^\wedge}^{T} (T-t)\{-\beta + \alpha(t-t^\wedge)[\beta(1-y) - \beta^\wedge(1-y^\wedge)]\}m(t)dt. \tag{26'}$$

If $\beta = \beta^\wedge$, then $I'(y) < 0$ and hence

$$\max I(y)\ (0 \leq y \leq 1) = I(0) = \beta^\wedge(1-y^\wedge)P(T-t^\wedge) + \beta^\wedge y^\wedge \int_{t^\wedge}^{T} (T-t)m(t)dt. \tag{27}$$

Let

$$\beta = \beta^\wedge \exp[c_\beta(t-t^\wedge)],\ d = c_\beta - \alpha y^\wedge.$$

Then under large d with respect to the main terms

$$I'(0) = (y\hat{}^{-1} - 1)/d^2 \exp[d(T-t\hat{})] > 0,\ I'(1) < 0.$$

Thus, $\max I(y)\ (0 \le y \le 1) = I(y^*),\ y^* > 0$.

For NES, we consider the problem (71, Ch. 2): $\max(T-t\hat{})\ (0 \le y \le 1)$ under the restriction $c(t) \ge c^0(t)$. Let $c^0(t) = c^0 = \alpha(1-y)P(t)$ and α, β, y, and P be constants. On the strength of (5, 6, Ch. 2)

$$c(t) = c^0,\ P(t) = P,\ m(t) = \alpha y P = m\hat{},\ a(t) = t - t\hat{}$$

and therefore

$$\max I8(y)\ (0 \le y(t) \le 1) = \max (T-t\hat{})\ (0 \le y(t) \le 1) = T(y) - t\hat{} = \infty. \tag{28}$$

Of course, the conclusion (28) is valid under the condition that any process of disintegration is absent here.

Indeed, let α, y, and P be the same, but $\beta = \beta\hat{}\exp(-dt)$, where $\beta\hat{}$ and d are positive constants. Then

$$m(t) = \alpha y P,\ c(t) = \alpha\beta\hat{}Py(1-y)\exp[-d(t-t\hat{})][1-\exp(-dt\hat{})]/d$$

and therefore under $c(t) \ge c^0 =$ constant,

$$t - t\hat{} \le \ln\{\alpha\beta\hat{}Py(1-y)[1-\exp(-dt\hat{})]/[dc^0]\}/d. \tag{29}$$

It follows from (29) that

$$\max (T-t\hat{})\ (0 \le y \le 1) = \ln\{1/4\alpha\beta\hat{}P[1-\exp(-dt\hat{})]/[dc^0]/d. \tag{30}$$

Let us consider one more example when P, and $m\hat{}$ are positive constants, $y\hat{} = 1$, $\alpha = \beta = xs$, $xt\hat{}^2 = 2$. We will find y, m, c, and a for which $I1(y) = \max$. In our case

$$I1 = I = I(y) = \int_{t\hat{}}^{T} c(t)dt = \int_{t\hat{}}^{T}[\int_{a(t)}^{t} \alpha(s)m(s)ds - m(t)]dt. \tag{31}$$

So, regarding the infinitesimal of the first order, the variations

$$dP = 0 = \int_{a+da}^{t}(m+dm)ds - \int_{a}^{t} m ds,\ dI = \int_{t\hat{}}^{T}\{\int_{a(t)}^{t}[\alpha(s)-\alpha(a)]dm(s)ds - dm(t)\}dt$$

$$= \int_{t\hat{}}^{T}\{\int_{t}^{\hat{}a(t)}[\alpha(t)-\alpha(a(s))]ds - 1\}dm(t)dt,$$

where $\hat{}a(t) = a^{-1}(t),\ t \in [t\hat{}, a(T)],\ \hat{}a(t) = T,\ t \in [a(T), T]$. Note that the last formula is obtained on the basis of rearrangement of the integration order. Let us put

$$\int_t^{{}^\wedge a(t)} [\alpha(t)-\alpha(a(s))]ds - 1 = 0,\ t \in [t^\wedge, a(T)]\ (\alpha(t) = xt). \tag{32}$$

It is not difficult to verify that the solution of this equation is $a(t) = t-t^\wedge$. Therefore

$$m(t) = m^\wedge = P/t^\wedge,\ y(t) = (t-t^\wedge)y(t-t^\wedge)/t = (t-kt^\wedge)/t,\ t \in [kt^\wedge,(k+1)t^\wedge],\ k = 1, 2, \ldots,$$

$$c(t) = xm^\wedge t^\wedge(t-t^\wedge),\ t \in [t^\wedge,\ T-t^\wedge]. \tag{33}$$

With regard to the infinitesimal of the second order in the vicinity of the y obtained on the segment $[t^\wedge, T-t^\wedge]$, the variation

$$dI = \int_{t^\wedge}^{T-t^\wedge} \{\int_{a(t)+da}^{a(t)} [\alpha(s)-\alpha(a(t))]dsm^\wedge dt < 0.$$

Thus the m, y, c in (33) are optimal on $[t^\wedge, T-t^\wedge]$. On the segment $[T-t^\wedge, T]$ we have

$$dI = \int_{T-t^\wedge}^{T} \{\int_t^T [\alpha(t)-\alpha(s-t^\wedge)]ds - 1\}dm(t)dt,\ dm = \int_{t-t^\wedge}^{t} [\alpha y-\alpha(t-t^\wedge)y(t-t^\wedge)]dmds +$$

$$\int_{t-t^\wedge}^{t} \alpha mdyds,$$

where $\int_t^T [\alpha(t)-\alpha(s-t^\wedge)]ds < 1$. We assume that

$$T = (R+1)t^\wedge,\ R > 1;\ dy \leq 0,\ s \in [Rt^\wedge,(R+1)t^\wedge],\ dy = 0,\ s \in [t^\wedge, Rt^\wedge],$$

and show that $dm \leq 0$. In our case

$$\alpha(t-t^\wedge)y(t-t^\wedge) = x(t-Rt^\wedge)$$

and therefore

$$dm + x(t-Rt^\wedge)\int_{Rt^\wedge}^{t} dmds = \int_{Rt^\wedge}^{t} \alpha ydmds + \int_{Rt^\wedge}^{t} \alpha mdyds$$

from which

$$dm = \int_{Rt^\wedge}^{t} (\alpha y\, dm + \alpha mdy)\{1 - x(t-Rt^\wedge)(t-s)\exp[-x(t-s)(t+s-2Rt^\wedge)/2]\}ds,$$

where expression in $\{\ \} > 0$ on the strength of the inequality $\exp(u) > 2u,\ u > 0$. Therefore $dm \leq 0$ and hence $dI \geq 0$. It means that the optimal

$$y = 0,\ s \in [Rt^\wedge,(R+1)t^\wedge], \tag{34}$$

and the optimal

$$m = \int^{Rt^\wedge}_{t-t^\wedge} xsy(s)m^\wedge ds = xm^\wedge[(R+1)t^\wedge - t][t-(R-1)t^\wedge]/2,\ a'(t) = m/m^\wedge,\ a(Rt^\wedge) = (R-1)t^\wedge,$$

$$c = xm^\wedge \int^{Rt^\wedge}_{t-t^\wedge} sds - m + x\int^{t}_{Rt^\wedge} sm(s)ds,\ t \in [Rt^\wedge, (R+1)t^\wedge]. \qquad (34')$$

The optimal (maximal) value of the functional I on the segment $[t^\wedge, (R+1)t^\wedge]$ is

$$I^* = xm^\wedge t^\wedge \int^{Rt^\wedge}_{t^\wedge} (t-t^\wedge)dt + \int^{(R+1)t^\wedge}_{Rt^\wedge} c(t)dt, \qquad (35)$$

where c is given in (34'). One can see that $I^* \to \infty$ with the speed R^2 in order for $R \to \infty$ though P is a constant.

3. EXISTENCE AND UNIQUENESS OF SOLUTIONS

Theorem 5. *Under the assumptions of Theorems* 2-5, Ch. 3 *the problems* (64-71, Ch. 2) *have the minimizing generalized solutions, the minimizing approximate U-solutions and the minimizing U-solutions, i.e., the ordinary solutions y* on the set of measurable functions U*.*

Proof. We give the proof for the problem (C.1, Ch. 2) by analogy with [10]. In the other cases the proof is similar. Let in notations of [9] $y = x = (m, c)$, $u = y$, $b = a$,

$$g_0(y, u, b) = -\int^{t}_{t^\wedge} c(t)dt,\ g_1(y, u, b) = \int^{t}_{a(t)} m(s)ds - P(t),\ 0 \le a(t) \le t, \qquad (36)$$

and the state equation (19) has the form

$$m(t) = \int^{t}_{0} f_1[t,s,m(s),y(s),a(t)]ds,\ c(t) = \int^{t}_{0} f_2[t,s,m(s),y(s),a(t)]ds,$$

$$f_1 = \begin{cases} \alpha(t,s)y^\wedge(s)m^\wedge(s), & a(t) \le s \le t^\wedge, \\ \alpha(t,s)y(s)m(s), & a^*(t) \le s \le T, \\ 0, & 0 \le s \le a(t), \end{cases} \quad f_2 = \begin{cases} \beta(t,s)[1-y^\wedge(s)]m^\wedge(s), & a(t) \le s \le t^\wedge, \\ \beta(t,s)[1-y(s)]m(s), & a^*(t) \le s \le T, \\ 0, & 0 \le s \le a(t), \end{cases} \quad a^*(t) = \max\,[t^\wedge, a(t)]. \qquad (37)$$

Let us accept

$$Y = C([t^\wedge, T]) \times C([t^\wedge, T]),\ B = \{a(t) \in C^1([t^\wedge, T]) \mid 0 \le a(t) \le T\}. \qquad (38)$$

As a control space U we consider the space ordinary controlling functions y: $[t^\wedge, T] \to \mathbb{R}$ and the space Σ of generalized controlling functions σ representing functions of the time with the values on the set of measures. Respectively, let us determine the admissible set of control values

$$R^*(t) = [0, 1] \subset \mathbb{R}, \tag{39}$$

the control functions

$$U^* = \{y \in U |\ y(t) \in R^*(t)\} \tag{40}$$

and the generalized control functions

$$\Sigma^* = \{\sigma(t, r) \in \Sigma | \int_{R^*(t)} \sigma(t, r) dr = 1. \tag{41}$$

Let V be the set of values for the state variable $x = (m, c)$ under $y \in R^*$ and $a \in B$, $V \subset \mathbb{R} \times \mathbb{R}$. These are the following facts:

(i) on the strength of continuity and monotonicity of the operator in the state equation (37) the set V is bounded and closed;

(ii) on the strength of the assumptions of Theorem 2 the set B is compact;

(iii) the function $g = (g_1, g_2)$ is continuous;

(iv) the set $A(\Sigma^*) \neq \varnothing$.

Furthermore, the following properties of the function $f(f_1, f_2)$, f: $[t^\wedge, T] \times [0, T] \times V \times R^* \times B \to \mathbb{R} \times \mathbb{R}$, are true:

(i) f is continuous over t under $(s, x, y, a) \in [0, T] \times V \times R^* \times B$;

(ii) f is continuous over (x, y, a) under $(t, s) \in [t^\wedge, T] \times [0, T]$;

(iii) f has a discontinuity under $s = a(t)$ but it is measurable by s under $(t, x, y, a) \in [t^\wedge, T] \times V \times R^* \times B$;

(iv) there exists an integrable function $\psi : [0, T] \to \mathbb{R}$ such that

$$\sup |f(., s, ., ., .)| ([t^\wedge, T] \times V \times R^* \times B) \leq \psi(s),\ s \in [0, T]. \tag{42}$$

Thus, all the conditions of Theorem 1 have been realized and therefore the problem C.1 (64, Ch. 2) has the minimizing solution $(m, c, \sigma, a) \in V \times \Sigma^* \times B$.

Let us prove now existence of the approximate U-solution. Let $(x^\wedge, \sigma^\wedge, a^\wedge)$ be the minimizing relaxed solution. It is easy to see that $x^\wedge$ is the unique solution of the Volterra-type equation $x = F(x, \sigma^\wedge, a^\wedge)$ and the equation $x = F(x, y, a^\wedge)$ has a solution for each $y \in U^*$ and sufficiently close to $\sigma^\wedge$ in the metric space Σ. It means that the conditions of Theorem 2 are valid.

Note that according to [9] the sequence y_j will converge to the measure σ if

$$\text{Lim} \int_0^T \varphi(s, y_j(s)) ds\ (j \to \infty) = \int_0^T ds \int_0^1 \varphi(s, r) \sigma(r) dr \tag{43}$$

for arbitrary $\varphi \in L([0, T] \times [0,1])$.

Let us prove, at last, the existence of the ordinary solution y on the set of measurable functions. Indeed, there are the following properties of the problem:

i) $g(x, \sigma, a) = g^\wedge(x, a)$, $g^\wedge(x, a)$: $X \times B \to \mathbb{R} \times \mathbb{R}$ for all $(x, \sigma, a) \in X \times \Sigma^* \times B$;

the condition ii) - iv) of Theorem 3 are obviously valid;

v) the set of functions $f^\wedge(r) = f(., s, x, r, a), f^\wedge: R \to L_1([t^\wedge, T])$, makes up a convex subset $\{f^\wedge(r) | r \in R^*(s)\}$ in the metric space of integrable functions $[t^\wedge, T] \to \mathbb{R} \times \mathbb{R}$ (it follows from the formula (36)-(40)) for all $(s, x, a) \in [0,T] \times V \times B$. Thus, for the problem considered Theorem 3 is valid on the strength of which our problem has the minimizing U-solution (x, y, a).

It should be noted that the set B in [9] is usually the convex subset of a finite vector space. But in Ch. V, VI, and VIII of [9] there are remarks about the validity of the theorems when B is, in particular, the convex compact subset of a space of continuous functions.

At any case (and to weaken the conditions of Theorem 5) we can give another proof based on reducing of the equation of motion to the form (see (22, Ch. 3))

$$m(t) = \int^{t^\wedge}_{a(t)} \alpha(t,s)y(s)m(s)ds + \int^{t}_{t^\wedge} \alpha(t,s)y(s)m(s)\}ds,$$

$$M(t) = \int^{t}_{0} m(s)ds,$$

$$a(t) = M^{1}[M(t) - P(t)],\ t \in [t^\wedge, T], \tag{44}$$

i.e., to the case of $n = 3$ in Theorem 1. If $a(t) \leq t^\wedge$ in (44), then we can establish easily the validity of all the conditions of Theorem 1. Otherwise, we have to consider (44) under $t \in [t^\wedge, t_1]$, $a(t) \leq t^\wedge$, then under $t \in [t_1, t_2]$, $a(t) \leq t_1$, and so on until we pass T. And also we have to prove that the operator F corresponding to the system (44) is continuous. This last property of F is almost obvious with regard to the natural restriction $\alpha(t,s)y^\wedge(s)m^\wedge(s) > 0$, $s \in [0,t^\wedge]$. □

Theorem 6. *Under the assumptions of Theorems* 6, 7, Ch.3 *the problems* (64-71, Ch.2) *have the minimizing U-solutions* x^*, y^*.

Proof. It is similar to the above, but the role of a state equation (19) will play the system of the equations in (51, Ch. 2) with respect to m, c and α. □

Theorem 7. *Under the assumptions of Theorem* 10, Ch. 3 *the problems* (64-71, Ch.2) *have the minimizing U-solutions* x^*, y^*, z^*, e^*, g^*, r^*, w^*.

Proof. It is similar to the above. □

Theorem 8. *In the particular case* $\beta(t,s) = k(t)\alpha(t,s)$, *where* k *is a continuous function, the problem* C.1 *has a minimizing relaxed solution* (m, c, σ, a) *such that for all* $s \in [t^\wedge, T]$ $\sigma(s)$ *is a measure supported on a set of at most* 2 *points of* $[0,1]$.

Proof. We can rewrite the relations (36), (37) in the form

$$g_0(y, u, b) = -\int_{r}^{t} c(t)dt,\ c(t) = k(t)\int_{a(t)}^{t} \alpha(t,s)m(s)ds - k(t)m(t),$$

$$g_1(y, u, b) = \int_{a(t)}^{t} m(s)ds - P(t),\ 0 \le a(t) \le t,\ m(t) = \int_0^t f[t,s,m(s),y(s),a(t)]ds,$$

$$f = \begin{cases} \alpha(t,s)y^\wedge(s)m^\wedge(s),\ a(t) \le s \le t^\wedge, \\ \alpha(t,s)y(s)m(s),\ a^*(t) \le s \le T, \\ 0,\ 0 \le s \le a(t), \end{cases} \tag{45}$$

It means that for our problem Theorem 4 takes place and also $l = n = 1$. □

Later we will show that in our case the measure σ supported on two points $r = y_{min}(t)$ (in particular, $r = 0$) and $r = 1$ on the boundary of $R^*(t)$. Then [2] we have the relaxed solution

$$\sigma(s, r) = p(s)\delta(r-1) + [1-p(s)]\delta[r-y_{min}(s)], \tag{46}$$

where δ is Dirac measure, and therefore

$$x(t) = F(x, \sigma, b) = \int_0^t \int_{R^*(s)} f[t, s, m(s), r, b(t)]\sigma(s, r)drds =$$

$$\int_0^t \{f[t, s, m(s), 1, b(t)]\, p(s) + f[t, s, m(s), y_{min}(s), b][1-p(s)]\}ds. \tag{47}$$

From (47), (37) and Theorem 5 it is easy to conclude that

$$y^*(t) = p(t) + y_{min}(t)[1-p(t)] \tag{48}$$

is the desired ordinary minimized solution. In the case $R^*(t) = [0,1]$ the optimal control $y^*(t) = p(t)$, which is the probability that the control will be equal to the value $r = 1$.

Let $g_0{}'_u(y, u, b) = g_0'(t)$ be the gradient of $g_0(y, u, b)$ and let $g_0'(t)$ be positive, negative for t belonging to the segments $T1$, $T2$ correspondingly for all admissible u. Then in (46) and (48) we obviously have $p(t) = 1$, $t \in T1$, and $p(t) = 0$, $t \in T2$. Thus, for determination of real structure of the solutions of optimization problems we need to determine and investigate the gradients of the respective specific functionals.

The theorems similar to Theorem 8 can be also established for other optimization problems.

4. ON STABILITY OF SOLUTIONS

Theorem 9. *Under assumption of s.*1, *the remained term R in the formula of the finite increment*

$$L(x+\delta x, y+\delta y, z+\delta z, t) - L(x, y, z, t) = \delta L(x, y, z, t) + R, \tag{49}$$

where L and δL are given by the relations' (6)-(11), *satisfies the condition*

$$|R| = o(\|\delta y\|_{L\infty}) \text{ or (what is the same) } |R|/\|\delta y\|_{L\infty} \to 0 \text{ for } \|\delta y\|_{L\infty} \to 0. \qquad (50)$$

Besides, regarding the main terms, the variation of the functional I (5) $\delta I = I'(y)\delta y$, *where* $I'(y)$ *is given by the formula* (14).

Proof. Under the assumption of s.1 there exist finite values $\|k\|_{C}$, $\|\Phi\|_{C}$, $\|p\|_{C}$, $s \in [0,T]$, $t \in [t\char`^,T]$, and also norms $\|x\|_C$ and $\|\varphi\|_C$ of the desired solutions $x(t)$ and $\varphi(t)$, $t \in [t\char`^,T]$, under any controls y, $y \in U$. On the strength of continuity of the relation $x(t)$ of $y(t)$ and finiteness of the segment $[t\char`^,T]$, the equality $\|\delta x\|_C = O(\|\delta y\|_{L\infty})$ or (what is the same) $\|\delta x\|_C/\|\delta y\|_{L\infty} < c$, where c is a certain constant, is valid. The remainder term R contains an analytical expression over products of variations δy and δx and depends on the given and desired functions [8], and also on $T\text{-}t\char`^$. Thus, $|R| = o(\|\delta y\|_{L\infty})$. The last conclusion of Theorem is trivial. □

It means that in the case of MM (5-11, Ch. 2) and under the assumptions of s. 1, we have the property of *stability of solutions* for the optimization problem C.1-C.8, Ch. 2.

Let us dwell on some of the problems C.1-C.8, Ch. 2 in more detail.

Lemma 1. *The gradient of the functional I with regard to MM* 5, 6, Ch. 2 *under the condition that P is given has the form*

$$I'(y) = -m(t)\int_t^{\char`^A(t)} [\alpha(s,t)\psi(s)+\beta(s,t)]ds,\ \char`^A(t) = a^{-1}(t),\ t\in [t\char`^,a(T)];\ T,t\in [a(T),T], \qquad (51)$$

where $\psi(t)$, $t \in [t\char`^,T]$, *is determined uniquely under any given* $y(t)$ *with the help of the equations* (5, 6, Ch. 2) *and the conjugate equation*

$$\psi(t) = \int_t^{\char`^A(t)} [\alpha(s,t)y(t) - \alpha(s,a(s))y(a(s))]\psi(s)ds - \int_t^{\char`^A(t)} \{\beta(s,t)[1-y(t)]-$$

$$\beta(s,a(s))[1-y(a(s))]\}ds. \qquad (52)$$

Lemma 2. *The gradient of the functional I with regard to MM* (5, 6, Ch. 2), *f is given, has the form*

$$I'(y) = -m(t) \int_t^{\char`^A(t)} \{[\alpha(s,t)\psi_1(s) + \beta(s,t)] + [\alpha(s,t) - \beta(s,t)]\psi_2(s)\}ds, \qquad (53)$$

where ψ_i, $i = 1, 2$, *are determined with the conjugate equations*

$$\psi_1(t) = \int_t^{\char`^A(t)} \{\alpha(s,t)y(t)\psi_1(s) -\beta(s,t)[1-y(t)]+ [\alpha(s,t)y(t)+\beta(s,t)(1-y(t))]\psi_2(s)\}ds,$$

$$\psi_2(t)[\alpha(t,a(t))y(a(t))+\beta(t,a(t))(1-y(a(t))] =$$

$$-\psi_1(t)\alpha(t,a(t))y(a(t))+\beta(t,a(t))(1-y(a(t)). \tag{54}$$

Lemma 3. *The gradient of the functional*

$$I1 = \int^{T}_{t^\wedge} \int^{t}_{b(t)} \beta(t,s)[1-y(s)]m(s)dsdt \tag{55}$$

with regard to MM (9, Ch. 2), P_1 *and* P_2 *are given, has the form*

$$I1'(y) = -m(t)\{\int^{\wedge A(t)}_{t} [\alpha(s,t) - \alpha(s,a(s))]\psi(s)ds + \int^{\wedge B(t)}_{t} [\beta(s,t) - \beta(s,b(s))]ds\},$$

$$^\wedge B(t) = b^{-1}(t),\ t \in [t^\wedge, b(T)];\ T,\ t \in [b(T),T], \tag{56}$$

the function ψ *being defined by*

$$\psi(t) = y(t)\int^{\wedge A(t)}_{t} [\alpha(s,t) - \alpha(s,a(s))]\psi(s)ds - (1-y(t))\int^{\wedge B(t)}_{t} [\beta(s,t) - \beta(s,b(s))]ds, \tag{57}$$

where a, b, and m are the solutions of the system considered (*see Theorem* 4, Ch. 3).

Lemma 4. *The gradient of the functional* (55) *with regard to MM* 9, Ch. 2, *f and P are given, has the form*

$$I1' = -m(t)\{\int^{\wedge A(t)}_{t} [\alpha(s,t)(\psi_1(s)+\psi_3(s))+\psi_2(s)]ds - \int^{\wedge B(t)}_{t} [\beta(s,t)(\psi_3(s)-1)+\psi_2(s)]ds\}, \tag{58}$$

the functions ψ_i, $i = 1, 2, 3$, *being defined by the equations*

$$\psi_1(t) = y(t)\int^{\wedge A(t)}_{t} \{\alpha(s,t)[\psi_1(s)+\psi_3(s)] + \psi_2(s)\}ds + [1-y(t)]\int^{\wedge B(t)}_{t} \{\beta(s,t)[\psi_3(s)-1] +$$

$$\psi_2(s)\}]ds, [\psi_1(t)+\psi_3(t)]\alpha(t,a(t)) + \psi_2(t) = 0,\ [\psi_3(t) -1]\beta(t,b(t)) + \psi_2(t) = 0, \tag{59}$$

where a, b, and m are the solutions of the system considered (*see Theorem* 5, Ch. 3).

Lemma 5. *The gradient of the functional I3 with regard to MM* (5, 6, Ch. 2), *c is given, has the form*

$$I3'(y) = -m(t)\int^{\wedge A(t)}_{t} [\alpha(s,t)\psi_1(s) - \beta(s,t)\psi_2(s)]ds, \tag{60}$$

where ψ_i, $i = 1, 2$, *are determined with the conjugate equations*

$$\psi_1(t) = \int^{\wedge A(t)}_{t} \{\alpha(s,t)y(t)\psi_1(s) + \beta(s,t)[1-y(t)]\psi_2(s) - 1\}ds,$$

$$\psi_1(t)\alpha(t,a(t))y(a(t)) + \psi_2(t)\beta(t,a(t))[1-y(a(t))] = 1. \quad (61)$$

The similar lemmas are also valid for the functionals *I2-I8* (Ch.2).

Lemma 6. *The gradient of the functional* I ($I = \int^T_{t^*} \int^t_{a(t)} \beta(t,s)[1-x(s)-y(s)]m(s)dsdt$) *with regard to MM* (51, Ch. 2), $\nu = 1$, $\mu = 1$, $c_1 = 0$, *P is given*, $\alpha_1 = \alpha(s) = \beta(t,s)$, *has the form*

$$I'(x,y) = (I'_x, I'_y),\ I'_x(x,y) = -m(t)\alpha(t)\int^{\wedge A(t)}_t [\psi_1(s) + 1]ds,\ I'_y(x,y) =$$

$$-m(t)\alpha(t)\int^{\wedge A(t)}_t [\psi_2(s) + 1]ds, \quad (62)$$

where the conjugate variables ψ_i, $i = 1, 2, 3$, *are determined with the relations*

$$\psi_1(t) + \int^T_t m(u)\int^{\wedge A(u)}_u [1-x(u)(\psi_1(s)+1) - y(u)(\psi_2(s)+1)]ds\}du = 0,$$

$$\psi_2(t) + \int^{\wedge A(t)}_t \{\alpha(t)[1-x(t)(\psi_1(s)+1) - y(t)(\psi_2(s)+1)] - \psi_3(s)\}ds = 0,$$

$$\alpha(a(t))[1-x(a(t))(\psi_1(t)+1) - y(a(t))(\psi_2(t)+1)] - \psi_3(t) = 0, \quad (63)$$

and also m, a, and α *are the solutions of the system* 51, Ch. 2.

Lemma 7. *The gradient of the functional* I ($I = \int^T_{t^*} \int^t_{a(t)} \beta(t,s)[1-x(s)-y(s)-z(s)-e(s)-g(s)-r(s)-w(s)]m(s)dsdt$) *with regard to MM* (55, Ch. 2), $\nu = 1$, $\mu = 1$, $c_1 = d_1 = e_1 = r_1 = 0$, $ai(t) = a(t)$, $i = 1,\dots, 8$, *P is given, has the form*

$$I'(x, y, z, e, g, r, w) = (I'_x, I'_y, I'_z, I'_e, I'_g, I'_r, I'_w),\ I'_x = -m(t)\int^{\wedge A(t)}_t [\beta(t) + \alpha(t)\psi_1(s)]ds,$$

$$I'_y = -m(t)\int^{\wedge A(t)}_t [\beta(t) + \alpha(t)\psi_2(s)]ds,\ I'_z = -m(t)\int^{\wedge A(t)}_t [\beta(t) + \alpha(t)\psi_3(s)]ds,$$

$$I'_e = I'_g = I'_r = I'_w = -m(t)\beta(t)\int^{\wedge A(t)}_t ds, \quad (64)$$

where

$$\psi_1(t) - \int^T_t\int^{\wedge A(u)}_u m(u)d(u,s)dsdu = 0,\ d(u,s) = x(u)\psi_1(s) + y(u)\psi_2(s) + z(u)\psi_3(s),$$

$$\psi_2(t) - \int^{\wedge A(t)}_t \{[\beta(1-x-y-z-e-g-r-w)]|_t - \alpha(t)d(t,s) - \psi_4(s)\}ds = 0,$$

$$\psi_3(t) + \int^{\wedge A(t)}_t [(1-x-y-z-e-g-r-w)m]|_t\, ds = 0,$$

$$-[\beta(1-x-y-z-e-g-r-w)]|_{a(t)} + \alpha(a(t))d(a(t),t) + \psi_4(t) = 0. \quad (65)$$

The proof of Lemmas 1-7 is almost trivial according to the material of s. 1 (see the relations (1)-(14)). It should be noted only that for the proof of Lemmas 6, 7 we need to use, in addition, the formulae of the type

$$\int_{t^\wedge}^{T} \psi_1(t) \int_{a(t)}^{t} \alpha(s)x(s)m(s)dsdt = \int_{t^\wedge}^{T} [\alpha(t^\wedge) + \int_{t^\wedge}^{t} \alpha'(u)du]x(t)m(t) \int_{t}^{\wedge A(t)} \psi_1(s)ds =$$

$$\int_{t^\wedge}^{T} \int_{u}^{T} x(t)m(t) \int_{t}^{\wedge A(t)} \psi_1(s)dsdt\, \alpha'(u)du + \int_{t^\wedge}^{T} \alpha(t^\wedge)x(t)m(t) \int_{t}^{\wedge A(t)} \psi_1(s)ds. \quad (66)$$

Theorem 10. *With regard to the main terms, the variations of the functionals I1 and I3 for different cases of MM* (5-9, 51, 55, Ch. 2) *are given by the formulae* $\delta I = I'(y)\delta y$, *where* $I'(y)$ *are given by the respective formulae* (51)-(65).

Proof. It is obvious from Lemmas 1-7 and the proof of Theorem 9. □

5. INVESTIGATION OF STRUCTURE OF SOLUTIONS

The first essential result on the properties of solutions of the problem C.1 (see Ch. 2) has been obtained by V.M. Glushkov and V.V. Ivanov [3]. The result consisted qualitatively in that for "small" T-$t^\wedge$ the desired $y(t)$ is minimally possible ($y(t) = y_{min}(t)$ by virtue of the restriction $a'(t) \geq 0$), but for "large" T-$t^\wedge$ the desired $y(t)$ may differ from the minimally possible on the larger initial part of the segment $[t^\wedge,T]$ and only on the smaller final part of $[t^\wedge,T]$ the desired $y(t)$ is minimally possible. The notions "small" and "large" depend on the values of the functions α and β, namely, the greater the functions in question, the closer to $t^\wedge$ is the boundary between "small" and "large" segments.

The obtained result was in essence an application of the fact that for "large" T-$t^\wedge$ growth of $m(s)[1-y(s)]$ overtakes diminution of $1-y(s)$. The result has obtained, in the sequel, the important qualitative very general interpretation: the record of an external function for any ES can be obtained only under the conditions of its sufficiently comfortable guarantee, that is, under the significant fraction of resources sent to internal needs of ES.

The *structure of solution* for the problem C.1 and other optimization problems has been investigated in detail mainly by Yu.P. Yatsenko [4, 5, 10]. Let us refer to the appropriate results below without detailed proofs (these proofs can be found in [4, 5, 10] and [9], Ch. 2). Some of those results will follow from Theorem 8 and also from the asymptotic properties of the solutions for some of the optimization problems.

Lemma 8. *If* $\alpha(t,s)$ *and* $P'(t)$ *belong to the class of Lipschitz over* t, $t \in [t^\wedge,T]$, *then on the segment* $[t^\wedge,T]$ *there exists the unique integrable function* $y_{min}(t) = 0$, $t \in \Delta \subset [t^\wedge,T]$: $a'_{min}(t) > 0$; $y_{min}(t) = y^{\cdot}(t) \geq 0$, $t \in [t^\wedge,T] \setminus \Delta$: $a'_{min}(t) = 0$, *where* $a_{min}(t)$ *is determined uniquely from the equation* 5, 6, Ch. 2, *under* $y = y_{min}$. *And also* $0 < y_{min}(t) < 1$, $t \in [t^\wedge,T]$, *if*

$$P''(t) < \alpha(t,t)P'(t) + \int_{t^\wedge}^{t} \alpha'_t(t,s)P'(s)ds + \int_0^{t^\wedge} \alpha'_t(t,s)y^\wedge(s)m^\wedge(s)ds, \tag{67}$$

and $y_{\min}(t) = 0$, *if* $P'(t) \le 0$ *or*

$$0 \le P'(t) \le \int_{z(t)}^{t^\wedge} \alpha(t,s)y^\wedge(s)m^\wedge(s)ds,$$

$$m^\wedge(z(t))z'(t) - \int_{z(t)}^{t^\wedge} \alpha(t,s)y^\wedge(s)m^\wedge(s)ds = -P'(t),\ z(t^\wedge) = 0. \tag{68}$$

We will assume later that the condition $0 \le y_{\min}(t) \le 1$ is valid.

Theorem 11. *There exists the instant* θ, $t^\wedge \le \theta < T$, *such that* $I'(y_{\min},t) < 0$ *under* $t \in (\theta,T)$ *and* $y^*(t) = y_{\min}(t)$. If $\theta > t^\wedge$, *then the segment* $[t^\wedge,\theta]$ *consists of subintervals, where either* $I'(y^*,t) > 0$ *and* $y^*(t) = 1$, *or* $I'(y^*,t) < 0$ *and* $y^*(t) = y_{\min}(t)$, *or* $I'(y^*,t) = 0$ *and* $y_{\min}(t) \le y^*(t) \le 1$, *and here, all these cases are possible depending on the given functions of the problem.*

The more refined results can be obtained in the special cases

$$A.\ \beta(t,s) = k(t)\alpha(t,s),\ \text{B. } a(T) \le t^\wedge \text{ (then } a^{-1}(t) = T \text{ and } y(a(s)) = y^\wedge(a(s))\text{)}. \tag{69}$$

Lemma 9. *Let* A. *or* B. *be valid and* $I'_y(a^*, s) = 0$, $s \in [t,{}^\wedge A(t)]$. *If* $a(t) > a^*(t)$, $a(u) \ge a^*(u)$, $u \in (t,T]$, *then* $I'_y(a(t), t) < 0$, *and if* $a(t) < a^*(t)$, $a(u) \le a^*(u)$, $u \in (t,T]$, *then* $I'_y(a(t), t) > 0$.

We give a sketch of the proof in the case of both conditions A. and B.
By (51) and (52) we have in the case of A.

$$\psi(t) = -k(t). \tag{70}$$

and in the case of B.

$$\psi(t) = -\int_t^T [\beta(s,t) - \beta(s,a(s))]ds = \int_t^T [\alpha(t,s) - \alpha(s,a(s))]\psi(s)ds, \tag{71}$$

So, in both cases we have

$$I(a) = \int_{t^\wedge}^{T} k(t)[\int_{a(t)}^{t^\wedge} \alpha(t,s)m^\wedge(s)ds + \int_{t^\wedge}^{t} \alpha(t,s)m(s)ds - m(t)]dt,$$

$$dI = I(a+da) - I(a) = \int_{t^\wedge}^{T} k(t)[\int_{a(t)+da(t)}^{a(t)} \alpha(t,s)m^\wedge(s)ds +$$

$$\int_{t^\wedge}^{t} \alpha(t,s)dm(s)ds - dm(t)]dt,\ P(t) = \int_{a(t)}^{t^\wedge} m^\wedge(s)ds + \int_{t^\wedge}^{t} m(s)ds,$$

$$dP = \int^{a(t)}_{a(t)+da(t)} m\hat{}(s)ds + \int^{t}_{t\hat{}} dm(s)ds = 0. \tag{72}$$

Using (70) - (72) and

$$\int^{t}_{t\hat{}} \alpha(t,s)dm(s)ds = \alpha(t,t)\int^{t}_{t\hat{}} dm(v)dv - \int^{t}_{t\hat{}} \alpha'_s(t,s)\int^{s}_{t\hat{}} dm(v)dvds,$$

$$\int^{T}_{t\hat{}} k(t)\int^{t}_{t\hat{}} \alpha'_s(t,s)\int^{s}_{t\hat{}} dm(v)dvdsdt = \int^{T}_{t\hat{}}\int^{T}_{t} k(s)\alpha'_t(s,t)\int^{t}_{t\hat{}} dm(v)dvdsdt,$$

$$-\int^{T}_{t\hat{}} k'(t)\int^{t}_{t\hat{}} dm(s)dsdt = k(T)\int^{a(T)}_{a(T)+da(T)} m\hat{}(s)ds + \int^{T}_{t\hat{}} k(t)dm(t)dt \tag{73}$$

obtained on the strength of integration by parts and inversion of the order of integration we have

$$dI = \int^{T}_{t\hat{}} k(t)\int^{a(t)}_{a(t)+da(t)} [\alpha(t,s) - \alpha(s,a(t))]m\hat{}(s)dsdt + k(T)\int^{a(T)}_{a(T)+da(T)} m\hat{}(s)ds. \tag{74}$$

The desired properties follow from (74) immediately.

<u>Theorem 12.</u> *For the problem* C.1 *in the cases* (44) *and* (69) *there exists the "best" function* $a^*(t)$, $t \in [t\hat{},T]$, *such that* $I'(a^*,t) = 0$, $t \in [t\hat{},\theta]$, $t\hat{} \leq \theta < T$, *and* $I'(a^*,t) < 0$, $t \in (\theta,T)$. *Depending on the values of a disagreement*

$$\Delta_0 = |a^*(t\hat{}) - a(t\hat{})|, \Delta_1 = |a^{*\prime}(t\hat{}) - a'(t\hat{})| \tag{75}$$

and a length of the planning interval T-$t\hat{}$ *it is possible that the optimal solutions* $a(t)$ *and* $y(t)$ *have the following behavior variants:*

(*i*) $y(t) = y_{\min}(t)$, $t \in [t\hat{},T]$;
(*ii*) $y(t) = 1$, $t \in (t\hat{},s_1)$, *$a(t)$ does not intersect $a^*(t)$*, $y(t) = y_{\min}(t)$, $t \in [s_1,T]$;
(*iii*) $y(t) = 1$, $t \in (s_{i-1},s_i)$, $y(t) = y_{\min}(t)$, $t \in [s_i,s_{i+1}]$, $i = 1,\dots, N-1$, $s_0 = t\hat{}$, $s_N = T$,
$a(t)$ intersects $a^*(t)$ N times, $N \geq 1$. (76)

The case of unlimited quantity of switching is possible. In this case under $s_i \to s\hat{}$, where $s\hat{}$ is a condensation point, the values $|s_i - s_{i-1}|$ are decreasing geometrically with a ratio $q < 1$.

<u>Theorem 13.</u> *For MM* (51, Ch. 2), $\nu = 1$, $\mu = 1$, $z = c_1 = 0$, $\alpha_1 = \beta_1 = \alpha(s)$, *if*

$$m\hat{} > 0, 0 < P'(t) \leq P'_{max} = \int^{t\hat{}}_{a(t)} \alpha\hat{}(s)y\hat{}(s)m\hat{}(s)ds,$$

$$m\hat{}(a(t))a'(t) - \int^{t\hat{}}_{a(t)} \alpha\hat{}(s)y\hat{}(s)m\hat{}(s)ds = -P(t), t \in [t\hat{},T], \tag{77}$$

and the segment $[t\hat{},T]$ *is sufficiently large, then the optimal*

$$x^*(t) = 1, t \in [t^\wedge, t_1], y^*(t) = 1, t \in [t_1, t_2], x^*(t) = y^*(t) = 0, t \in [t_2, T], \quad (78)$$

and

$$t_1 - t^\wedge \asymp T - t^\wedge, t_2 - t_1 \asymp (T - t^\wedge)^{1/2}, T - t_2 \asymp 1. \quad (79)$$

The economic interpretation of Theorem 13 is the following. To achieve the maximum of summarized output of the consumer product for the sufficient large period T-$t^\wedge$ it is necessary that all new WP have to be directed on the largest initial part $[t^\wedge, t_1]$ of $[t^\wedge, T]$ to the branch of Science (S), after this on some internal part $[t_1, t_2]$ of $[t^\wedge, T]$ to the branch of production of means of production (A) and at last only on the end period T-t_2 to the branch of the consumer product production (B). In addition, $t_1 - t^\wedge = O(T - t^\wedge)$, $t_2 - t_1 = O([T - t^\wedge]^{1/2})$, $T - t_2 = O(1)$. Reduction of the maximum possible period of the consumer product production from $T - t^\wedge$ to $T - t_2$ is compensated by essential growth of new WP efficiency.

Theorem 14. *If the condition* A. (69) *is fulfilled;* $y_{min}(t) < 1, t \in [t^\wedge, T]$*;* $\beta(t,s)$ *is a monotone increasing function of s;* α *and P are slowly varying functions (i.e.,* α'_t, α'_s, *and* $P'(t) \ll 1$) *and the solution* y^* *of the problem* C.1 *has a finite number of switches, then this solution is unique.*

Theorem 15. *If the conditions* A. *and* B. *are fulfilled;* $\beta(t,s)[1-y^\wedge(s)]$ *and* $\alpha(t,s)y^\wedge(s)$ *are monotone increasing functions of s, then the problem* C.1 *has not more than one solution.*

Theorem 16 (on the turnpike or the attracting locus [1]). *Let the following conditions:*

$$i)\ \beta(s) = b^s,\ b > 1, \text{ or } \beta(s) = s^k,\ k > 0;\ ii)\ \alpha(t,s) = k(t)\beta(s), \quad (80)$$

where k is rapidly enough decreasing function of t, be fulfilled. Then for the problem C.1 *there exists the "best" function* $a^0(t)$ *("turnpike") such that* $I'_y(a^0) = 0$, *the solution of the problem* C.1 $a^*(t) \to a^0(t)$ *under* $t, T \to \infty$ *and also the behavior of* $a^0(t)$ *under* $t \to \infty$ *does not depend on the initial values* $m^\wedge(t)$, $y^\wedge(t)$, $t \in [0, t^\wedge]$, *and is determined only by functions* α *and* β.

6. CASE OF NON-UNIQUENESS OF SOLUTIONS

Let us consider now a case of *non-uniqueness of solutions* for optimization problems on the example of the same problem C.1, Ch.2.

Let α, $m^\wedge$, $y^\wedge$, P' be constants, $\beta = \beta(s)$. Then on the base of (51), (52) supposing $a(t) \le t^\wedge$, $t \in [t^\wedge, T]$, from $I' = I'(y, s) = 0$, $t \in [t^\wedge, \theta]$, $t^\wedge < \theta < T$, it follows that

$$\alpha \int_s^T \psi(t)dt = -\beta(s)(T-s),\ \psi(s) = (1-y^\wedge)\int_s^T [\beta(a(t)) - \beta(s)]dt,\ t \in [t^\wedge, \theta];$$

$$\psi(t) = \alpha[y(t)-y^\wedge]\int_t^T \psi(s)ds - \beta(t)(T-t)[1-y(t)] + (1-y^\wedge)\int_t^T \beta(a(s))ds,\ t \in [\theta,T];$$

$$m' = \alpha(y-y^\wedge)m + P'\alpha y^\wedge,\ m^\wedge = \alpha y^\wedge m^\wedge t^\wedge,\ [t^\wedge-a(t)]m^\wedge = P - \int_{t^\wedge}^t m(s)ds,\ t \in [t^\wedge,T];$$

$$I(y) = \int_{t^\wedge}^T \{m^\wedge(1-y^\wedge)\int_{a(t)}^{t^\wedge} \beta(s)ds + \int_{t^\wedge}^t [1-y(s)]\beta(s)m(s)ds\}dt. \tag{81}$$

Note that any equation of the form $f(t) = 0$, $t \in [t^\wedge,\theta]$, where f is a differentiable function, is identical to the equation $f'(t) = 0$, $f(t^\wedge) = 0$ or $f'(t) = 0$, $f(\theta) = 0$. Therefore under $t^\wedge = 1$, $\alpha y^\wedge = 1$, $y^\wedge = 1/2$, the relation (81) is identical to

$$\beta(a(s)) = -u'' - u',\ u = \beta(s)(T-s),\ u'(1)+u(1) = \int_1^T \beta(a(s))ds;\ m(t) = P' + m^\wedge a'(t),$$

$$y(t) = 1/2+[m'(t)-P']/[2m(t)],\ t \in [1,\theta],\ a(1) = 0,\ a'(1) = [m^\wedge-P']/m^\wedge \geq 0, \tag{82}$$

where the value θ is determined by the condition $I'(y,\theta) = 0$, $I'(0,t) < 0$, $\theta < t < T$. We have, besides (82),

$$2\int_\theta^T \psi(s)ds = -u(\theta),\ \psi(t) = -\int_t^T \psi(s)ds - \beta(t)(T-t) + 1/2\int_t^T \beta(a(s))ds,\ \theta < t < T. \tag{83}$$

Let us put, for example, $\beta(s) = \beta(\theta)[1+3(s-\theta)]$, $s \in [\theta,T]$,

$$\beta(s) = s,\ s \in [0,1/2];\ \beta(s) = s-1/2,\ s \in [1/2,1];\ \beta(a(s)) = -u'' - u' = 1/2(1-e^{-(t-1)}),$$

$$u(s) = \beta(s)(T-s),\ u'(1)+u(1) = \int_1^T \beta(a(s))ds,\ s \in [1,\theta];\ \beta(a(s)) = 1/2(1-e^{-(\theta-1)}). \tag{84}$$

It is not difficult to verify that the equation (84) with respect to $a(s)$ has the following two solutions

$$a_1(s) = 1/2(1-e^{-(s-1)}),\ a_2(s) = 1-1/2e^{-(s-1)},\ s \in [1,\theta], \tag{85}$$

and besides

$$u(t) = 1/2(T-t)(1-e^{-(t-1)}) + C(e^{-(t-1)}-e^{-(T-1)}),\ t \in [1,\theta],\ C = 1/2(e^{T-1}-e^{T-\theta}) + 1/2(T-\theta-1)e^{T-\theta},$$

$$\psi(t) = [1/4(1-e^{-(\theta-1)})-\beta(\theta)(1+3(T-\theta+2)](1-e^{t-T}) + 6\beta(\theta)(t-T), \tag{86}$$

where θ is the solution of the equation

$$f(\theta) = \beta(\theta)[1+6(T-\theta)] +$$

$$[1/2(1-e^{-(\theta-1)})-2\beta(\theta)(1+3(T-\theta+2)](T-\theta+e^{-(T-\theta)}-1)/(T-\theta) = 0. \tag{87}$$

It is easy to see that $f(T) = \beta(T) > 0, f(T\text{-}1) < 0$. It means that the desired θ meets the relation $T\text{-}1 < \theta < T$. We show that $a_1(t)$ and $a_2\hat{}(t) = a_m(t),\ t \in [1,\theta\hat{}];\ a_2\hat{}(t) = a_2(t),\ t \in [\theta\hat{},T]$, where $a_m(t)$ corresponds to $y_{max}(t) = 1,\ t \in [1,\theta\hat{}]$, deliver local maximums to the functional I. The function $a_m(t)$ and value $\theta\hat{}$ can be determined by the formulae:

$$a_m(t) = (2\text{-}1/[3(T\text{-}1)])(e^{t-1}\text{-}1)\text{-}2(1\text{-}1/[3(T\text{-}1)])(t\text{-}1), a_m(\theta\hat{}) = a_2(\theta\hat{}),\ a_m'(\theta\hat{}) = a_2'(\theta\hat{}).$$

It is not difficult to see that under large T we have $1 < \theta\hat{} < 2$. Then, we have

$$\delta I = \int^T_\theta \{-m(t)\int^T_t [2\psi(s)+\beta(t)]ds\}dt\text{-}\int^T_1 [\beta'(a(t))(\delta a(t))^2/2\text{-}\int^t_1 \delta y(s)\delta m(s)\beta(s)ds]dt.$$

$$\delta m(t) = 2\int^t_1 (y(s)\text{-}1/2)\delta m(s)ds + 2\int^t_1 m(s)\delta y(s)ds. \tag{88}$$

Therefore

$$\delta m(t) = 2\int^t_1 m(s)\exp\{2\int^t_s [y(u)\text{-}1/2]du\}\delta y(s)ds \tag{89}$$

and

$$\int^t_1 \delta y(s)\delta m(s)\beta(s)ds \geq m(\beta)m(m)(\int^t_1 \delta y(v)dv)^2/2 \geq 0. \tag{90}$$

Since $\int^T_t [2\psi(s)+\beta(t)]ds \geq 0,\ t \in [\theta,T]$, and besides $\beta'(a_i(t)) > 0,\ i = 1, 2$, we have

$$\delta I(y_1),\ \delta I(y_2) \leq 0,\ y_1(t) = 0,\ t \in [0,T];\ y_2(t) = 1,\ t \in [1,\theta\hat{}];\ y_2(t) = 0,\ t \in [\theta\hat{},T]. \tag{91}$$

It means that $y_1(t)$, $y_2(t)$ and the respective $a_1(t)$, $a_2\hat{}(t)$ deliver local maximums to I.

We can obtain also the case when $I(y_1) = I(y_2)$ [4, 5].

7. ASYMPTOTIC PROPERTIES OF SOLUTIONS

Let us consider some *asymptotic properties* of the relations (51)-(65) as $t \to T$ or $t, T \to \infty$. If $t \to T$, then $\hat{}A(t) = T$ for $a(T) \leq t \leq T$ and $\psi(t)$ in (52) tends to 0. So, in (51) under $\alpha(s,t)\psi(s) > -\beta(s,t),\ s \in [t,T],\ I1'(y) < 0$. It means that the desired y which maximizes $I = I1$ has to be minimum possible in the end of $[t\hat{},T]$ or on the whole segment $[t\hat{},T]$ if $T\text{-}t\hat{}$ is sufficiently small.

It is easy to see that the same property takes place for the relations' (53)-(57).

Under the condition of Lemma 5 and $t \to T$ we have

$$I3'(y) \approx m(t)\int^{\hat{}A(t)}_t \beta(s,t)/\{\beta(s,a(s))[1\text{-}y(a(s))]\}ds > 0, \tag{92}$$

that is the desired y that minimizes $I3$ has to be also a possible minimum in the end of $[t\hat{},T]$ or on the whole $[t\hat{},T]$ if $T\text{-}t\hat{}$ is sufficiently small. In a similar way, under the condition of Lemma 6 and $t \to T$ we have $\psi_1(t),\ \psi_2(t) \to 0$ and therefore $I1'_x,\ I1'_y < 0$ in the end of $[t^\wedge, T]$. It means that the desired x and y which maximize $I1$ have to be minimum in the end of $[t^\wedge, T]$.

At last, under the conditions of Lemma 7 and $t \to T$ we have $\psi_i(t) \to 0$, $i = 1, 2, 3, 4$, and hence $I1'_j < 0$, $j = x, y, z, e$, in the end of $[t^\wedge, T]$, that is the desired x, y, z, and e have to be minimum in the end of $[t^\wedge, T]$.

Let us show that $y(t) = 0$ cannot be the solution of the problem C.1 under certain conditions. Let $y(t) = 0$, $P(t) \neq 0$, $t \in [t^\wedge, T]$, in MM (5, 6, Ch. 2). Then $a(t) < t^\wedge$. Indeed, if $a(t) \geq t^\wedge$, then $m(t) = 0$ and $P(t) = \int_{a(t)}^{t} m(s)ds = 0$, but it is impossible. Let

$$0 < y^\wedge(s), \alpha(t,s) \sim \alpha\exp[-c_1(t-s)],\ \beta(t,s) \sim \beta\exp[-c_2(t-s)],\ c_1, c_2 > 0,\ \alpha > c_1,\ t \to \infty. \quad (93)$$

From (93) and (51), (52) it follows that under t, $T-t \to \infty$ ($y(t) = 0$, $t \in [t^\wedge, T]$) $-\psi_2(t) \gtrsim -\alpha m(y^\wedge)\int_t^T \exp(-c_1 s)[-\psi_2(s)]ds + \beta/c_2$, from which $-\psi_2(t) \gtrsim \beta/c_2$,

$$I'(y) \sim m(t)\int_t^T \{\alpha\exp[-c_1(s-t)][-\psi_2(s)] - \beta\exp[-c_2(s-t)]\}ds \gtrsim m(t)\beta/c_2(\alpha/c_1 - 1) > 0.$$

It means that $y(t) = 0$, $t \in [t^\wedge, T]$, cannot be the solution of the problem C.1.

Let more $x(t) = y(t) = 0$, $t \in [t^\wedge, T]$, in MM (51, Ch. 2), $\alpha = \beta$, $\mu = \nu = 1$, $c_1 = 0$. Then $a(t) < t^\wedge$ and under t, $T-t \to \infty$ from (62), (63) it follows that

$$I'_x = -m(t)\alpha(t)\int_t^T [\psi_1(s)+1]ds,\ I'_y = -m(t)\alpha(t)\int_t^T [\psi_2(s)+1]ds,\ \psi_1(t) = -\int_t^T (T-s)m(s)ds,$$

$$\psi_2(t) = \int_t^T \psi_3(s)ds - \alpha(t)(T-t),\ \psi_3(t) = -\alpha(a)x(a)[\psi_1(t)+1] - \alpha(a)y(a)[\psi_2(t)+1] + \alpha(a), \quad (94)$$

from which $I'_x \geq m(t)\alpha(t)\int_t^T [\min(m)(T-s)^2/2 - 1]ds \geq 0$. Apparently, $I'_x(r) = 0$ implies $r = T$. Under

$$0 < x^\wedge(t),\ y^\wedge(t) \leq 1,\ \alpha^\wedge(t) > 0,\ t \in [0, t^\wedge],$$

we have

$$\psi_2(t) = -\int_t^T \alpha^\wedge(a(s))y^\wedge(a(s))\psi_2(s)ds + f(t),\ f(t) = f_1(t) - f_2(t),\ f_1(t) =$$

$$-\int_t^T \{x^\wedge(a(s))\alpha^\wedge(a(s))[\psi_1(s)+1] + \alpha^\wedge(a(s))[1 - y^\wedge(a(s))]\}ds,\ f_2(t) = \alpha(t)(T-t). \quad (95)$$

Under $T-t$, $t \to \infty$,

$$f_1(t) \times (T-t)^3,\ f_2(t) \times (T-t)(t-t^\wedge).$$

This relation follows from

$$\alpha'(t) = \int^{t^\wedge}_{a(t)} \alpha^\wedge(s)x^\wedge(s)m^\wedge(s)ds < \text{constant}.$$

So, on the greater part of $[t^\wedge, T]$ the asymptotic behavior of $f(t)$ is determined by the function $f_1(t)$. Besides,

$$\psi_1(t) \asymp -(T-t)^2,\ I'_x > 0,$$

but

$$\psi_2(t) = -\int^T_t f'(s)\exp[-\int^s_t \alpha^\wedge(a(r))y^\wedge(a(r))dr]ds > 0,\ I'_y < 0. \tag{96}$$

Let $(t-t^\wedge) \asymp (T-t)^2$. Then the mutual influence of $f_1(t)$ and $f_2(t)$ is compensated. And what is more, on the segment $[T^\wedge, T]$, $T^\wedge = T-(T-t^\wedge)^{1/2}$, the asymptotic behavior of $f(t)$ is determined by the function $f_2(t)$ and also

$$-\psi_2(t) \asymp (t-t^\wedge)(1-e^{-(T-t)}) > 0,\ I'_y > 0,\ t\in (T^\wedge, T^{\wedge\wedge}) \subset [t^\wedge, T]. \tag{97}$$

It means that $x^*(t)$ for the problem C.1 is more than 0 on $[t^\wedge, r]$, $r \asymp T$, and $y^*(t)$ is more than $y(t) = 0$ on $(T^\wedge, T^{\wedge\wedge})$, where $T^{\wedge\wedge} \asymp T$, $T^{\wedge\wedge} - T^\wedge \asymp (T-t^\wedge)^{1/2}$ (see Theorem 13).

8. ADDITIONS AND REMARKS

The similar problems of investigation for multivariate MM are remained still open. We dwell here on two problems: optimization by using a certain other control functions and property of the functionals in the vicinity of the optimal internal control functions. The first problem is especially important for NES since the distributions *x, y, z*- types can be pre-assigned and we have to use other control functions, for example, the functions *u, v*-type described in Ch. 2 (see (56), (57), Ch. 2). With regard *u* we can consider the same problems as above with the same results, if we have, for example,

$$m(t) = \int^t_{a(t)} \alpha(t,s)u(s)y(s)m(s)ds,\ c(t) = \int^t_{a(t)} \beta(t,s)[U-u(s)][1-y(s)]m(s)ds,$$

$$0 < u \le U. \tag{98}$$

The reason for such kind of relations can be scantiness of external resources $f(t)$. Otherwise, we will have simpler MM than (98), namely

$$m(t) = \int^t_{a(t)} \alpha(t,s)u_1(s)y(s)m(s)ds,\ c(t) = \int^t_{a(t)} \beta(t,s)u_2(s)[1-y(s)]m(s)ds,$$

$$0 \le u_i \le U_i,\ i = 1,\ 2 \tag{99}$$

In (99) the control functions u_1 and u_2 can be created independently for optimization of functionals which depend on *m, c*. As to the function *v*, with the replacement of *m* by *m/v* we will have actually the same MM and optimization problems as above.

One more possibility for optimization of NES is creating their "prehistory" in a such manner that their following behavior would be pre-assigned. If, for example, in MM (98) α, β, and uy are given, and $m(t)$, $a(t)$ are pre-assigned for $t \in [t^\wedge, T]$, $a(t) < t^\wedge$, then we can determine the prehistory $m^\wedge(t)$, $t \in [0, t^\wedge]$, solving the equation

$$-\alpha(t,a)u(a)y(a)m^\wedge(a)a' + \int_a^{t^\wedge} \alpha'_t(t,s)u(s)y(s)m^\wedge(s)ds =$$

$$[m(t) - \int_{t^\wedge}^{t} \alpha(t,s)u(s)y(s)m(s)ds]'. \tag{100}$$

The second problem is important for the numerical solution of the optimization problems. If it appears that near that t for which the gradient of functional equals to zero this *functional is strongly convex* (or *concave*), then the well-known numerical methods will be convergent in the same vicinity by the rate of geometric progression. We show here that the functional I (see (64), Ch. 2) is strongly concave in the vicinity of internal optimal values of the control function y under both conditions A. and B. (69) and the usual condition $\alpha'_s(t,s)m^\wedge(s) > 0$. Indeed, let $I'(y,t) = 0$, $t \in [d,e] \subset [t^\wedge, T]$. Then under both conditions we have by analogy with (72)- (74)

$$dI(y) = I(y+dy) - I(y) = (I''(y)dy, dy) = \int_d^e k(t) \int_{a(t)+da(t)}^{a(t)} [\alpha(t,s) - \alpha(s,a(t))]m^\wedge(s)dsdt$$

$$= -\int_d^e k(t)\alpha'_s(t,a(t))m^\wedge(a(t))(da)^2/2dsdt < 0, \tag{101}$$

to within the second order of da and hence dy, since da is an increment corresponding to dy by MM (5, 6, Ch. 2). The desired property follows from (101) immediately.

The similar results take also place under more general conditions and for some other optimization problems [10].

REFERENCES

1. Ashmanov S.A., Introduction to Mathematical Economics, M: Nauka, 1984, 296 pp. (in Russian).
2. Balakrishnan A.V., Applied Functional Analysis, M.: Nauka, 1980, 384 pp.
3. Glushkov V.M., Ivanov V.V., Modeling of Optimization of Work Places Distribution Between the Branches of Production A and B, Kibernetika, N 6, 1977, 117-131 (in Russian).
4. Glushkov V.M., Ivanov V.V., Yatsenko Yu.P., On Uniqueness of Solutions of One Optimization Problem, DAN SSSR, 256, N 4, 1981, 802-806 (in Russian).
5. Gluskov V.M., Ivanov V.V., Yatsenko Yu.P., Analytical Investigation of One Class of Dynamic Models, II, Cybernetics, Plenum Publish. Corp., 18, N 3, 1982, 390-401.
6. Pontryagin L.S., Boltyanskii V.G., Gamkrelidze R.V., and Mischance E.F., The Mathematical Theory of Optimal Processes, Wiley (Inter-science), 1962, 391 pp.
7. Tikhonov A.N., Vasil'eva A.B., Sveshnikov A.G., Differential Equations, M: Nauka, 1980, 232 pp. (in Russian).
8. Vasil'ev F.P., Extreme Problems Solution Methods, M: Nauka, 1981, 400 pp. (in Russian).
9. Warga J., Optimal Control of Differential and Functional Equations, Academic Press, 1972, 531 pp.
10. Yatsenko Yu. P., Integral Models of Systems with Controllable Memory, K.: Naukova dumka, 1991, 219 pp. (in Russian).

PART II:

OPTIMAL NUMERICAL METHODS

CHAPTER 5

SOLUTION OF VOLTERRA EQUATIONS WITH PRE-ASSIGNED ACCURACY

A *complete* or total absolute *error* Δ for the approximate solution of Volterra integral equations of the second and the first kind is considered. This error is less or equal to the sum of errors of three types because of: input data inaccuracy and incompleteness (*inherent error*), approximate algorithms (*method error*), and realization of algorithms on computers (*round-off error*). The *optimal numerical methods* with *pre-assigned accuracy* allowing us to obtain $\Delta < \varepsilon$, where $\varepsilon > 0$ and pre-assigned, are determined. The minimal in order numbers of necessary basic computers' operations under realization of the methods and the respective algorithms are estimated. And also namely those equations and methods for which the numbers of operations are admissible in practice under rather small ε are investigated.

1. VOLTERRA EQUATIONS OF THE SECOND KIND

Let us consider *Volterra integral equations* (V.i.e.) *of the second kind*:

$$x(t) = \int_0^t f(t,u,x(u))du, \ t \in [0,T], \ x(0) = 0, \tag{1}$$

where $f(t)$ is a function of the class F (considered below) and $x(t)$ is the desired solution.

1.1. Inherent errors

Let X be a linear normalized space of functions determined on $[0,T]$ with the norm $\|.\|_t$, $0 < t \leq T$, $\|.\|_T = \|.\|$, such that

$$\|x_1x_2\|_t \leq \text{vrai max } |x_1(u)| \, \|x_2\|_t \, (0 \leq u \leq t), \quad \|\int_0^t x(u)du\| \leq \int_0^t \|x\|du, \ x_1, x_2 \in X. \tag{2}$$

Lemma. *In the cases of* $X = L_p = L_p([0,T])$, $1 \leq p < \infty$, *and* $X = C([0,T])$ *the properties* (2) *are valid.*

Proof. Validity of the first one is obvious. The proof of the second one in the case $X = C$ is almost trivial. In the case $X = L_p$ the second inequality is the consequence of the generalized inequality of Minkowski (see [6]):

$$\{\int_0^t |\int_0^t x(v,u)du|^p dv\}^{1/p} \leq \int_0^t \{\int_0^t | x(v,u)|^p dv\}^{1/p} du, \; 1 \leq p < \infty, \tag{3}$$

in which we need to put $x(v,u) = x(v\text{-}u)$, $x(v\text{-}u) = 0$ for $u > v$. □

We assume that the equation (1) has the unique solution $x \in X$ for each $f \in F$ and that $F \subset X$. When f runs F the respective solution of equation (1) runs the determined set $X^\wedge$. Namely the whole $X^\wedge$ is the approximate solution of the equation (1) without loss of information. However, for easing of the subsequent applications the approximate solution is usually the determined $x^\wedge \in X^\wedge$ corresponding to the determined $f^\wedge \in F$. Then the *inherent* or hereditary *error* sought for is

$$\Delta_I = \sup \|x - x^\wedge\|_t \,(x \in X^\wedge),\; 0 < t \leq T. \tag{4}$$

Denoting

$$\Delta f = \int_0^t [f(t,u,x(u)) - f^\wedge(t,u,x^\wedge(u))]du,\; \Delta x = x(t) - x^\wedge(t), \tag{5}$$

we easily find

$$\Delta x = \int_0^t [f(t,u,x(u)) - f(t,u,x^\wedge(u))]du + \Delta f, \tag{6}$$

from which under the condition that f belongs to the class of *Lipschitz functions* over x:

$$\|f(.,x_1) - f(.,x_2)\| \leq L\|x_1 - x_2\|, \tag{7}$$

on the strength of the Lemma and the well-known inequality by Chaplygin we have

$$\|\Delta x\|_t \leq L\int_0^t \|\Delta x\|_u du + \|\Delta f\|_t \tag{8}$$

and also

$$\|\Delta x\|_t \leq L\int_0^t \|\Delta f\|_u \exp[L(t-u)]du + \|\Delta f\|_t \leq \max \|\Delta f\|_u \,(0<u<t)e^{Lt},$$

$$0< t \leq T,\; \Delta_I \leq \sup \|\Delta f\| \,(f \in F)e^{LT}. \tag{9}$$

Thus, we have proved the following.

Theorem 1. *Under the condition that the function f in* (1) *complies with* (7) *and under the condition* (2) *for the inherent error* Δ_I *the estimates* (4) *and* (9) *are valid.*

Note, that these estimates are well known in the space *C*.
Let instead of (7) the condition

$$L'(x_1 - x_2) < f(.,x_1) - f(.,x_2) < L''(x_1 - x_2),\ -L < L',\ L'' < L, \tag{7'}$$

be given. Then

$$L'\int_0^t \Delta x du + \Delta f < \Delta x < L''\int_0^t \Delta x du + \Delta f,$$

$$L'\int_0^t \Delta f e^{L'(t-u)} du + \Delta f < \Delta x < L''\int_0^t \Delta f e^{L''(t-u)} du + \Delta f,$$

$$|\Delta x| < \max\,[|L'|\int_0^t |\Delta f| e^{L'(t-u)} du,\ |L''|\int_0^t |\Delta f|\, e^{L''(t-u)} du] + |\Delta f| \tag{8'}$$

from which

$$\Delta_1 \le \max\,(2, \exp[\max\,(L', L'')T])\max\,|\Delta f|\ (0 < u < T). \tag{9'}$$

In particular,

$$\Delta_1 \le 2\max\,|\Delta f|\ (0 < u < T),\ \ \max\,(L', L'') \le 0. \tag{9''}$$

Thus, it also takes place

Theorem 1'. *Under the condition of Theorem* 1 *with the replacement of* (7) *by* (7') *for the inherent absolute error* (4), *the estimates* (9') *and* (9") *are valid.*

It should be noted that under (7')

$$L'\int_0^t x(u)du + f_1(t) \le x(t) \le L''\int_0^t x(u)du + f_1(t),\ f_1(t) = \int_0^t f(t,u,0)du, \tag{10}$$

from which we can estimate $|x(t)|$ from both above and below in several cases. This can give more precise estimations of the hereditary relative error regarding (9') and (9").

Let the constants a and b be determined such that

$$f^+(t,u,x) = \sup f(t,u,x)\ (f \in F) > ax + b,\ a > 0, \tag{11}$$

and besides

$$L^+\delta x < f(.,x) - f(.,x^+) < L^{+\prime}\delta x,\ \delta x = x - x^+,\ x^+(t) = \int_0^t f^+(t,u,x^+(u))du. \tag{7''}$$

Then

$$x^+(t) > a\int_0^t x^+(u)du + bt,\ \delta f = \int_0^t [f(t,u,x^+(u)) - f^+(t,u,x^+(u))]du \leq 0 \tag{12}$$

and also

$$|x^+(t)| > |b|/a|1\text{-}e^{-at}(1+at)|e^{at},$$

$$L^+\int_0^t \delta f \exp(L^+(t\text{-}u))du + \delta f < \delta x < L^{+\prime}\int_0^t \delta f \exp(L^{+\prime}(t\text{-}u))du + \delta f, \tag{13}$$

from which

$$|\delta x| < \max(|L^+|,|L^{+\prime}|)\int_0^t |\delta f| \exp(L^+(t\text{-}u))du + |\delta f|,$$

$$\|x\|_t < \max(2,\exp(L^+t))\max\|\delta f\|_u\ (0<u<t) \tag{14}$$

and more

$$\|x/x^+\|_t < a/|b|e^{-at}1/|1\text{-}e^{-at}(1+at)|\max(2,\exp(L^+t))\max\|f\|_u\ (0<u<t) \tag{15}$$

Thus, we have

Theorem 2. *Under the condition* (2), (7"), *and* (11) *for the inherent relative error the estimate* (15) *is valid.*

We will show, at last, how the least possible value

$$\Delta^* = r^*_x = \inf[\sup\|x - x^\wedge\|\ (x \in X^\wedge)]\ (x^\wedge \in X^\wedge) \tag{16}$$

can be estimated. We assume that $f \in F$ has everywhere a continuously differentiable f'_x and that the derivative f'_x runs F' while f runs F. For an arbitrary set F with the norm $\|.\|$ we denote by f^* *Chebyshev center* and by r^*_f *Chebyshev radius* of F:

$$r^*_f = \inf\sup\|f_1 - f\|\ (f \in F) = \sup\|f^* - f\|\ (f \in F), \tag{17}$$

where lower bound succeeds over all the elements f_1 for which $\|f_1 - f\|$ have sense.

Let f^*, $f'_x{}^*$, and x^* be Chebyshev centers for F, F', and $X^\wedge$ respectively.

Theorem 3. *Under the condition that*

$$\|f^* - f_*\| = o(\sup\|f\text{-}f^*\|\ (f \in F)),\ f_*{}'_x = f'_x{}^* \tag{18}$$

and F is symmetrically relative to its Chebyshev center f^*, the solution x_* of the *equation*

$$x(t) = \int_0^t f^*(t,u,x(u))du \tag{19}$$

is asymptotically optimal by accuracy, i.e.,

$$\|x^* - x_*\| = o(\sup \|x - x^*\| \; (x \in X^\wedge)) = o(r^*_x), \; r^*_x \to 0. \tag{20}$$

Proof. We have

$$x - x_* = \int_0^t [f(t,u,x) - f^*(t,u,x_*)]du =$$

$$\int_0^t \{[f_*(t,u,x) - f_*(t,u,x_*)] + [f^*(t,u,x) - f_*(t,u,x)] + [f(t,u,x) - f^*(t,u,x)]\}du =$$

$$\int_0^t f_*{}'_x(x_*)(x-x_*)du + \int_0^t [f(t,u,x)-f^*(t,u,x)]du + o(\|x-x_*\|), \tag{21}$$

after which the matter is reduced to the linear case that is almost trivial. □

Consequently, from Theorem 3 it follows that any approximate *solution* of the equation (19) with higher order accuracy than r^*_f will be *asymptotically optimal* by *accuracy*. It will be the same result if f^* is given with the sufficient accuracy.

For the methods, algorithms, and software for determinations of optimal by accuracy approximations of f^* from several classes see [9, 10] and the next Ch. (6).

1.2. Method errors

To begin with, we will solve the equation (1) on the segment $[0,\Delta T]$, $\Delta T < T$. We consider two methods of the *approximately iterative* (AI)-type [4]: the first one (M_1) for the case when f does not have smoothness of high degree and the second one (M_2) when f is an analytical function.

Let $f(t,u,x)$ applies with (7) and has ith, $i = 0, 1, 2$, derivatives over all the arguments that are less by modules than the respective given constants $Li = Li(f)$ (in a domain when the solution exists). Then we take as the approximate solution of the equation (1) the approach to

$$x_N(t) = \Sigma^{N-1}_{\nu=0} (x_{\nu+1} - x_\nu); \; x_{\nu+1} = \int_0^t f(t,u,x_\nu(u))du, \; x_0 = 0, \tag{22}$$

where the integrals are computed by the formula of the trapezium [2].

Applying the well-known steps, we have on $[0,\Delta T]$

$$\|x_1 - x_0\|_t = \|f_0\|_t, \; f_0 = \int_0^t f(t,u,0)du,$$

$$\|x_2 - x_1\|_t \leq Lt\max\|f_0\|_u\,(0 < u < t),\ldots,\ \|x_{v+1} - x_v\|_t \leq (Lt)^v/v!\,\max\|f_0\|_u\,(0 < u < t),\ldots \tag{23}$$

from which the error

$$\Delta_2(M_1,t) = \|x - x_N\|_t \leq \sum{}^{\infty}_{v=N}\ \|x_{v+1} - x_v\|_t \leq \max\|f_0\|_u(0 < u < t)(Lt)^N/N![1 + Lt/(N+1) +$$

$$(Lt)^2/(N+1)(N+2)+\ldots] < \max\|f_0\|_u(0 < u < t)(2\pi N)^{-1/2}(Lte/N)^N[1+$$

$$Lt/(N+1)+(Lt)^2/(N+1)(N+2)+\ldots],\ 0 \leq t \leq \Delta T. \tag{24}$$

Let us divide the segment $[0,\Delta T]$ into M parts by the points

$$t_j = j\Delta t,\ \Delta t = \Delta T/M,\ j = 1,\ldots,M, \tag{25}$$

and put

$$x\hat{}_{v+1j} = \Delta t\{[f(t_j,0,0)+f(t_j,t_j,x\hat{}_{vj})]/2+\Sigma^{j-1}{}_{s=1}\,f(t_j,t_s,x\hat{}_{vs})\} =$$

$$\int {}'f(t_j,u,x\hat{}_v(u))du\ (u \in [0,\ t_j]), \tag{26}$$

where for an arbitrary function f a function $'f$ denotes the polygon:

$$'f(u) = [f(t_s)(t_{s+1} - u) + f(t_{s+1})(u - t_s)]/\Delta t,\ u \in [t_s, t_{s+1}], \tag{27}$$

where $x\hat{}_v(u)$ is a step function with the values $x\hat{}_{vs}$ for $u \in [t_s, t_{s+1}]$. It is well known [2] that

$$\int |f - 'f|du\ (u \in [0,\ t_j]) < L2(f)/12(\Delta t)^2 t_j. \tag{28}$$

Therefore

$$|x\hat{}_{v+1j} - x_{v+1}(t_j)| \leq |x_{v+1}(t_j) - \int {}'f(t_j,u,x_v(u))du\ (u \in [0,\ t_j])| +$$

$$|\Sigma^{j-1}{}_{s=0}\int [{}'f(t_j,u,x_v(u)) - {}'f(t_j,u,x\hat{}_v(u))]du\ (u \in [t_s,\ t_{s+1}])| <$$

$$2(f(.,x_v))/12(\Delta t)^2 t_j + Lt_j \max|x\hat{}_{vs} - x_{vs}|\ (1 \leq s \leq j). \tag{29}$$

It is not difficult to show that

$$df(.,.,x_v(u))/du = \partial f/\partial u + \partial f/\partial x\, dx_v/du;$$

$$d^2f(.,.,x_v(u))/du^2 = \partial^2 f/\partial u^2+\partial^2 f/\partial x^2(dx_v/du)^2+\partial f/\partial x d^2x_v/du^2+2\partial^2 f/[\partial u\partial x]dx_v/du. \qquad (30)$$

So, we should estimate $(x_v)'$ and $(x_v)''$ for $v = 1,..., N$-1. We have

$$(x_{v+1})' = f(t,t,x_v(t)) + \int_0^t \partial f/\partial t(t,u,x_v(u))du,$$

$$(x_{v+1})'' = \partial f/\partial t + \partial f/\partial u(u=t) + \partial f/\partial x\,(x_v)'+ \partial f/\partial t(t,t,x_v(t)) + \int_0^t \partial^2 f/\partial t^2(t,u,x_v(u))du, \qquad (31)$$

and therefore

$$(x_{v+1})' < L0+L1\Delta T,\ (x_{v+1})'' < L1(3+L0+L1\Delta T)+L2\Delta T =$$

$$3L1+L0L1+(L1^2+L2)\Delta T. \qquad (32)$$

Combining the obtained estimations, we easily find $L2(f(.,x_v)) < d$,

$$d = L2+L2L0^2+2L0L2+3L1^2+L0L1^2+(2L0L1L2+L1^3+3L1L2)\Delta T+L2L1^2\Delta T \qquad (33)$$

and also

$$\Delta x_{v+1} < dt_j/12(\Delta t)^2 + Lt_j\Delta x_v;\ \Delta x_v = \max |x\hat{}_{vs} - x_{vs}|\ (1 \le s \le j), \qquad (34)$$

from which

$$\Delta x_v < dt_j/12(\Delta t)^2(1-(Lt_j)^v)/(1-Lt_j) < dt_j/6(\Delta t)^2,\ Lt_j < L\Delta T < 1/2. \qquad (35)$$

Now regarding (24), it is not difficult to obtain

$$\Delta_2(M_1,t_j) = |x(t_j) - x\hat{}_{Nj}| \le |x(t_j) - x_N(t_j)| + |x_N(t_j) - x\hat{}_{Nj}| \le$$

$$L0t_j(2\pi N)^{-1/2}(e/2N)^N \{1+1/[2(N+1)]+...\} + dt_j/6(\Delta t)^2,\ Lt_j \le 1/2,\ 1 \le j \le M. \qquad (36)$$

As the result we have the following.

Theorem 4. *The estimate (33) and (36) give the error of the method (22)-(27). In addition, the number of f-computation operations does not exceed* $(M+1)(M+2)N/2$.

It should be noted that in a similar way we can estimate the error and the number of operations if we apply the more precise formula by Simpson (involving 4th derivation of f). It is convenient to have the approximate solution as the respective splice [3, 9, 10, 12] instead of the array of $x^\wedge_{Nj}$. Under the condition of smaller smoothness of f, one can create the method of numerical realization of (22) based on the rational approximation of f [2, 8] or the *method of Monte-Carlo* [12].

Let $x^\wedge(t)$ be the approximate solution of the equation (1) and let us consider the case $t \in [k\Delta T,(k+1)\Delta T]$, $k = 1,\dots,R$-1, $R\Delta T = T$. Let us also put

$$x_k(t) = \int^{k\Delta T}_0 f(t,u,x^\wedge(u))du + \int^{t}_{k\Delta T} f(t,u,x_k(u))]du,\ x_{kv}(t) = \int^{k\Delta T}_0 f(t,u,x^\wedge(u))du +$$

$$\int^{t}_{k\Delta T} f(t,u,x_{kv-1}(u))du,\ x_{k0}(t) = \int^{k\Delta T}_0 f(t,u,x^\wedge(u))du,\ v = 1,\dots, N; \tag{37}$$

$$x^\wedge_{kvj} = x^\wedge_{k0j} + \int {}^{t}f(t_{kj},u,x^\wedge_{kv-1}(u))du\ (u \in [k\Delta T,\ t_j]),\ x^\wedge_{k0j} = \int^{k\Delta T}_0 {}^{t}f(t_{kj},u,x^\wedge(u))du,$$

$$t_{kj} = k\Delta T + j\Delta t,\ j = 1,\dots, M,\ \Delta t = \Delta T/M,\ v = 1,\dots, N, \tag{38}$$

where $x^\wedge_{kv}(u)$ is the stepwise function which is equal to $x^\wedge_{kNs}$ on the segment $[t_{ks}, t_{ks+1}]$ and $x^\wedge(t)$ is the respective polygon of t, $t \in [t_j, t_{j+1}]$.

By analogy with (33) and (36), from (37) and (38) it follows

$$\Delta x = x(t) - x^\wedge(t) = x(t) - x_k(t) + x_k(t) - x_{kN}(t) + x_{kN}(t) - x^\wedge(t) = \int^{t}_0 [f(t,u,x(u)) -$$

$$f(t,u,x^\wedge(u))]du + r_k(t),\ r_k(t) = \int^{t}_{k\Delta T} [f(t,u,x^\wedge(u)) - f(t,u,x_{kN}(u))]du + \int^{t}_{k\Delta T} [f(t,u,x_{kN}(u))$$

$$-f(t,u,x_k(u))]du + x_k(t) - x_{kN}(t) + x_{kN}(t) - x^\wedge(t), \tag{39}$$

from which by the virtue of (7') and (9"), where $r_k(t)$ plays the the role of Δf, we have

$$\Delta_2(M_1) = \max\ |x(t)-x^\wedge(t)|\ (t \in [0,T]) \le 3[kL0\Delta T/N! + kd\Delta T/6(\Delta t)^2] + x''(t)|/8(\Delta t)^2 \le$$

$$3L0T/N! + dT/2(\Delta t)^2 + 1/8[3L1 + L0L1 + (L1^2+L2)\Delta T](\Delta t)^2,\ j = 1,\dots, M;\ k = 1,\dots, R\text{-}1;$$

$$R\Delta T = T;\ \Delta t = \Delta T/M. \tag{40}$$

Thus, it establishes the following

Theorem 4'. *The method* (37), (38) *error is given by the estimate* (33) *and* (40). *In addition, the number of f-computations does not exceed* $(M+1)(M+2)NR/2$.

Let us consider now the second case when $f(z,v,w)$ is an analytical function of 3 complex-valued arguments z,v,w in the domain

$$D = Er \times Er \times D(H), \tag{41}$$

where Er is the closed domain bounded by the *ellipse of Zhukovsky*

$$\partial Er = \{(x,y)|x = \Delta T/2 + a_r \cos t,\ y = b_r \sin t,\ t \in [-\pi, \pi]\},$$

$$a_r = \Delta T(r+r^{-1})/4,\ b_r = \Delta T(r-r^{-1})/4,\ r > 1, \tag{42}$$

and the domain $D(H)$ is the closed circle with the radius H and the center 0.

Let us consider another *approximately iterative* (*AI*) *method* [4] setting

$$w(z) = \int_0^z f(z,s,w(s))ds,\ w_v(n;z) = \int_0^z L_n(f(.,w_{v-1}(n;.),s))ds =$$

$$\sum_{i=0}^n f(z,x_i,w_{v-1}(n;x_i))\int_0^z p_i(s)ds, w_v(n;x_j) = w_{vj} = \Delta T/2\sum_{i=0}^n a_{ij}f(x_j,x_i,w_{(v-1)i});$$

$$j = 0,\dots,\ n;\ v = 0,\dots,\ N;\ a_{ij} = 2/\Delta T\int p_i(s)ds\ (u \in [0,\ x_j]) = e_i/n[1 - c_j + c_i/2(1 -$$

$$c_{2j}) + \Sigma_{v=2}^n e_v c_{iv}(c_{j(v-1)}/(v-1) - c_{j(v+1)}/(v+1) - 2/(v^2-1))];\ e_0 = e_n = 1/2;\ e_j = 1, j = 1,\dots,\ n-1;$$

$$c_k = \cos(k\pi/n);\ x_j = x_j(n) = \Delta T/2(1 - \cos(j\pi/n)). \tag{43}$$

<u>Theorem 5</u>. *Let in the equation*

$$w(z) = \int_0^z f(z,s,w(s))ds \tag{44}$$

the function f be analytical in the domain $D^\wedge = Er^\wedge \times Er^\wedge \times D(H^\wedge)$, $\Delta T^\wedge > 0$, $H^\wedge > 0$, $r^\wedge > 1$. *Then if under* $r \in [1, r^\wedge]$ *and natural numbers n and N we take the number H* $\le H^\wedge$, $\Delta T \le \Delta T^\wedge$, *such that the inequality*

$$H > (a_r + \Delta T/2)(\|f\|_{C(D^\wedge)} + L\Delta_{2,\Delta T}),\ a_r = a_r(\Delta T) = \Delta T/4(r+r^{-1}),$$

$$D = Er \times Er \times D(H), \tag{45}$$

is fulfilled, where $\Delta_{2,\Delta T}$ *denotes the right part of the inequality*

$$|w(z) - w_N(n;z)| \le LO(f)\{(\|L_n\|+1)r(1+r^{-2n-2})/[(1-q)(r^\wedge - r)](r/r^\wedge)^n(a_r + \Delta T/2) +$$

$$L^{-1}e^q q^{N+1}/(N+1)!\},\ \|L_n\| < 2/\pi \ln n + 1,\ q = L(a_r + \Delta T/2),\ LO(f) = \|f\|_{C(D^\wedge)}, \tag{46}$$

L is Lipschitz constant for f over the argument w (in *D*), *then the polynomial* $w_N(n;z)$

created by the AI-algorithm (43) *approximates the solution* $w(z)$ *of the equation* (44) *in the domain Er such that the inequality* (46) *is fulfilled.*

The proof is similar to that of Theorem 2 in the article [4].
From Theorem 5 and the mode of the proof of Theorem 4', it follows the following

Theorem 5'. *If the segment* $[0,T]$ *is divided on the sub-segments* $[(k\text{-}1)\Delta T, k\Delta T]$, $L\Delta T < 1/2$, $k = 1,..., R$, *and for every of these sub-segments the condition of Theorem* 5 *takes place, then the error* $\Delta_2(M_2)$ *for* $[0,T]$ *under the piecewise polynomial approximation* (43)*-type is estimated by the relation*

$$\Delta_2(M_2) < \exp(LT)4\Delta_{2,\Delta T},\ L\Delta T < 1/2. \tag{47}$$

In addition, the total number f-computation does not exceed $RN(n+1)^2$.

Let us explain the origin of the constant 4: under $L\Delta T = 1/2$, $r_k < k\Delta_{2,\Delta T}$ we have

$$\Delta_2(M_2) < \Delta_{2,\Delta T}\Sigma^{R-1}{}_{k=1} \exp((R\text{-}k)L\Delta T)k < e^{LT}\Delta_{2,\Delta T}\int{}^{\infty}{}_{0}\, xe^{-x/2}dx = 4e^{LT}\Delta_{2,\Delta T}.$$

1.3. Round-off errors

Let us denote f_l for f computed on a computer, where l is the number of binary digits in an adder of the computer. Not infrequently the estimate

$$\Delta_3(f) = |f\text{-}f_l| < C_f(|f|+1)2^{-l} \leq C_f(L0+1)2^{-l}, \tag{48}$$

is valid [2], where the constant C_f is not very large. Let us see the additional error of the solution that can result the error of (48)-type.

Theorem 6. *Under the condition* (7'), (9"), *and* (48) *the round-off error for the approximate solution* (38) (*in the floating point system*)

$$\Delta_3(M_l) < \{2.12L0[(M+1)R+2]^2 + 4C_f(L0+1)(M+1)R\}2^{-l},\ 2^{-l}(M+1)R < 0.1. \tag{49}$$

Proof. Let us apply to the scheme (38) the well-known estimate [16] of the type

$$|\Sigma^{m}{}_{k=1}[(a_k)_l - a_k]| < \Sigma^{m}{}_{k=1}|a_k||E_k| < 1.062^{-l}\max|a_k|$$

$$(1 \leq k \leq m)(m+2)^2/2,\ 2m < 0.1. \tag{50}$$

As a result, the maximal round-off error along one step (see also (26)) of an iteration

$$\Delta_{3l} < 1.062\ ^{-l}L0[(M+1)R+2]^2/2 + \Delta_3(f)(M+1)R;\ 2^{-l}(M+1)R < 0.1. \tag{51}$$

By the virtue of convergence of the iterative process (38) over v with the similarity ratio $L\Delta T < 1/2$, the summarizing error along one step does not exceed $2\Delta_{3l}$ [15, 16]. This error along subsequent steps will be as a hereditary error. So, by the virtue of (9"), the accumulated error does not exceed $4\Delta_{3l}$. □

Theorem 7. *The round-off error for the approximate solution* (43)

$$\Delta_3(M_2) < 1/(1\text{-}q)[n+3)^2/2L0 \max |a_{ij}| 1.062^{-l} + \Delta_3(a_{ij})L0 + \Delta_3(f) \max |a_{ij}|(n+1)],$$

$$(n+2)2^{-l} < 0.1;\ \Delta_3(a_{ij}) < (n+7)\ln n/n\ 1.062^{-l}\Delta_3(\cos)\ln n;\ \Delta_3(\cos) < 2^{-l+1};$$

$$|a_{ij}| < 3\ln n/n;\ q = L(a_r + \Delta T/2) < 1. \tag{52}$$

The proof can be obtained by analogy with the proof of Theorem 6 on the strength of (43) and (49) and the well-known estimates [16] of the round-off error for any arithmetical operation $*$:

$$(a * b)_l = (a * b)(1 + e_x),\ |e_x| < 2^{-l}. \tag{53}$$

From Theorem 7 and mode of the proof of Theorem 6, it follows also the next

Theorem 7'. *Under the condition of* (7) *and Theorem* 5' *the round-off error for the piecewise polynomial approximation (with regard only to the main terms)*

$$\Delta_3(M_2, T) < e^{LT}4\Delta_3(M_2, \Delta T) = L0e^{LT}2^{-l+3}4n\ln n/(1 - q),$$

$$(n+2)R2^{-l} < 0.1,\ L\Delta T \leq 1/2. \tag{54}$$

1.4. Analysis of the total errors

Connecting the estimates (9"), (40) and (49) we obtain the estimation of the complete absolute error (in Chebyshev metric) for the approximate solution of V. i. e. (1) by the method M_l (37), (38):

$$\Delta(M_l) < 2max|\Delta f|\ (0 < u < T) + 3TL/N! + dT/2(\Delta t)^2 + 1/8[3L1 + L0L1 +$$

$$(L1^2 + L2)\Delta T](\Delta t)^2 + \{2.12L0[(M+1)R+2]^2 + 4C_f(L0+1)(M+1)R\}2^{-l},$$

$$2^{-l}(M+1)R < 0.1. \tag{55}$$

In a similar way, connecting the estimates (9), (46), (47), and (54), we obtain the estimation of the complete error for the solution of V. i. e. (1) by the AI-method M_2 (43)-type (with regard only to the main terms and $q \leq 1/2$, $r/r^\wedge \leq 1/2$):

$$\Delta(M_2) < \exp(LT)\{\max|\Delta f|\ (0<u<T) + 4L0[\pi \ln n2^{-n} + eL^{-1}/(2^N N!)] + 2^{-l+4}L04n\ln n\},$$

$$2^{-l}(n+2)R < 0.1. \tag{56}$$

From (55), one can see that in order that $\Delta(M_1)$ has the pre-assigned order of decreasing ε, it is sufficient that $|\Delta f|$, R/M^2, R/N^N, and $(MR)^2 2^{-l}$ have the order of ε. The number operations M^2NR of the first method will be logarithmically equivalent to the number of input data M^2R for $\varepsilon \to 0$, $1 > \varepsilon > 0$, if, for example, $N = \ln(1/\varepsilon)$, $M^2R = 1/\varepsilon$. In addition, if $\ln R$ has the order of $\ln(1/\varepsilon)$, then l also has the order of $\ln(1/\varepsilon)$. Thus, both hypotheses [1] will be fulfilled for the first method. However, it is clear that the conditions of this method efficiency, especially the conditions (9"), are very restrictive.

From (56), it follows that in order $\Delta(M_2)$ has the pre-assigned order of decreasing ε it is sufficient that $e^R|\Delta f|$, $e^R \ln n/2^n$, $e^R/2^N N!$, and $e^R n\ln n/2^l$ have the order of ε. Under fixed R the number of digits l has also the order of $\ln(1/\varepsilon)$, i.e., the second method will be also stable according to the hypotheses [1]. However, it is clear that in practice we have to use additional R digits, i.e., we have to compute under increased digitalization. The hypothesis about logarithmic equivalence between the numbers of input data and the operations does not take place for the second method. This fact prompts us that there is the determined reserve for improving of the second method. Indeed, applying, under the same input data, instead of the simple iteration the method that combines the simple iteration and the *method of Newton* [2], in the respective estimation of the complete error we will have the term of the form $1/2^{2N}$ instead of $1/(2^N N!)$. After which the property of logarithmic equivalence will take place.

It should be noted that any method and the respective *algorithm* for which the hypotheses [1] are valid may naturally be called *optimal* by accuracy and number of operations. We note also the certain *reserves* for raising of *accuracy*.

Decreasing of inherent errors can be achieved at the expense of the solution of the restoration of input data on the base of measuring systems output problems that are also reduced to the solution of i. e. of Volterra-type and Wiener-Hopph-type [8, 9].

Decreasing of method errors can be achieved [2] at the expense of accounting of different values of the estimating constants $Li(f)$, $i = 0, 1, 2$, on each sub-segment $[k\Delta T, (k+1)\Delta T]$, $k = 0, \ldots, R-1$, and the special dividing of these sub-segments. Applying instead of Δt the steps Δt_k such that the estimations of the errors for different sub-segments were identical, we can minimize the method error (under given number of f-computations).

Decreasing of round-off errors can be achieved [15] at the expense of the special techniques of computation, for example, the summation should be realized by way of pair-wise addition of the results of addition. Besides, a probability character of round-off error [15] should be considered.

2. VOLTERRA EQUATIONS OF THE FIRST KIND

2.1. Two main approaches

Let us consider *V. i. e. of the first kind*

$$Ax = \int_0^t a(t,u,x(u))du = f(t),\ t \in [0,T]. \tag{57}$$

The first approach consists in reduction (57) to the V. i. e. of the second kind

$$x(t) = a^{-1}[- \int_0^t a'_t(t,u,x(u))du + f'(t)],\quad a(x) = a(t,t,x(t)), \tag{58}$$

by way of formal differentiation under the condition of the constant sign of $a'_x(t,t,x(t))$. It can be shown that all the methods of s.1 can be applied to (58) without principal variations. If, as usually occurs, instead of $a(t,u,x(u))$ and $f(t)$ the respective classes A and F of their approximate value $a^\wedge(t,u,x(u))$ and $f^\wedge(t)$ are given, then under the transition from (57) to (58) the important problem of numerical differentiation or determination of the respective classes A' and F' is aroused.

For several new results about the solutions of this problem see [7, 13, 14]. For our purposes it is sufficient to note that we can determine the classes A and F such that A' and F' have been investigated.

It is clear that the method of reduction (57) to (58) is not universal and it is still necessary to have an approach for the direct solution of (57) as V. i. e. of the first kind, which is an *ill-posed problem* [7, 8, 9, 13, 14].

2.2. Quasi-solution method

Let us consider in detail the well-known *method of quasi-solution* [7] in connection with (57) if a priori we know the set M of the solution correctness such that on this set the operator A has the *reciprocal operator* A^{-1} for which the estimate of the *continuity module*

$$w(A^{-1},d) = \sup \|A^{-1}(f1) - A^{-1}(f2)\| \ (\|f1-f2\| < d, f1, f2 \in A \cap M) \tag{59}$$

is given, where $\|.\|$ is the sign for the norm of linear normalized Banach or Hilbert space.

The general scheme now consists in solving the problem

$$I(x) = \|Ax - f\| = \min\ (x \in M),\ A \in A,\ f \in F, \tag{60}$$

which is reduced to the convex programming for the convex set M and convexity of $I(x)$ in the domain $D(x_0)$: $I(x) < I(x_0)$, where x_0 is the given initial approximation for the solution of the problem (60).

If x^*_M is the exact solution of the problem (60), x^* is the desired solution, $x^* \in M$, and x_{Md} is the approximate solution of the problem (60), i.e.,

$$\|Ax_{Md} - f\| < \|Ax^*_M - f\| + d,\ d > 0,\ x_{Md} \in M, \tag{61}$$

where d, in general case, is the *numerical error* for the solution of the problem (60) (does not exceed the sum of the method error and round-off error), then

$$\|Ax^* - Ax_{Md}\| < \|Ax^* - f\| + \|Ax^*_M - f\| + d < 2\|Ax^* - f\| + d, \tag{62}$$

and hence

$$\|x^* - x_{Md}\| < w(A^{-1}, 2\|Ax^* - f\| + d). \tag{63}$$

But

$$\|Ax^* - f\| \leq d(A,M) + d(F), \tag{64}$$

where

$$d(A,M) = \sup\ [\sup \|A1x - A2x\|\ (A1,\ A2 \in A)]\ (x \in M),\ d(F) = \sup \|f1 - f2\|\ (f1,\ f2 \in F). \tag{65}$$

By substituting (64), (65) in (63), we have the following

Theorem 8. *The complete error of the quasi-solution method for the approximation solution* (61) *of the problem* (60)

$$\Delta = \|x^* - x_{Md}\| < w(A^{-1}, 2[d(A,M) + d(F)] + d). \tag{66}$$

Note that the similar estimate (without regarding of the round-off error) is well known to specialists. Determined value d can be obtained from the finite-dimensional approximation of the solution of the problem (60) [2].

2.3. Estimate of continuity modules for reciprocal operators

It is necessary to know the function $w(A^{-1},d)$ for the effective application of (66). There are plenty of scientific works devoted to the estimation of $w(A^{-1},d)$ (see, for example, [7, 13]). We dwell here on a simple estimate that is well known [7] for a linear operator.

Let the *derivative by Fresher* of the operator A in the point $x^\wedge$, $x^\wedge \in D(x_0)$,

$$A'x = \int{}^{t}_{0} a'(t,u)x(u)du, \ a'(t,u) = a'_x(t,u,x^\wedge), \tag{67}$$

has the property

$$\|Ax - Ax^\wedge - A'x\| < L2\|x - x^\wedge\|, \tag{68}$$

where $L2$ is a given constant, and introduce the equation

$$A1x_1 = (A')^*A'x_1 = \int{}^{T}_{0} k(t,u)x_1(u)du = (A')^*f = f_1(t), \ k(t,u) = \int{}^{T}_{0} a'(v,u)a'(v,t)dv$$

$$= \int{}^{u}_{t} a'(v,u)a'(v,t)dv \ (u < t), \ -\int{}^{T}_{u} a'(v,u)a'(v,t)dv \ (u > t),$$

$$(A')^*f = \int{}^{t}_{0} a'(u,t)f(u)du = f_1(t). \tag{69}$$

Besides, let the *set of correctness*

$$M(x^\wedge) = \{x \,|\, x = A1v, \ \|v\| \le R\} \tag{70}$$

Then [7]

$$w(A1^{-1},d) < (2Rd)^{1/2}. \tag{71}$$

Assuming $\|Ax_1 - Ax_2\| < d$, under condition that $x_1, x_2 \in M \subset S(x^\wedge,r)$, where S is the sphere in Hilbert space H with center $x^\wedge$ and radius r, on the strength of (68) and (71) we have

$$\|A'x_1 - A'x_2\| = \|A'x_1 + Ax^\wedge - Ax_1 - (A'x_2 + Ax^\wedge - Ax_2) + Ax_1 - Ax_2\| < d + r/L2,$$

$$\|A1(x_1 - x_2)\| \le \|A'\|(d + r/L2), \ \|x_1 - x_2\| < (2R\|A'\|(d + r/L2))^{1/2}. \tag{72}$$

By virtue of arbitrariness of r, from here and (70) it follows that

$$w(A^{-1},d) < (2\sup \|A'\|Rd \ (x^\wedge \in D(x_0))^{1/2} < (2LRd)^{1/2} \tag{73}$$

on the set of correctness

$$M = \cap M(x^\wedge)\ (x^\wedge \in D(x_0)). \tag{74}$$

Thus, we have the following

Theorem 9. *The continuity module* $w(A^{-1})$ *satisfies the inequality* (73) *in the domain* $D(x_0)$ *under the condition that the relations' (67), (69), (70), and (74) determine the set of correctness for the operator A.*

Note that $M \neq \varnothing$ under the sufficient large R and small $d(D(x_0))$, where $d(.)$ is the diameter of domain ..

3. ON REGULARIZATION FOR LINEAR EQUATIONS

Let us consider the linear equation:

$$A' = A,\ A1 = A^*A,\ f_1 = A^*f,\ x_1 = x,\ M(x^\wedge) = M, \tag{75}$$

and introduce the equation (see [8])

$$A_a x^\wedge_a = (aI + A^\wedge 1)x^\wedge_a = f^\wedge_1,\ x^\wedge_a = 1/af^\wedge_1 + 1/a \int^t_0 k(t,u)x^\wedge_a(u)du,\ a > 0, \tag{76}$$

where I is the unit operator, $A^\wedge 1$ and $f^\wedge_1$ are given estimates of $A1$ and f_1 . We solve (76) by the *least square method*, i.e., we take as the approximate solution

$$x_n = x_{na} = \Sigma^{n-1}_{k=0} a_k T_k(2/Tt\text{-}1),\ T_k(u) = \cos(k \arccos u), \tag{77}$$

and we find unknown coefficients a_k by way of the solution of the linear algebraic system

$$\Sigma^{n-1}_{k=0} a_k(A_a T_k, T_j) = (f^\wedge_1, T_j),\ j = 0,\dots, n\text{-}1, \tag{78}$$

where for any f_1 and f_2

$$(f_1, f_2) = 1/T \int^T_0 f_1(t) f_2(t)dt. \tag{79}$$

It follows from the results [2, 8, 15] that the inherent error for the solution of the equation (57) or (75)

$$\Delta_1 = \|x\text{-}x^\wedge_a\| < aR + dI(\|f\| + 3\|A\|)(R\|A1\| + aR)/a = (a + C_1 dI/a)R,\ dI = d(A) + d(F); \tag{80}$$

the least square method error

$$\Delta_2 = \|x\hat{}_a - x_{n\,a}\| < (a+\|A1\|)/a\; d_n(x\hat{}_a), \tag{81}$$

where $d_n(x\hat{}_a)$ is the value of the best approximation for $x\hat{}_a$ by algebraic polynomials in the power n - 1; the round-off error under realization of the least square method on a computer

$$\Delta_3 = \|x_{na} - x\hat{}_{nal}\| < (a+\|A1\|)/a\; n2^{-l}\|x_{n\,a\,l}\|, \tag{82}$$

where l is the number of digits in a computer's summarizer (we assume here that the elements of (79)-type are computed exactly, otherwise additional errors as hereditary ones can be included in the estimation (80)).

Noting that $d_n(x\hat{}_a)$ in the presence of rth derivatives of $f_1(t)$ and $k(t,u)$ (by t) can be estimated in the form $d_n(x\hat{}_a) < C_2R/(an^r)$, where C_2 can be found effectively by the virtue of (76), and connecting the estimations (80)-(82), we obtain (with the accuracy of the main terms)

$$\Delta < R(a+C_1dI/a+C_2/(an^r)+C_3/an2^{-l}). \tag{83}$$

Let us put $C_2/(an^r) = C_3n2^{-l} = C_1dI$. Then (83) can be rewritten in the form

$$\Delta < R(a+3C_1dI/a). \tag{84}$$

Let the parameter of regularization $a = (3C_1dI)^{1/2}$, which minimizes the estimate Δ (84). As a result

$$\Delta < 2R(3C_1dI)^{1/2};\; C_2/n^r = 3^{1/2}(C_1dI)^{3/2},\; 2^{-l} = C_1/(C_3n)dI. \tag{85}$$

We summarize the following

Theorem 10. *The complete error of the method* (75)-(79) *is estimated by the relation's* (85), *which allow us to determine necessary parameters of the method and the computer for the solution of the linear equation* (69), (75) *with pre-assigned accuracy* (*under the condition of sufficient smallness of* dI).

It should be noted that for the method under consideration the hypotheses [1] will be also fulfilled if the algebraic system (78) is solved by the special algorithm [9] combining the iterative process with the approximation of the matrix of this system by a band matrix with the center along the main diagonal.

In conclusion, we note that the considered results without principal difficulties can be transfered to the systems of V. i. e. of the second and first kind. For a few different results of other mathematicians see [3, 11].

REFERENCES

1. Bakhvalov N.S., On Properties of Optimal Methods for the Solution of the Problems of Mathematical Physics, ZVMiMF, 10, N 3, 1970, 555-568 (in Russian).
2. Bakhvalov N.S., Numerical Methods, M.: Nauka, 1973, 632 pp. (in Russian).
3. Brunner H.A., Survey of Recent Advances in the Numerical Treatment of Volterra Integral and Integro-Differential Equations, J. of Comput. and Appl. Math., v. 8, N. 3, 1982, 213-229.
4. Dzyadyk V.K., Approximately Iterative Methods of Approximation by Polynomials for the Solution of the Cauchy Problem, ZVMiMF, 26, N. 3, 1986, 357-372 (in Russian).
5. Gear C. W., Numerical Initial Value Problems in Ordinary Differential Equations, Prentice-Hall, 1971, 253 pp.
6. Hardy G.H., Littlewood J.E., Polya G., Inequalities, Cambridge University Press, 1934, 314 pp.
7. Ivanov V.K., Vasin V.V., Tanana V.P., The Theory of Linear Ill-Posed Problems and Its Applications, M.: Nauka, 1978, 206 pp. (in Russian).
8. Ivanov V.V., The Theory of Approximate Methods and Their Application to the Numerical Solution of Singular Integral Equations, Nordhoff Int. Publ., 1976, 330 pp.
9. Ivanov V.V., Numerical Methods for New Class of Models of Developing Systems, in book: Computing Process and Systems, 3, M.: Fizmatgiz, 1985, 166-179 (in Russian).
10. Ivanov V.V., Tvalodze Yu.G., Zhuzhunashvili A.Sh., On the Solution of Volterra Integral Equations with Pre-assigned Accuracy, M.: VINITI, 1886-89, 1989, 33 pp. (in Russian, see also in English UCD/CCM Report No. 102, 1997, 1-12).
11. Linz P., Analytical and Numerical Methods for Volterra Equations, SIAM Studies in Appl. Math., 1985, 227 pp.
12. Mikhaylov G.A., Optimization of Monte-Carlo Weight Methods, M.: Nauka, 1987, 147 pp. (in Russian).
13. Morozov V.A., Regular Methods for the Solutions of Ill-Posed Problems, M.: Izd-vo MGU, 1974, 360 pp. (in Russian).
14. Tikhonov A.N., Arsenin V.P., Methods for the Solution of Ill-Posed Problems, M.: Nauka, 1974, 223 pp. (in Russian).
15. Voevodin V.V. Computational Foundation for Linear Algebra, M.: Nauka, 1977, 304 pp. (in Russian).
16. Wilkinson D.X., Algebraic Problem of Eigen-Values, M.: Nauka, 1970, 564 pp. (in Russian).

CHAPTER 6

REDUCTION TO VOLTERRA-TYPE EQUATIONS

It is shown that many problems of applied mathematics can be reduced to the solutions of Volterra-type integral and integro-differential equations. So, the numerical methods for V. i. e. developed in Ch. 5 can be used for the numerical solution of wide class of problems.

In addition, a general procedure for the optimal by accuracy finite approximation of initial data of a given class is considered.

1. MATHEMATICAL PROGRAMMING PROBLEMS

Highly general and important problems are the *mathematical programming* problems of the form

$$f(x) = \inf (x \in X), X = \{x: e_i(x) = 0,\ i = 1,\dots, l;\ p_j(x) \leq 0,\ j = 1,\dots, m\}, \tag{1}$$

where the functions f, e_i, and p_j are continuously differentiable in $\mathbb{R}^n$ ($\mathbb{R}^n$ is n-dimensional Euclidean space). Problems of the solution of linear or nonlinear operator equations of the form $Ax = y$, where the operator A and element y are given, are easily reduced to (1) by means of the solution of the minimization problem for the norm of the residual $f(x) = \|Ax-y\|$ and a certain finite-dimensional approximation of the desired element x. In a similar way, the more general problems of the solution of equations regarding several restrictions (see, for example, the method of quasi-solution in Ch. 5) are also easily reduced to (1).

We consider how (1) is reduced to the problem of minimization of functions without restrictions, i.e., in all the space $\mathbb{R}^n$ [3, 5], and how the last problem is reduced to the solution of V. i. e..

Let

$$F(x,q) = [f(x) - q]_+^{s+1} + g(x), \tag{2}$$

where q is a real parameter; s is an integer; $y_+ = \max\{0, y\}$ for arbitrary y; and g is a function with the property

$$g(x) = 0\ (x \in X),\ > 0\ (x \notin X). \tag{3}$$

The examples of g(x) are

$$g(x) = \Sigma^{l}_{i=1} \; |e_i(x)| + \Sigma^{m}_{j=1} \; [p_j(x)]_+ \tag{4}$$

and

$$g(x) = \Sigma^{l}_{i=1} \; e_i^2(x) + \Sigma^{m}_{j=1} \; [p_j(x)]_+^s. \tag{5}$$

Let the values of $f(x)$ belong to the segment $[q^-, q^+]$, $q^- = \inf f(x) \; (x \in X) < \infty$.

Lemma 1. *If a settled value* $q\hat{} \geq q^-$ $(q\hat{} < q^-)$, *then the value of the global minimum of* $F(x,q)$ *in* $\mathbb{R}^n$ *is equal to* 0 (> 0).

Lemma 2. *Let the function f,* e_i, *and* p_j *be convex. Then the function* (2), *where g is given by* (4) *or* (5) *is also convex.*

Thus, in the condition of this Lemma we have the well-known *convex programming problem.*

Lemma 3. *The function* (2), *where g is given by* (5), *will be s-time differentiable if each of the function f,* e_i, *and* p_j *has the same property.*

The proof of Lemmas 1-3 is obvious.

For analytical functions instead of (2) one can introduce, for example,

$$F(x,q,r) = [f(x) - q]^2 + \Sigma^{l}_{i=1} e_i^2(x) + \Sigma^{m}_{j=1} [p_j(x) + r_j^2]^2, \; r = (r_1, \ldots, r_m). \tag{6}$$

Lemma 4. *The function* $F(x,q,r)$ *is analytical over all the arguments and the desired solution* $x^* = x(q^*)$ *of the problem* (1) responds to the least $q = q^*$ *for which*

$$F(x^*, q^*, r^*) = \inf F(x, q^*, r) \; (x, r \in \mathbb{R}^{n+m}) = 0. \tag{7}$$

The proof of Lemma 4 is also obvious.

Let us consider the following algorithm:

i) create point $x_m = x(q_m)$ for which $|F(x_m, q_m) - \inf F(x, q_m) \; (x \in \mathbb{R}^n)| < \varepsilon$, $m = 0, 1, \ldots$;

ii) if $F(x_m, q_m) > \varepsilon$, then $q_{m+1} = q_0 + h2^m$, where h and q_0 are arbitrary numbers, $h > 0$, for example, $h = 1$ and $q_0 = 0$;

iii) if $0 \leq F(x_m, q_m) \leq \varepsilon$, then $q_{m+1} = (q_{m-1} + q_m)/2$;

iv) if $F(x_{m+1}, q_{m+1}) > \varepsilon$, then $q_{m+2} = (q_{m+1} + q_m)/2$ and if $F(x_{m+1}, q_{m+1}) < \varepsilon$, then $q_{m+2} = (q_{m+1} + q_{m-1})/2$;

v) stop if $q_m - q_{m-1} \le \varepsilon$.

Theorem 1. *Let the conditions*

$$|f(x)-f(x^\wedge(x))| \le M(f')|x-x^\wedge(x)|,\ |g(x)-g(x^\wedge(x))| \ge m(g')|x-x^\wedge(x)|,$$

$$x \in \{x: 0 < g(x) < 2\varepsilon\},\ M(f') = \max|f'|,\ m(g') = \min|g'|, \tag{8}$$

take place, where $x^\wedge(x)$ *is the nearest to the x point of the set* $\{x: g(x) = 0\}$, *and let us consider for the simplicity the case* (4) *and* $s = 0$. *Then after the algorithm's stopping we have*

$$|q_m - \inf f(x)\ (x \in X)| \le 2\varepsilon(1 + M(f')/m(g')). \tag{9}$$

Besides, the number of the value $f(x)$ *computation does not exceed*

$$N = 2\max(0, \log_2[|q^- - q_m|/h]) + \log_2 h/\varepsilon. \tag{10}$$

Proof. The relation (10) is obvious. If $q_m \ge q^-$, then $q_m - q^- \le q_m - q_{m-1} \le \varepsilon$. If $q_m \le q^-$, then on the strength of (8)

$$q^- - q_m = q^- - f(x^\wedge(x)) + f(x^\wedge(x)) - f(x_m) + f(x_m) - q_m < 2\varepsilon(1 + M(f')/m(g')). \tag{11}$$

In a similar way we can receive the estimates in the other cases of the functions f and g and under the more general condition $|g(x) - g(x^\wedge(x))| > \nu|x - x^\wedge(x)|$, $\nu > 0$, $x \in \{x: 0 < g(x) < 2\varepsilon\}$. □

Thus, we have the reduction of the initial problem (1)-type to the finite sequence (with usually not large number) of the minimization problems in $\mathbb{R}^n$.

Let us consider now the reduction of the last problem to the solution of V.i.e..

Lemma 5. *Any finite solution of the systems of V.i.e. of the form*

$$x(t) = x_0 - \int_0^t f'(x(u))du,\ f'(x) = \operatorname{grad} f(x),\ x(0) = x_0,\ t \ge 0, \tag{12}$$

has the property $f'(x(\infty)) = 0$, *i.e.,* $x(\infty)$ *is a local minimum of* $f(x)$.

The proof is obvious.

Theorem 2. *Let* $x(t)$ *be the solution of* (12) *and*

$$\|x_0 - x(t)\| \le d,\ t \ge 0,\ \cos(F'(x), x_0 - x(\infty)) \ge c > 0,\ F(x) - F(x^*) \le \|F'(x)\|\|x - x^*\|. \tag{13}$$

Then

$$F(x(T)) - F(x(\infty)) \le \varepsilon,\ T \ge d^2/(c\varepsilon). \tag{14}$$

Proof. On the strength of (13) we have

$$d \ge \|x_0 - x(t)\| \ge \int_0^t \|F(x(u))\cos(F'(x(u)), x_0 - x(\infty))\| du \ge c\int_0^t \|F'(x(u))\| du, \tag{15}$$

from which

$$\int_0^t \|F(x(u))\| du \le d/c. \tag{16}$$

So, there exists the least $t^\wedge$, $t^\wedge \le d/(c\varepsilon)$, such that $\|F'(x(t^\wedge))\| \le \varepsilon/d$. Indeed, otherwise

$$\int_0^{t^\wedge} \|F(x(u))\| du > \varepsilon t^\wedge/d \ge d/c. \tag{17}$$

Therefore, $F(x(t^\wedge)) - F(x^*) \le \varepsilon$. It remains to note that $F(x(T)) - F(x^*) \le F(x(t^\wedge)) - F(x^*)$ for $T \ge t^\wedge$. □

Theorem 3. *Let*

$$\|F'(x)\| \le M,\ (F'(x), x - x(\infty)) \ge m\|x - x(\infty)\|^2. \tag{18}$$

Then

$$\|x(t) - x^*\| \le (\varepsilon/M)^{1/2},\ F(x) - F(x^*) \le \varepsilon,\ T \ge 1/(2m)\ln[M\|x_0 - x^*\|/\varepsilon]. \tag{19}$$

Proof. On the strength of (18) we have

$$1/2 d/dt\|x - x^*\|^2 = ((x - x^*)', x - x^*) = (F(x) - F(x^*), x - x^*) = -m\|x - x^*\|^2,$$

$$F(x) - F(x^*) \le M\|x - x^*\|^2, \tag{20}$$

from which (19) follows rather easily. □

Thus, the matter reduces to the solution of Volterra-type equations of the form (12) on the finite segment [0,T], which is necessary for the methods of Ch. 5.

2. SOME OTHER PROBLEMS

The problem of the solution of *operational equations* of the form

$$Ax = y, \tag{21}$$

where usually two elements are to be given and third is to be found, reduces to the mathematical programming with the help of the equivalent problem of minimization of the norm of the residual:

$$\|Ax - y\| = \min (x, y, A \in X \times Y \times A), \tag{22}$$

where X, Y, and A are given sets, and a finite approximation (see the next s. 3) of all the elements in (22). In particular, by means of (21) and (22) the solution of the boundary problems of mathematical physics can be reduced to the mathematical programming and by the same token to the solution of Volterra equations.

The important problem is the selection of the *base system of* the *elements* $e_1,..., e_n$ with the help of which we need to find the solution of the equation $Ax = y$. The "good" system is any one that gives a "good" approximation, for example, for the desired solution x under not "large" n. In practice, the sense of the solution prompts us usually to select the "good" system. In theory, there are many well-known "good" systems for some classes of functions [2, 3] and the certain new, rather effective ones (see, for example, [1, 7]).

It should be noted that there is easier way of reduction to Volterra-type equations in some particular cases. We show this on the example of two boundary problems for the Boltzmann equation. Professor Luis H. Garcia-Rubio posed this example. It was arising in connection with certain applications of the Boltzmann equation to the problems of chemical engineering [8]. In the BI-spherical system of coordinates these problems reduce to the plane case.

Let for the simplicity sin = s, cos = c, sinh = sh, cosh = ch, $y(u,v) = y$. Then we have the following nonlinear equation:

$$\nabla y = (chu\text{-}cv)^3/(a^2sh^2u^\wedge sv)[\partial/\partial u(sv/(chu\text{-}cv)\partial y/\partial u)+\partial/\partial v(sv/(chu\text{-}$$

$$cv)\partial y/\partial v)] = kshy, \tag{23}$$

where ∇ is the operator of Laplace in the BI-spherical coordinates, $2a(chu^\wedge - 1)$ is the distance of the closest approach between two symmetric with respect to the origin spheres of the radius a, and k is a Debye screening parameter.

From (23) it follows that

$$\partial^2 y/\partial u\partial v = ka^2sh^2u^\wedge\int^v{}_0\, ss/sv[((chu\text{-}cv)/(chu\text{-}cs)^3)'_u shy(u,s) +$$

$$+ (chu\text{-}cv)/(chu\text{-}cs)^3 chy(u,s)\partial y(u,s)/\partial u]ds + \int^v{}_0\, ss/sv\{((chu\text{-}cv)shu/(cu\text{-}$$

$$cs)^2)'_u\partial y(u,s)/\partial u + [(chu\text{-}cv)shu/(chu\text{-}cs)^2\text{-}((chu\text{-}cv)/(chu\text{-}cs))'_u]\partial^2 y(u,s)/\partial u^2 -$$

$$(chu\text{-}cv)/(chu\text{-}cs)\partial^3 y(u,s)/\partial u^3\}\mathrm{d}s, \tag{23'}$$

In the case of the bounds

$$\partial y/\partial u = 0,\ u = 0;\ \partial y/\partial v = 0,\ v = 0, \pi;\ y = y\hat{},\ u = u\hat{} \tag{24}$$

we can find the solution in the form

$$y = \int^{v}{}_{0} \int^{u\hat{}}{}_{u} rs(\pi\text{-}s)w(r,s)drds + y\hat{}, \tag{25}$$

where w can be found from (23) or (23'). The reason is all the bounds (24) are fulfilled for (25).

Similarly, in the case of the bounds

$$(chu\hat{} - cv)\partial y/\partial u = kb\sigma^*,\ u = u\hat{},\ b = ashu\hat{}, \tag{24'}$$

instead of $y = y\hat{}$, we can find the solution in the form

$$y = y(0,0) + kb\sigma^* u^2/[(chu\hat{}\text{-}cv)2u\hat{}] + \int^{v}{}_{0} \int^{u}{}_{0} r(u\hat{}\text{-}r)s(\pi\text{-}s)z(r,s)drds, \tag{25'}$$

where z can be found from (23). On the basis of the known structure of the solutions, we should instead of r take $r^{1/2}$, and similarly instead of $u\hat{}\text{-}r$, s, and $\pi\text{-}s$.

Determination of the desired functions w and z can be reduced to the solution of Volterra-type equations. For example, to find w in the case of the problem (23'), (24), we have

$$\partial^2 y/\partial u\partial v = -uv(\pi\text{-}v)w(u,v),\ \partial y/\partial u = -\int^{v}{}_{0} us(\pi\text{-}s)w(u,s)ds,$$

$$\partial^2 y/\partial u^2 = \int^{v}{}_{0} s(\pi\text{-}s)[w(u,s) - u\partial w(u,s)/\partial u]ds,$$

$$\partial^3 y/\partial u^3 = -\int^{v}{}_{0} s(\pi\text{-}s)[2\partial w(u,s)/\partial u + u\partial^2 w(u,s)/\partial u^2]ds. \tag{26}$$

By substituting (26) in (23'), we obtain the *Volterra-type* integro-differential *equation* that can be solved by the method of simple iteration. In addition, we have to apply formulae of squaring for analytical functions with selection of singularities and with regard to the ill-posed problem of numerical differentiation [3].

Of course, this is only general idea. Further, more detailed, consideration will be necessary, if we want to develop this approach.

3. ON OPTIMAL BY ACCURACY METHODS FOR THE APPROXIMATION OF FUNCTIONS

Reduction of many other problems to (1) and hence to (12) is based on the *approximation* of initial data and desired results with the help of objects that are completely determined by finite number of parameters, finite information.

In practice, we really have only finite information about each applied problem of continuous or infinite-type. This is the main reason each problem is naturally imbedded in the class of problems with the same information about their initial data.

The main *postulate* of the so-called "*Information-based complexity theory*" [10] is the following. If we have any problem with at least two possible solutions under given information then there does not exist an algorithm that allows us to find any solution with confidence. We can find only the approximate solution with accuracy that is equal or greater than 1/2 of distance between the different possible solutions.

This finite information naturally determines the classes of data and results with the same information. It is important that the desired approximation was optimal by accuracy under given information.

We consider here the so-called *method of bound functions* [4] for the construction of the *optimal by accuracy approximation.*

Let a class functions F be given. We determine two bound functions for the F:

$$f^+(x) = \sup f(x)\ (f \in F), f^-(x) = \inf f(x)\ (f \in F). \tag{27}$$

Then it is obvious that the *optimal by accuracy approximation* $f^*(x)$ of functions from the class F and the error $\rho^*(x)$ of the approximation we can find with the help of the formulae:

$$f^*(x) = [f^+(x) + f^-(x)]/2, \quad \rho^*(x) = [f^+(x) - f^-(x)]/2. \tag{28}$$

The functions f^* and ρ^* are called *Chebyshev center and Chebyshev radius* of F respectively.

As example of application of this method, we consider the following class of functions:

$$F \equiv C^n{}_{1,L,N,\varepsilon}: |f(x) - f(y)| \le L\|x-y\|_1 = L\max |x_i - y_i| (i = 1,\dots, n),\ x, y \in \pi: [0,1]^n,$$

$$|f(x^j) - f^\wedge_j| \le \varepsilon_j,\ x^j = (x_1{}^j,\dots, x_n{}^j),\ j = 1,\dots, N, \tag{29}$$

where $L, N, n, x_i{}^j, f^\wedge_j, \varepsilon_j, i = 1,\dots, n; j = 1,\dots, N,$ are given real numbers (N, n are integers).

<u>Theorem 4.</u> *The optimal by accuracy approximation and the respective error on the class* $C^n{}_{1,L,N,\varepsilon}$ *are given by the formulae* (22), *where*

$$f^{+}(x) = \min \{f^{\wedge}_j + \varepsilon_j + L\|x - x^j\|_1\} \; (j = 1,\dots, N),$$

$$f^{-}(x) = \max \{f^{\wedge}_j - \varepsilon_j - L\|x - x^j\|_1\} \; (j = 1,\dots, N). \tag{30}$$

In addition,

$$\rho^*(x) \leq \max |\varepsilon_j| \; (j = 1,\dots, N) + L/N^{1/n}. \tag{31}$$

For the proof, see [3].

Note that the error $\rho^*(x)$ depends on $(2 + n)N + 3$ magnitudes here. It is not difficult to construct examples when $\rho^*(x) = 0$ for arbitrary values of L and ε_j, $j = 1,\dots, N$.

The similar results can be obtained in many other classes of functions [3, 4, 9, and 10]. In particular, the optimal by accuracy approximation of functions with unknown in advance singular points is considered in [4].

A priori given estimates of L and/or ε_j, $j = 1,\dots, N$, can be strongly overstated. To improve them we can solve the problem (the *method of quasi-solution*)

$$L \simeq \max |f(x^j) - f(x^k)|/\|x^j - x^k\|_1 \; (j, k = 1,\dots, N, j \neq k) = \min \tag{32}$$

under given $f^{\wedge}_j$, ε_j, $j = 1,\dots, N$, and observance of the relation (23) or the problem (the *method of residual principle*)

$$\varepsilon = \max |\varepsilon_j| \; (j = 1,\dots, N) = \min \tag{33}$$

under given L and $f^{\wedge}_j$, $j = 1,\dots, N$, and observance of the relation (23). It is clear that the problem (26) and (27) can be reduced to the well-known *linear programming problem* [6].

REFERENCES

1. Daubechies I., Orthonormal Bases of Compactly Supported Wavelets, Communication in Pure and Applied Mathematics, 41, 1988, 909-996
2. Dzjadyk V.K., Ivanov V.V., On Asymptotic and Estimates for the Uniform Norms of the LaGrange Interpolation Polynomials Corresponding to the Chebyshev Nodal Points, Analisis Mathematica, 9, 1983, 85-97.
3. Ivanov V.V., Methods of Computation. Guidance, Kiev: Naukova dumka, 1986, 584 pp. (in Russian).
4. Ivanov V.V., Optimal by Accuracy Approximation of Functions with Unknown in Advance Singular Points, Proceeding of MI AN USSR, 180, 1987, 120-121 (in Russian).
5. Ivanov V.V., Lyudvichenko V.A., On One Method of Unconditional Optimization for the Solution of Mathematical Programming Problems, Kibernetika, N 52, 1977, 1-8 (in Russian).
6. Karmanov V.G., Mathematical Programming, M.: Nauka, 1975, 272 pp. (in Russian)
7. Kaiser G., A Friendly Guide to Wavelets, Birkhäuser, 1994, 300 pp.
8. Orazem M.E. et al., Development of Physic-Chemical Models for Electrochemical Impedance Spectroscopy, Electrochimica Acta, 38, 1993, 1903-1911.
9. Traub J.F., Wozniakowski H., A General Theory of Optimal Algorithms, Academic Press, 1980, 341 pp.
10. Traub J.F., Wozniakowski H., Perspective on Information-Based Complexity, Bulletin of the AMS, 26, N 1, 1992, 29-52.

CHAPTER 7

SOME COMPLEMENTS

1. ON ADAPTIVE ALGORITHMS OF OPTIMIZATION

We consider adaptive algorithms of optimization on the example of the problem:

$$I = \int_{t^\wedge}^{T} c(t)dt = \int_{t^\wedge}^{T} \int_{a(t)}^{t} \beta(t,s)[1-y(s)]m(s)dsdt = \max\ (y) \tag{1}$$

under preserving the relations

$$m(t) = \int_{a(t)}^{t} \alpha(t,s)y(s)m(s)ds,\ P(t) = \int_{a(t)}^{t} m(s)ds,\ a'(t) \geq 0,$$

$$0 \leq y_{\min}(t) \leq y(t) \leq 1,\ t \geq t^\wedge \geq 0, \tag{2}$$

where the functions α, β, P, y_{min} as well as all the elements on the prehistory $[0,t^\wedge]$ are given, and the conditions of Theorem 2, Ch. 3 and Lemma 8, Ch. 4 are held.

1.1. Subsidiary algorithm

Regarding the structure of the control y (Ch. 4), we introduce the *subsidiary algorithm* A the idea of which:

$$y_{r+1}(t) = \begin{cases} 1,\ I'(y_r, t) > 0, \\ y_0(t) = y_{\min}(t),\ t \in [t^\wedge, T], \\ y_{\min}(t),\ I'(y_r, t) \leq 0,\ r = 0, 1, \ldots, \end{cases} \tag{3}$$

where I' is the gradient of the functional I. To guarantee convergence, the algorithm A is complicated by introducing the following steps:

i) Determination of the set

$$Er = \{t\colon I'(y_r, t)[y_r(t) - (1 + y_{\min}(t))/2] < 0\}. \tag{4}$$

If $Er = \varnothing$, then $y_r(t)$ is the desired solution.

ii) Determination of the interval $er = (t_r, t_r^\wedge)$ of a constant sign of $I'(y_r,t)$, $er \subset Er$,

$$\int_{er} |I'(y,t)|dt = \max. \tag{5}$$

iii) Determination of rth correction

$$h_r(t) = \begin{cases} 1, I'(y_r, t) > 0, \\ y_{\min}(t), I'(y_r, t) \leq 0, \ t \in er. \end{cases} \tag{6}$$

iv) Determination of the r+1th approximation

$$y_{r+1}(t) = \begin{cases} h_r(t), \ t \in er, \\ y_r(t), \ t \in [t^\wedge, T] \backslash er, \end{cases} \tag{7}$$

and computation $I(y_{r+1})$.

v) If for a few k the values $I(y_{k+1}) - I(y_k) < \varepsilon$, then stop, otherwise, go to i).

The magnitude ε is a parameter of the algorithm A. If we construct the sequence ε_n, $\varepsilon_n \to 0$, $n \to \infty$, the algorithm A will produce the sequence

$$\{y_i\}^{\infty}_{i=1} = \cup \{y_{kn}\} (k = 1, \ldots, r_n; \ n = 1, \ldots). \tag{8}$$

Theorem 1. *For every i there exists* $\varepsilon^\wedge_i > 0$ *such that under* $\varepsilon_i < \varepsilon^\wedge_i$ *we have* $I(y_{i+1}) > I(y_i)$, *i.e., the algorithm* A *allows us to construct a maximizing sequence* $\{y_i\}$, $i = 1, \ldots$.

The proof follows from the fact in [6, Ch. 4] that under a small ε_i the variation $\delta y = y_{i+1} - y_i$ is a collection of the needle-shaped variations and the expression for the respective variation of the functional has the form

$$\delta I = \int^{T}_{t^\wedge} \{-m(t) \int^{\wedge A(t)}_{t} [\alpha(s,t)\psi(s) + \beta(s,t)]ds\} \delta y(t)dt, \tag{9}$$

where the function ψ is determined by the conjugate equation (52, Ch. 4).

Theorem 2. *Let* y^* *be the solution of the problem* (1), (2). *Then*

$$\lim I(y_i) \, (i \to \infty) = I(y^*) = \max I(y) \, (y \in U), \ U = \{y: y \in L_\infty[t^\wedge, T], \ y_{\min}(t) \leq y(t) \leq 1\}. \tag{10}$$

The proof follows from the structure of the desired solution (see Ch. 4) and the fact that the functions from U can be approximated with arbitrarily accuracy by the piecewise constant functions $U^\wedge$. It is clear that $y_i \in U^\wedge$. Therefore $\{y_i\}$ is the maximizing approximate solution on the strength of the proof of Theorem 1, Ch. 4.

Theorem 3. *If the optimal control* y^* *is a cluster set in* U *under all* $t \in [t^\wedge, T]$ *and has a finite number of switching points N, then the algorithm* A *converges by the functional with an exponential rate. Besides, starting from a certain R,*

$$d_i = I(y^*) - I(y_i) < q^i,\ q < 1 - (1 - \|y_{\min}\|)/N,\ i > R. \qquad (11)$$

Proof. We have

$$d_i = \int^T_{t^\wedge} \{[|I'(y^*,t)| - I'(y^*,t)][y_i(t) - y_{\min}(t)] + [I'(y^*,t) + |I'(y^*,t)|][1 - y_i(t)]\}dt/2. \qquad (12)$$

On the strength of Theorems 1, 2 under sufficiently large i the structure of $y_i(t)$ near the boundary values of $y^*(t)$ and the values of $I'(y_i,t)$ are closed to the optimal ones. Therefore, regarding the main term, the increment

$$\Delta_i = I(y_{i+1}) - I(y_i) \geq d\int_{ei} |I'(y_i,t)|dt,\ d = 1 - \|y_{\min}\| > 0, \qquad (13)$$

and on the strength of the structure of the algorithm A (its second step)

$$d_i \leq \int_{Ei} |I'(y^*,t)|dt \leq N\int_{ei} |I'(y_i,t)|dt. \qquad (14)$$

Thus, we have

$$d_{i+1} = d_i - \Delta_i \leq d_i[1 - d/N]. \qquad (15)$$

The estimate (11) follows easily from (15). □

1.2. Adaptive algorithms

Let us consider the *adaptive algorithm* A^ consisting of the following steps:

i) Application of the algorithm A until for a few sequential i

$$|I(y_{i+1}) - I(y_i)| < \varepsilon \text{ and/or } g_i = t_i^\wedge - t_i < \delta. \qquad (16)$$

Let t_{ik}, $k = 1,..., N_i$, is the ordered from the left to the right set of switching $y_i(t)$, $t \in [t^\wedge, T]$. If $\delta_{ik} = t_{ik+1} - t_{ik} > \delta$ for all $k = 1,..., N_i - 1$, then y is the approximate solution of the problem. Otherwise, go to the next step ii).

Let all the values $\delta_{ik} \leq \delta$ on the segments $[t_j, tj^\wedge]$, $j = 1,..., L$, and let

$$D = \cup\ [t_j, tj^\wedge]\ (j = 1,..., L). \qquad (17)$$

ii) Determination of the desired control $y(t)$, $t \in D$, by the standard method of the gradient's project

$$y_{r+1}(t) = y_r(t) + \alpha_r I'(y_r,t),\ r = 1, 2,..., \qquad (18)$$

where α_r are selected by a method [21] which can provide convergence with the exponential rate and with regard to the restriction $y_{\min}(t) \le y_{r+1}(t) \le 1$.

The δ and ε are parameters of the algorithm A^. As above, the algorithm A^ provides the infinite sequence

$$\{y_i\}^{\infty}_{i=1} = \cup \{y_{kn}\} \ (k = 1,..., r_n;\ n = 0,1,..., ;\ \delta \to \delta/2^n,\ \varepsilon \to \varepsilon/2^n). \tag{19}$$

Theorem 4. *If* 1) *the problem has the unique solution* (y^*, a^*, m^*); 2) $\alpha(t,s) = k(t)\beta(t,s)$ *and* $a^*(T) \le t$^; 3) $a^{*\prime}(t) > 0$, $\beta'_s(t,s) > 0$, *then for* $\delta \to 0$ *the algorithm* A^ *converges by the functional I with the exponential rate, i.e., for a certain R,*

$$I(y^*) - I(y_i) < q^i,\ q < 1,\ i > R. \tag{20}$$

Proof. In the general case the segment $[t$^$,T]$ consists of three sets: C, on which the number of switches is finite and $t_{ik+1} - t_{ik} > \delta$; D, on which $y^*(t)$ is intrinsic in U; F, on which $t_{ik+1} - t_{ik} \le \delta$ and the number of switching points can be infinite. The algorithm A^ = A for $t \in C$ and has the exponential rate of its convergence. For $t \in D$ functional $I(y)$ is strongly concave under $y = y^*$ (see s. 8, Ch.4). So, the rate of convergence here is also exponential [21]. As to F its measure has the order δ and in asymptotic, when $\delta \to 0$, does not affect the rate of convergence of $I(y_i)$. □

It should be noted that the respective proof of Theorem 1-4 with the more details and more general conditions one can find in [10], Ch. 4.

2. APPROXIMATION OF GENERALIZED FUNCTIONS

The solutions of integral equations are not infrequently existing in the so-called classes of *generalized functions* [3, 4, 18, 19]. In order to find out approximations to such generalized solutions on the basis of, for example, the least square method, it is necessary to have instances of the *complete systems* of generalized and easily computable *functions*. One of the such systems is constructed and justified below.

2.1. Preliminary consideration

Let $C^1([0,T])$ be the set of continuously differentiable on $[0,T]$ functions $x(u)$ for which

$$x(0) = x'(0) = 0,\ (x, y)_1 = \int_0^T x'(u)y'(u)du,\ \|x\|_1 = \left(\int_0^T |x'(u)|^2 du\right)^{1/2} \tag{21}$$

and W^1 be the *completion of* the *set* $C^1([0,\mathrm{T}])$ in the norm by (21). Let else $x \in L_2([0,T])$ and $y \in W^1$, where $L_2 = L_2([0,T])$ is the space of square integrable functions on $[0,T]$. Then we denote W^{-1} the completion of the space L_2 in the norm

$$\|x\|_{-1} = \sup \left| \int_0^T x(u)y(u)du \right| / \|y\|_1 \ (y \in W^1, y \ne 0). \tag{22}$$

It is known [3, 19] that Dirac δ-function can be constructed as element of W^{1} and that for $x \in W^{-1}$

$$\|x\|_{-1} = \sup|(x, y)|/\|y\|_1 \ (y \in W^1, y \neq 0), (x, y) = \lim \int_0^T x_i(u)y(u)du \ (i \to \infty), \quad (23)$$

where $x_i \in L_2$ and $\|x-x_i\|_{-1} \to 0$ as $i \to \infty$.

Let D be the *operator of generalized differentiation* (the derivative is regarded in the sense of Sobolev [19]). The operator D transforms the space W^1 to L_2 and the conjugate operator D^* transforms the space L_2 to W^{-1}, moreover the equalities

$$\|Dy\|_2 = \|y\|_1, \quad \|D^*x\|_{-1} = \|x\|_2 \quad (24)$$

are valid. We also have the following relations

$$(x_1, x_2)_{-1} = (Jx_1, Jx_2)_1 = \int_0^T dJx_1/du dJx_2/du du, \ \|x\|_{-1} = \|Jx\| = \left(\int_0^T |dJx/du|^2 du\right)^{1/2},$$

$$dJx/du = -\int_0^u x(v)dv, \quad (25)$$

where $x_1, x_2 \in W^{-1}$; $Jx_1, Jx_2 \in W^1$. The operator J under $x \in L_2$ has the form

$$Jx = \int_0^T G(u,v)x(v)dv, \ G(u,v) = t\text{-}u \ (v < u), \ t\text{-}v \ (v > u); \quad (26)$$

and under $x \in W^{-1}$ we have

$$Jx = \lim Jx_i, \ i \to \infty, \ x_i \in L_2; \ \|x-x_i\|_{-1} \to 0, \ i \to \infty. \quad (27)$$

Let $x \in L_2$ and let *Steklov function*

$$x_h(u) = 1/2h \int_{u-h}^{u+h} x(v)dv, \ h > 0, \quad (28)$$

We use the following properties of this function:

$$\|x-x_h\|_2 < w_2(x,h), \ \|x_h'\|_2 < 1/h w_2(x,h), \ w_2(x,h_n) = \sup \left(\int_0^T |x(u+v)-x(u)|^2 du\right)^{1/2}, \quad (29)$$

where under each v, $|v| < h_n$, the integral is computed with the respect to that part of the segment $[0,T]$ which contains the both u and $u + v$.

2.2. Construction of a complete orthonormalized system in the space W^{-1}

Let us divide the segment $[0,T]$ into n segments $[kh,(k+1)h]$, $k = 0,\dots, n-1$, $h = T/n$, and denote

$$e_{k,n}(u) = \delta(u-kh),\ u \in [0,T],\ k = 0,\dots, n-1. \tag{30}$$

Theorem 5. *Let $x \in W^{-1}$, $X \in L_2$, and $D^*X = x$. There exists the numbers a_{kn} such that*

$$\|x - \sum_{k=0}^{n-1} a_{kn}\delta(u-kh)\|_{-1} < (1+2^{-1/2})w_2(X,h),\ n = 1, 2,\dots . \tag{31}$$

Proof. On the strength of (24) and the fact that Dirac δ-function is the generalized derivative of the respective step function we have (for $x \in W^{-1}$, $X \in L_2$)

$$d = \|x - \Sigma_{k=0}^{n-1} a_{kn}\delta(u-kh)\|_{-1} = \|D^*(X - \Sigma_{k=0}^{n-1} a_{k,n}\theta(u-kh)\|_{-1} =$$

$$\|X - \Sigma_{k=0}^{n-1} a_{k,n}\theta(u-kh)\|_2 = (\Sigma_{j=0}^{n-1} \int_{jh}^{(j+1)h} |X(u) - \Sigma_{k=0}^{n-1} a_{k,n}\theta(u-kh)|^2 du)^{1/2}. \tag{32}$$

Using Steklov function X_h, we find the numbers $a_{k,n}$, $k = 0,\dots, n-1$, from the condition

$$\Sigma_{j=0}^{n-1} \int_{jh}^{(j+1)h} |X(u) - \Sigma_{k=0}^{n-1} a_{k,n}\theta(u-kh)|^2 du = \Sigma_{j=0}^{n-1} \int_{jh}^{(j+1)h} |X(u) - X_h(jh)|^2 du. \tag{33}$$

For $jh \le u \le (j+1)h$, $j = 0,\dots, n-1$, we have

$$\Sigma_{k=0}^{n-1} a_{k,n}\theta(u-kh) = X_h(jh);\ \Sigma_{k=0}^{j} a_{k,n} = X_h(jh)\ j=0,\dots, n-1, \tag{34}$$

from which

$$a_{k,n} = X_h(kh) - X_h((k-1)h),\ \ k = 1,\dots, n-1,\ a_{0,n} = X_h(0). \tag{35}$$

Let us estimate the value d:

$$d = (\Sigma_{j=0}^{n-1} \int_{jh}^{(j+1)h} |X(u) - X_h(jh)|^2 du)^{1/2} \le$$

$$\|X - X_h\|_2 + (\Sigma_{j=0}^{n-1} \int_{jh}^{(j+1)h} |X_h(u) - X_h(jh)|^2 du)^{1/2}. \tag{36}$$

Since

$$|X_h(u) - X_h(jh)|^2 = |\int_{jh}^{u} X_h'(u)du|^2 \le (u-jh)\int_{jh}^{(j+1)h} |X_h'(u)|^2 du,\ \int_{jh}^{(j+1)h} |X_h(u) -$$

$$X_h(jh)|^2 du \le \int_{jh}^{(j+1)h} [(u-jh)\int_{jh}^{(j+1)h} |X_h'(v)|^2 dv]du = h^2/2 \int_{jh}^{(j+1)h} |X_h'(u)|^2 du, \tag{37}$$

we have

$$(\Sigma_{j=0}^{n-1} \int_{jh}^{(j+1)h} |X_h(u) - X_h(jh)|^2 du)^{1/2} \le h/2^{1/2}(\int_0^T |X_h'(u)|^2 du)^{1/2} = h/2^{1/2}\, \|X_h'\|_2. \tag{38}$$

Using inequality (36), we have

$$d = \|X-\Sigma^{n-1}_{k=0} a_{k,n}\theta(u-kh)\|_2 \leq \|X-X_h\|_2 + h/2^{1/2}\|X_h'\|_2 \leq (1+2^{-1/2})w_2(X,h)\to 0,\ h\to 0. \square \quad (39)$$

Theorem 6 [4]. *The system of the functions*

$$e\hat{}_{1,n}(u) = 1/(T-u_{1,n})^{1/2}\delta(u-u_{1,n}),\ e\hat{}_{k,n}(u) = (T-u_{k-1,n})^{1/2}/((T-u_{k,n})(u_{k,n}-u_{k-1,n}))^{1/2}\delta(u-u_{k,n}) -$$

$$(T-u_{k,n})^{1/2}/((T-u_{k-1,n})(u_{k,n}-u_{k-1,n}))^{1/2}\delta(u-u_{k-1,n}),\ u_{k,n} = kh,\ k=2,3,\ldots;\ n=1,2,\ldots; \quad (40)$$

k, n are relatively prime numbers, is the orthonormalized in the norm of W^{-1}.

Proof. We first show that the system of the *functions* $e_{k,n}$, $k = 1,\ldots, n$, are *linearly independent*. By contradiction, for example, $\sum^{n}_{k=1} a_k e_{k,n}(u) = 0$, $a_n \neq 0$. Then $e_{k,n}(u) = \Sigma^{n-1}_{k=1} d_k e_{k,n}(u)$ or $\delta(u-u_{k,n}) = \Sigma^{n-1}_{k=1} d_k(u-u_{k,n})$, $d_k = a_k/a_n$. From here for any continuous function $f(t)$, it follows $f(u_{n,n}) = \Sigma^{n-1}_{k=1} d_k f(u_{k,n})$. Let $f(u_{n,n}) \neq 0$ and $f(u_{k,n}) = 0$, $k = 1,\ldots, n-1$. It is impossible on the strength of (42). Therefore $e_{k,n}(u)$, $k = 1,\ldots, n$, are linearly independent. We construct now the *system of orthonormalized functions* by the well-known *Schmidt's algorithm*. The first element of the system has the form $e\hat{}_{1,n} = e_{1,n}/\|e_{1,n}\|_{-1}$. Since

$$\|e_{1,n}\|_{-1} = \|\delta(u-u_{1,n})\|_{-1} = \|J\delta(u-u_{1,n})\| = (\textstyle\int^T_0 |-\int^u_0 \delta(v-u_{1,n})dv|^2\, du)^{1/2} = (T-u_{1,n})^{1/2} \quad (41)$$

we have $e\hat{}_{1,n} = \delta(u-u_{1,n})/(T-u_{1,n})^{1/2}$. The second element is computed by the formula

$$e\hat{}_{2,n} = x_2/\|x_2\|_{-1},\ x_2 = e_{2,n} - (e_{2,n}, e\hat{}_{1,n})_{-1}\, e\hat{}_{1,n}. \quad (42)$$

Let us define the scalar product $(e_{2,n}, e\hat{}_{1,n})_{-1}$:

$$(e_{2,n}, e\hat{}_{1,n})_{-1} = 1/(T-u_{1,n})^{1/2} \textstyle\int^T_0 \theta(u-u_{2,n})\theta(v-u_{1,n})du = (T-u_{2,n})/(T-u_{1,n})^{1/2}. \quad (43)$$

Since $x_2 = \delta(u-u_{2,n}) - (T-u_{2,n})/(T-u_{1,n})\delta(u-u_{1,n})$, we have

$$\|x_2\|_{-1} = (\textstyle\int^T_0 |-\theta(u-u_{2,n}) + (T-u_{2,n})/(T-u_{1,n})\theta(u-u_{1,n})|^2 du)^{1/2} =$$

$$[(T-u_{2,n})(u_{2,n}-u_{1,n})/(T-u_{1,n})]^{1/2}. \quad (44)$$

Thus, we prove the formula (40) for $k = 2$. Developing a similar computation for the elements $e\hat{}_{k,n} = x_k/\|x_k\|$, we obtain the desired formula (40) for any k. □

We show that the application of (40) results in easily computable functions (under the condition that the initial functions are easily computable). Let the linear operator

$Rf = \int_a^b r(t,u)f(u)du$ is given, where $r \in C \times L_2$. Then

$$Re^\wedge_{1,n} = 1/(T-u_{1,n})^{1/2}\, r(t,u_{1,n});\ Re^\wedge_{k,n} = (T-u_{k-1,n})^{1/2}/[(T-u_{k,n})(u_{k,n}-u_{k-1,n})]^{1/2}\, r(t,u_{k,n}) -$$

$$(T-u_{k,n})^{1/2}/[(T-u_{k-1,n})(u_{k,n}-u_{k-1,n})]^{1/2}\, r(t,u_{k-1,n}),\ u_{k,n},\ u_{k-1,n} \in [a,b],\ k = 2, 3,...;$$

$$n = 1, 2,... \quad (45)$$

From Theorem 6 it follows that the system of functions (45) is R-complete in L_2.

3. ON OPTIMAL METHODS FOR THE SOLUTION OF APPLIED PROBLEMS ON COMPUTERS

The method and the results of s. 3, Ch. 6 can be expanded for many other applied problems: solutions of algebraic, differential and integral equations, minimization of functions and functionals, etc. (see [6, 7, 20]).

After reduction of an initial problem or from the beginning, when we have a certain finite problem for its solution on a computer, the questions about round-off errors and numbers of computation operations are arising. We can answer to these questions in a reasonable way following the results of [1, 2, 6, 7], and Ch. 5.

We dwell here on some results concerning numbers of the necessary main operations on computers.

For linear algebraic systems in the form

$$\sum_{i=1}^{n} x_i(Ae_i, Ae_j) = (f, Ae_j),\ j = 1,..., n, \quad (46)$$

which appear by the well-known *method of least square* for the solution of the linear operator equation $Ax = f$, we can economize the number of basic operations at the expense of approximately tape structure of the respective matrices $\{(Ae_i, Ae_j)\}$. As a result [6], we can achieve n^2 operations instead of n^3 in order, where n is the number of the equations. The same result can be obtained for many other direct methods of mathematical physics.

For the problem of quadratic *Boolean programming*, i.e., determination of at least one vector x for which

$$f^* = \sup f(x)\ (x \in B^n),\ f(x) = \sum_{i=1}^{n} x_i(1 - \sum_{j=i+1}^{n} a_{ij}x_j),\ x = (x_1,..., x_n),$$

$$x_i,\ a_{ij} \in B = \{0,1\},\ j = i+1,..., n;\ i = 1,..., n,\ n > 1, \quad (47)$$

where B^n is the set of all the vertices of n-dimensional cube, it is known that we can achieve n^m operations if $a_{ij} = 0$ for $|i-j| > m$, where m does not depend on n. This result can be probably generalized on any applied mathematical programming problem, for the correlation among variables is decreasing fast depending on their distances.

The problem (47) in general case is the example of the so-called *NP-complete problems* [1, 2, 15] which are very difficult for the solution under large n. There is the conjecture that this problem does not belong to the class P, i.e., it cannot be solved with the help of a polynomial number of basic operations.

We hope that special machines non-algorithmic-type, with unlimited parallel modes, can be constructed for the effective solution all these problems in practice.

Investigation in the field of algebraic-based complexity theory also gives the many important results (see, for example [1]). In particular, it was proved [12] that the problem of linear programming belongs to the class P and the polynomial-time effective algorithms were constructed for its solution [10].

New trends in theoretical developments and computational methods, connecting, on the one hand, with the properties of the symmetries of solitons and other solutions of evolution equations and, on the other hand, with the possibilities of symbolic computation, one can find in [5, 11].

In general, let we need to solve an applied problem $P(I)$ with the help of an algorithm $A(X)$ on a computer $C(Y)$, where I, X, and Y are sets of parameters of the problem, algorithm and computer respectively. Then, in practice, we have deal with the so-called complete error E, necessary time T, and memory M of the computer,

$$E = E(I, X, Y),\ T = T(I, X, Y),\ M = M(I, X, Y). \tag{48}$$

The optimal by E algorithm $A(X)$ is important, in particular, because with insufficient accuracy, we should try to find reserve in improving of parameters I and/or Y, but not the algorithm.

The important *reserve* of improving the *accuracy* of initial data $I = I(t)$, $t \in [0, R]$, of an applied problem can be realized [6, 7, 17] by solving but again the Volterra-type equation of the first kind in the form

$$\int_{t-t^\wedge}^{t} K(t,s) I^\wedge(s)\,ds = I(t),\ t \in [0, R], \tag{49}$$

where $I^\wedge(s)$ is the desired vector of more precise data, $K(t,s)$ is the matrix of impulse transient functions of the respective measuring devices (which is presupposed known with the sufficient accuracy), $t^\wedge$ is the estimate from below for the time of end of all the transient processes.

For improving of the round-off error we can carry out computations with the additional accuracy (with extra binary digit bits).

The usual mode for improving of the time $T(I, X, Y)$ is the computation in a parallel way of the collection of the independent sub-problems representing the initial applied problem.

Improving $M(I, X, Y)$ implies usually making worse $T(I, X, Y)$ and vice versa.

4. ON SOFTWARE

The software appropriated to the problems of ES or development modeling is contained in [7, 8, 13]. The peculiarity of the software [7] consists in that as a rule the so-called estimating subroutines are contained in it, side by side with the ordinary subroutines resolving principal standard problems of computation mathematics. The former subroutines estimate the number of the necessary basic computer operations, the required computer memory capacity, and different kind of errors accompanying a process of the solution of applied problems on a computer.

The software [8, 13] is based on [7] and consists from the subroutines for the solution of the above mentioned problems of MM identification, for modeling of the problems of development prediction in general and in the form "if . . ., then . . .", and for modeling of some ES optimization problems.

The appropriate software developed in the USA can be found in [14, 16].

REFERENCES

1. Akho A., Khopkroft D., Ul'man D., Construction and Analysis of Computing Algorithms, M.: Mir, 1979, 536 pp. (in Russian).
2. Garey M.R., Johnson D.S., Computers and Intractability. A Guide to the Theory of NP-Completeness, W.H. Freeman and Co., 1979, 416 pp.
3. Gel'fand I.M., Shilov G.E., Spaces of Basic and Generalized Functions, M.: Fizmatgiz, 1958, 306 pp. (in Russian).
4. Gladkivskaya O.V., Investigation of the Identification Problems for One Class of Dynamic Models, Synopsis of Thesis, Kiev University, 1990, 15 pp.
5. Hereman W., New Trends in Theoretical Developments and Computational Methods, in the book: CRC Handbook of Lie Group Analysis of Differential Equations, 3, 1996, 367-413.
6. Ivanov V.V. et al., Characteristics of Problems, Algorithms, and Computers in Complex of Programs of Computation Mathematics, Reprint 84-36, Kiev: Glushkov Inst. of Cybernetics, 1984, 54 pp. (in Russian).
7. Ivanov V.V. et al., Complex of Routines POM-1 (Programs of Optimal Methods), Glushkov Inst. of Cybernetics, Kiev: 1985, 1250 pp. (Dep. in GOS FAP SSSR, N 50860000156) (in Russian).
8. Ivanov V.V. et al., Package of the Programs for Economic Systems Development Prognosis with Regard to the Technical Progress, Upravl. sistemy i mashiny, N 5, 1985, 100-102 (in Russian).
9. Ivanov V.V., Yatsenko Yu.P., Adaptive Algorithms of Optimization in the Integral Macro-Economic Models, DAN SSSR, N 5, 1986, 1053-1058 (in Russian).
10. Iri M., A Proof of the Polynomiality of the Iri-Imai Method, Technical Reports, Faculty of Engineering, University of Tokyo, 1991, 1-23.
11. Ito M., SYMCD-a REDUCE Program for Finding Conserved Densities of Systems of Nonlinear Evolution Equations, Comput. Phys. Commun. 79, 1994, 547-554.
12. Khachiyan L.G., Polynomial Algorithms in Linear Programming, ZVMiMF, 20, 1980, 51-68 (in Russian).
13. Mikhalevich V.S., Ivanov V.V. et al., Complex of Routines MDS-1 (Modeling of Developing Systems), Glushkov Inst. of Cybernetics, Kiev: 1986, 711 pp. (Dep. in GOS FAP SSSR, N 50880000105) (in Russian).
14. More J.J., Wright S.J., Optimization Software Guide, SIAM, 1993, xii+153 pp.
15. Pardalos P.M., Thoai N.V., Introduction to Global Optimization, Kluwer Academic Publishers, 1995, 318 pp.
16. Press W.H., Vetterling W.T., Teukolsky S., Flannery B.P., Numerical Recipes in C Numerical Recipes in Fortran The Art of Scientific Computing, Cambridge Univ. Press, 1992, C Version, 1992 pp.; Fortran Version, 992 pp.; C Example Book, 352 pp.; Fortran Example Book, 352 pp.
17. Provencher S.W, CONTIN: a General Purpose Constrained Regularization Program for Inverting Noisy Linear Algebraic and Integral Equations, Computer Physics Communication 27, North-Holland, 1982, 229-242.
18. Ramm A.G., Theory and Applications of Some New Classes of Integral Equations, Springer, 1980, 343 pp.
19. Sobolev S.L., Equations of Mathematical Physics, M.: Nauka, 1966, 443 pp. (in Russian).
20. Sukharev A.G., Optimal Search of Extreme, M.: Izdatel'stvo MGU, 1975, 100 pp.
21. Vasil'ev F.P., Numerical Methods for the Solution of Extreme Problems, M.: Nauka, 1980, 518 pp. (in Russian).

PART III:

INTRODUCTION TO APPLICATIONS

CHAPTER 8

RECONSTRUCTION OF ECONOMY CONTROL (by academician Glushkov)

V.M. Glushkov's concepts of the scientific systematic analysis [3] and development [3, 4] (see also [7, Ch. 2]) generalized in the monograph [5] can explain many urgent issues of any country's economy control reconstruction.

1. ON THE SCIENTIFIC SYSTEMATIC APPROACH

According to the approach, the effective solution of the reconstruction problems of country's economy control is based on: definition of the control objects, control body, and devices and their parameterization; structuring and elaboration of the models and their investigation; elaboration and ground of the goals, control algorithms, and appropriate means; estimation of the efficiency of the reconstruction process and effective realization of the latter.

A new control technology project elaborated in details and a detailed plan of the actions to convert the control technology from the existing to the new one are the necessary conditions for the efficiency of organizational reconstruction.

Very important issue is the organization of the perfecting process for the control systems. It is necessary that along with the acting system of organizational control we have the acting system of perfecting of the former system (and itself). Projection of the reconstruction of organizational control is complicated not less than projection of e g airplanes, but airplanes are projected not those specialists who control them. "...unfortunately, it is prolonged to hold the opinion that the systems of organizational control can be among other things projected and launched in the work by the managerial staff themselves. This fact has become a serious obstacle on the way of perfecting the national economy control systems on the actually scientific basis." ([5], p. 403). By V.M. Glushkov estimations, the employment of the reserves of the USSR's economy control improvement would allow them to achieve frontiers of every two past five-year plans for one under the same inputs.

2. PRINCIPAL NOTIONS OF THE DEVELOPMENT CONCEPT

The economic system on the whole and its separate subsystems (control objects and control bodies) as well as various other artificial systems created by human beings and natural systems are the instances of the so-called developing or evolutionary systems (ES). They all have the sequence of the typical common properties, namely, existence of: the prehistory of development; the subsystem of perfecting of the whole system (including itself); the functioning part of the system creating useful products and the obsolete part of the system; the new created elements of the system, usually more effective, and the old elements, usually less effective; the internal and external functions of the system; the distribution of the given resources between these functions; the resources, incoming from outside; etc.

The main element of the economy as ES is the working place (WP) which is determined by definition as totality of labor functions and their ware: material, energetic, information (including software and hardware) and personnel. The main characteristics of WP are indices of its functioning efficiency.

The main conditions and sources of ES development are availability of external resources and existence of the subsystem of development or perfection of the whole system including (and this is the principal source!) the perfection subsystem itself.

Thus, *self-development* or self-perfection is the core of the development *conception* ("concept of S-D or S-P").

The descriptions of ES models in detail and their investigation see above (PART I). Here we would like to stress the necessity of strengthening the services for perfecting of enterprises, detailing of their functions and raising of their efficiency. For this, it is necessary, in particular, to have the list of all the labor functions fulfilled, their wares and the respective indices of efficiency as a part of the basic indices of activity for every enterprise (or joined one). It is required for robotization of production and the solutions of many different problems of product quality and labor productivity raising. Furthermore, it is necessary also to create the labor function theory and the automated construction of means of production for realization of the new labor function with the help of the determined basis collection of labor functions. It is especially required for the so-called FAP (flexible automated production). At last, it is appropriated to have the plans of perfecting of WP on WP themselves for all enterprises and to conduct the work in this direction during a certain part of the total working time. All these in more detail see in [6] and Ch. 14, s. 3.

3. STRUCTURE AND FUNCTIONS OF ECONOMY AS ES

The structure and functions can be represented Fig.1, Ch. 1, where A is the perfecting subsystem for all ones (among them for itself) or the subsystem of new WP production; B is the subsystem of consumer goods and services production; f is the generalized resource, incoming into ES from outside in the unit of the time t; m is the number of the new WP in the time unit; c is the quantity of consumer goods and services in the time unit; G is the total quantity of the "waste" or obsolete WP; y is a relative share of WP distribution between A and B; α is the index of efficiency of WP functioning in the subsystem A; β is the index similar to α but for the subsystem B.

It should be noted that these symbols and relations for our MM are not common ones in the economics. The subsystem A is usually called as the system of production of means of production. We include in the subsystem B not only things of personal consumption but also the whole non-industrial consumption. It includes, in particular, all kind of armaments and all kind of public care (means of transportation, broadcasting, TV, etc.). And also, referring the production of a final product to A or B we refer to the same subsystem the production of all the intermediate products which are necessary for the manufacturing of the final product. The important qualitative conclusion follows from here that average complexity of manufacture into the subsystems A and B is approximately equal. It allows us to simplify the models in the case of macroeconomic applications.

4. BASE MM

The respective simplest, minimal, or *base MM of economy* as ES is

$$m(t) = \int_{a(t)}^{t} \alpha(t,s)y(s)m(s)ds,\ c(t) = \int_{b(t)}^{t} \beta(t,s)[1-y(s)]m(s)ds,\ f(t) = m(t) + c(t),$$

$$P(t) = \int_{a(t)}^{t} y(s)m(s)ds + \int_{b(t)}^{t} [1-y(s)]m(s)ds,\ G(t) = \int_{0}^{a(t)} y(s)m(s)ds +$$

$$\int_{0}^{b(t)} [1-y(s)]m(s)ds,\ 0 \le y \le 1;\ 0 \le a(t),\ b(t) < t;\ t \ge t\hat{}\ \ge 0, \qquad (1)$$

where $t\hat{}$ is the start of modeling; $m(t)$ is the rate of quantity of new WP production or productive capacity at the instant t; ym is a share of m which goes to the subsystem A of restoration of WP; $c(t)$ is the rate of quantity of the new external product at the instant t; $(1-y)m$ is a share of m which goes to the subsystem B of $c(t)$ production; $f(t)$ is the external resources of ES (in the same units as m and c); $a(t)$ and $b(t)$ are temporal boundaries for liquidation of the obsolete WP; $[\min[a(t\hat{}), b(t\hat{})], t\hat{}]$ is the ES development prehistory (all the elements of the model are usually given on it); $P(t)$ is the total quantity of functioning WP at the instant t; $G(t)$ is the total quantity of obsolete WP at the instant t; $\alpha(t,s)$ is the efficiency index for functioning of the subsystem A along the channel ym -- m, i.e., the number of the units of $m(t)$ created in the unit of the time starting from t per one unit of $y(s)m(s)$; $\beta(t,s)$ is the efficiency index for functioning of the subsystem B along the channel $(1-y)m$ --c, which is similar to α; the efficiency indices α and β may depend on all the elements of the model.

With position of economics the first *equation* in (1) is that of *fund growth*, the second is the *equation* of *consumption*, the third is the *balance equation* for input and output resources, the fourth is the *balance equation* for functioning labor power, and the fifth is the *equation* for obsolete or preserving in reserve *labor power*. As the unit for measurement of productive capacity we shall accept one WP (averaged). We presuppose that at the each given instant of the time t there is the unique newest, more effective technology m_t which are employed in the capacity $m(t)$. So, the function $\alpha(t,s)$ represents number of WP in the technology m_t created in the unit of the time per one WP (averaged) in the technology m_s. This function is usually increasing over s and decreasing over t. Because of finiteness of the set of different kind of technology both functions α and β are step functions. However, under small steps these functions can be approximated in the points of discontinuity with continuous functions.

The ordinary statement of problems with regard to MM (1) is as follows: exogenous elements α, β, f, and P to be given, distribution y to be found from the solution of an optimization problem (see, for examples, the problems C.1-C.8, Ch. 2) and m, a, b, c, and G to be found from the system (1) (usually on the pre-assigned temporal segment $[t^\wedge, T]$).

The important particular case of (1)

$$a(t) = b(t),\ da(t)/dt \geq 0. \tag{2}$$

The second relation in (2) means that closed down capacity is not opened for using. In the case (2) usually α, β, P (or G), and y are given and m, a, c, f, and G (or P) are found regarding the relations (1) and (2).

We can set down different optimization problems with regard to MM (1) or (1), (2). One of the statement is the problem minimization of the *labor power* under the condition that the planning period T and consumption $c(t)$ are given:

$$\int_{t^\wedge}^{T} P(t)dt = \min\ (y). \tag{3}$$

From general mathematical positions it is not difficult both to attach much more generality to the relations (1) and to make essentially more detailed models (see Ch.2). So, in (1) α, β, and y can be considered as matrix functions and m, c, P, and G as vector functions. In this connection, 1 is the unit matrix, 0 is the null matrix, and $0 \leq y \leq 1$ means the same inequalities for all respective components of the matrix y. The economic interpretation of these models and their investigation can be multi-sector analysis of economy of a separate country or economy of our planet.

Rigorous mathematical investigation of the (1), (2)-types models and (3)-type optimization problem is given in the PART I of this book.

5. MM OF INDICES OF EFFICIENCY OF α, β-TYPES

We consider some *MM of* α. *MM of* β will be similar.

The index α in the form

$$\alpha(t,s) = \alpha(s)(1 + cs)/(1 + ct) \tag{4}$$

can be obtained due to the known estimate of a constant c (can be found on the verifying phase of prediction, usually $c \approx 0.5$) and the array of approximated values of $\alpha(s)$. The last ones can be determined with the help of the prehistory of ES from the equations

$$m^*(t) = m(t)(1+ct) = \int_{a(t)}^{t} \alpha(s)y(s)m^*(s)ds,\ N(t) = \int_{a(t)}^{t} n(s)ds,$$

$$N^*(t) = \int_0^t n(s)ds,\ P_1(t) = \int_{a(t)}^{t} y(s)m(s)ds,\ P_2(t) = \int_{a(t)}^{t} [1-y(s)]m(s)ds, \tag{5}$$

under the condition that the capital stock $N(t)$, the inflow of funds per year $n(s)$, and the numbers of the personnel in the subsystem A $P_1(t)$ and the subsystem B $P_2(t)$ are given. Indeed, let $N^{*-1}(t)$ be the *reciprocal* function or *operator* for $N^*(t)$. Then

$$a(t) = N^{*-1}[N^*(t) - N(t)], \tag{6}$$

and

$$m(t) = P'(t) + m(a(t))a'(t),\ P(t) = P_1(t) + P_2(t),\ y(t) = [P_1'(t) + m(a(t))y(a(t))$$

$$a'(t)]/m(t),\ \alpha(t) = [m^{*\prime}(t) + \alpha(a(t))y(a(t))m^*(a(t))a'(t)]/[y(t)m^*(t)]. \tag{7}$$

One can see from (7) that we have to use the technique of the solution of the *ill-posed problem* of numerical differentiation (see Ch. 5, 6) to find the desired functions.

In the general case the desired function $\alpha(t,s)$ can be obtained on the basis of the least square method and various other methods (see Ch. 2) and [7, 8]. In particular, on the basis of detailing of the structure of ES, Ch. 2, s. 6.2., regarding the function α in the form similar to (3), we have

$$\alpha'(t) = \int_{a(t)}^{t} \alpha(t,s)x(s)m(s)ds,\ 0 \le x \le 1 - y, \tag{8}$$

with the substitution of $1\text{-}x\text{-}y$ for $1\text{-}y$ in (1). It corresponds to the selection in the subsystem A the branch of science that creates the new technology characterized by the function α. And also xm corresponds to a share of WP of scientists.

6. SOME IMPORTANT PROPERTIES OF ECONOMY AS ES

The investigation of economic system as ES results in the proof of the following important properties (see Ch. 3 and 4):

i) *For the establishment of "the record" of any external function of economic system during long periods of the time, for example, maximization of the output consumer goods and services or minimization of the labor input under the given level of consumption* (see (3)), *the most part of the resources must go to the internal needs of the system, i.e., on the perfecting subsystem A of ES.*

ii) Let us distinguish in the subsystem A of the economic system the branch of the Science that creates new technologies and in turn raises the above mentioned indices of efficiencies of the WP functioning (see MM (51), Ch.2).

Then for the establishment of "the record" of the external function of the economic system the Science has the highest priority (for distribution of the resources) and the remained part of A goes next and the subsystem B goes thereupon.

iii) *Any economic system can utilize rationally not more than determined quantity of the external resources f.*

iv) *Optimization of functioning of the economic system results inevitably in liquidation of the obsolete WP and therefore necessity of releasing workers to train for new professions and necessity of utilization of the economic system "waste".*

v) *The most important factors for acceleration of economic development and controlling of growth of specialists' number in the subsystems of A-type are raising of efficiency of WP in these subsystems, and creation and intrusion of various means of organization techniques, methods of scientific organization of labor, personal professional computers, automated WP for decision makers, etc.*

7. ON THE ROLE OF CENTRAL BODIES OF CONTROL

Let ES1 be a central *body of control* as ES and ES2 be a controlled ES. The generalized structure of the bonds between ES1 and ES2 has the form of Fig.2, Ch.1, where the values fi, mi, ci, yi, and Gi, $i = 1$ for ES1, and $i = 2$ for ES2, have the same sense as the values f, m, c, y and G in the case of Fig.1, Ch.1, and in the models (1) above.

From here it follows that the main role of central bodies of control (system ES1) relative to isolated businesses (system ES2) consist of

A. Establishment of bilateral bonds between ES1 and ES2 for the problems statement, control of their decisions and interchange of the results.

B. Determination of the appropriate resources $f2$, $f1 + f2 = f$.

C. Forming of the government orders like $m1''$, $c1''$ and $m2''$, $c2''$.

D. Determination of methods for utilization of $G1''$ and $G2''$.

E. Control of fulfillment of the self-compensatory condition

$$c2 \geq c1'' + m1'' + d, \tag{9}$$

where d is the fee for central bodies' supporting.

Of course, central bodies of control have much more number of functions not bound directly to ES2. We can construct *MM* of ES1 and ES2 *interaction* by analogy with (1)-(7) regarding the structure of two ES interactions (Fig 2, Ch.1). Besides, we should add the following bonds between both systems:

$$m1 = m1' + m1'' + m2'',\ m2 = m2' + m2'' + m1'',\ c1 = c1' + c1'' + c2'',$$

$$c2 = c2' + c2'' + c1'',\ f = f1 + f2,\ f1 + G2'' = m1 + c1,\ f2 + G1'' = m2 + c2. \quad (10)$$

Applying game theory to the problem of optimization of the interactions between ES1 and ES2, we receive at least two possibilities. Let the optimization problem consists in both

$$f_1(y1,y2),\ f_2(y1,y2) = \min\ (y1,\ y2), \quad (11)$$

where f_1, f_2 are purposeful functionals for ES1, ES2 correspondingly. Then the first possibility is determination of the so-called coordinated optimum, i.e., any *Pareto point* $(y^P1,\ y^P2)$ for which the set

$$\{(y1,\ y2)/([f_1(y1,y2) < f_1(y^P1,y^P2)] \wedge [f_2(y1,y2) \leq f_2(y^P1,y^P2)]) \vee$$

$$([f_1(y1,y2) \leq f_1(y^P1,y^P2)] \wedge [f_2(y1,y2) < f_2(y^P1,y^P2)])\} = \varnothing, \quad (12)$$

and, maybe, determination of the controls $y^P{}^\wedge 1$, $y^P{}^\wedge 2$ for which

$$f_1^{P\wedge} = f_1(y^P{}^\wedge 1,y^P{}^\wedge 2) = \min f_1(y^P1,y^P2), \quad (13)$$

where min is taken by all the *Pareto points* (12).

The second possibility is determination of the controls $y^\wedge 1$, $y^\wedge 2$ for which

$$f_1{}^\wedge = f_1(y^\wedge 1,y^\wedge 2) = \min f_1(y1,y2)\ (0 \leq y1,\ y2 \leq 1) \quad (14)$$

and determination of the controls y^*1, y^*2 for which

$$f_2(y^*1,y^*2) = \min f_2(y1,y2)\ ((y1,y2) \in \{y1,y2 \,|\, f_1(y1,y2) = f_1{}^\wedge\}) \quad (15)$$

or better

$$f_2(y^*1,y^*2) = \min f_2(y1,y2)\ ((y1,y2) \in \{y1,y2 \,|\, f_1(y1,y2) \geq f_1{}^* - ad\}, \quad (16)$$

where *ad* is "admittance" from a leader to a follower.

It is not bed to associate a central body and an isolated enterprise as a model of interrelation between society and its separate members, i.e., every member of society (every enterprise) obtains as a rule the determined task from society (central body), but for realization of the task one may usually use not all potentially available worker time.

So, it is possible to dispose the remainder of the time according to the proper judgment, in particular, to make use of it for determined initiative (regarding existing stimuli). The question arises about the optimal relation of the times. For the solution of the question it should weigh as always additional effects from realization of additional orders and losses from the complication of conditions for the creative activity. The model (1)-(15)-types allows us to consider this problem both on qualitative and quantitative level.

8. ON THE FORMATION OF COUNTRY'S DEVELOPMENT

V.M. Glushkov gave much attention to various mechanisms of the formation of society development with short- and long- term goals. By his opinion [3, 5], the desired mechanism has to consist of series iterates with the determined temporal steps on each of which the following actions take place:

i) Inquiry of social opinion for determination and correction of goals.

ii) Modeling of the consequences under assumption of acceptance of those or other goals regarding the neo-sphere factors (see Ch. 1 and the next Ch. 9).

iii) Information for community about the results of the modeling.

iv) Actual acceptance of the goals and their realization (after stabilization of the results of i) in the rotation of i), ii), iii) and i).

Let us dwell shortly on kinds and *methods of prediction* for achievement of those or other goals according to Glushkov [3-5].

At first, this is a demographic prediction, estimating forward, during two-three decades, variation of the country's population quantity and its composition by sex, age, national belonging, and so on. The second kind of prediction is that of international conditions and military-political situation trends. The third kind of prediction is that of science technical chances to guarantee social demands.

When the complex system of economic development goals has been determined the main control problem is to guarantee the most rapid achievement of the goals. In the condition of men resources' scantiness the criterion of the most rapid achievement of the goal practically accords to that of the most rapid growth of the productivity of labor, moreover not abstract productivity, but needed for realization of the goals standing before society. For the each complex goal the plan of its achievement has to be made up. This plan can be represented the most conveniently as network diagrams. After reconciliation and optimization of the plans in aggregate indices the process of their further detailing is starting.

The description in detail of the methods of prediction and control of discrete processes, in particular, STP, and the examples of application of the methods for prediction and control of social processes are given in [3, 5].

The problems, arising on the way of the complex solution of economic development, are proposed to overcome with the help of the general automated system for which courses of construction are also described in detail in [4].

9. SOME CONCLUSION OF A PRACTICAL FEATURE

The certain recommendations of a practical feature have been done in s.1. Let us consider some problems in economics from a position of the theory of ES.

The first problem is a reasonable relation between the industrial sphere and the rest ones and the connected problems of increasing or decreasing of administrative machinery, numbers of scientific workers, physicians, etc. It follows from the general theory that the non-industrial sphere, in particular, a relative number of scientific workers and physicians has to increase. The reason is that the final product of the non-industrial sphere is new WP, in particular, new technology mastering which industrial sphere can give the important new product and much more common products than under the simple quantitative growth of output. In a similar way, it follows that a relative share of administrative machinery has to increase if according to the theory of ES the main function of administrative machinery is creation of new, more effective WP.

The second problem is advisability of stimulating those or other kind of production. The usual approach is to stimulate, for example, production of critical commodities that are not enough. Such type of decision is superficial. It does not affect deeper nonlinear regularity of ES functioning. In particular, resolving decision can be stimulating the growth of labor productivity that is responsible for labor productivity of the production of the deficient output.

The third problem is an interconnection between the theory of development and market problems. Of course, the market and the inquiry of social opinion correct and stimulate directions and purposes of systems development. But the best ways of achieving of originating purposes can be found from the ES theory.

10. ON THE SOFTWARE FOR MM OF ECONOMY AS ES

This software [12] was elaborated in Glushkov Institute of Cybernetics, Kiev, the Ukraine. Its considerable part is the part of the software [7, 8, and 13, Ch. 7].

It exploits the models (1)-(3)-types (two- and nine-product models) and also the standard economic indices (national income, output of consumer goods, average productivity, rate of accumulation, manpower, capital stock, basic funds, profit, etc.), the new indices characterizing the control of reproduction of basic funds (quantity of new WP created in current year, quantity of obsolete WP eliminated in current year, growth of productivity of new WP, etc.), and the indices characterizing the control of science (investments in science, quantity of scientists, delays of penetration of results, coefficient of science efficiency, etc).

It can be used for multi-variant prediction of the main aggregate indices of economic systems development, for the solution of optimization problems, for training users to work with the new economic models, and as business game "Multi-variant prediction of the main aggregate macro-economic indices".

11. CONCLUSION

A. V.M. Glushkov's memoirs can serve as foundation for scientific approaches to the solution of various problems of economy, especially STP.

B. Comparison between V.M. Glushkov's methodology and his results and today discussions and propositions for the solution of economic reform problems (see, for examples, [1, 2, 9]) only emphasizes the advantage of Glushkov's ideas and concepts. In those works there is only a weak reference to the systematic approach significance, nothing is about the role of the subsystem of A-type and there are almost no propositions about the STP control mechanism. The similar self-organization approach for modeling of economics under technical change is developed later in [10, 11].

C. The traditional, classical forms of dynamic macro-models are based on using systems of ordinary differential equations. Such models are not entirely convenient, by the reason that the real macroeconomic systems are described by non-smooth and even discontinuous functions. What is more, the classical forms are poorly suited for the description of dynamics of turning over and final rejecting obsolete capacities from being used. Here we have introduced the description of dynamic mathematical models with the help of the system integral equations, which allow us to reduce the lacks.

D. These results and techniques can serve as theoretical groundwork for the complex projects of various countries' economics control general plan.

It should be emphasized that experience of application of MM of ES under consideration shows that there are no such complicated urgent problems of economy for which this technique would be useless. By analogy with the economy on the whole we can consider its separate brunches [6], any enterprise (Ch. 14, s. 3), and the other fields of human activity (Ch.9).

REFERENCES

1. Aganbegyan A.G., The Soviet Economics. View in the Future, M.: Ekonomika, 1988, 256 pp. (in Russian).

2. General Economic Perspective Until 2000 y., UN ESE E8.AP, 32, 1989, 292 pp.

3. Glushkov V.M., Introduction in ASC (Automated System of Control), Kiev: Tehnika, 1972, 310 pp. (in Russian).

4. Glushkov V.M., Macro-economic models and concepts of construction of GSAS (General State Automated System), M.: Statistika, 1975, 160 pp. (in Russian).

5. Glushkov V.M., Grounds of Paper-off Information, M.: Nauka, 1982, 552 pp. (in Russian).

6. Ivanov V.V., Conception of Development of Machine-Construction in the Ukraine, Kiev: Glushkov Inst. of Cybernetics, Reprint 92-16, 1992, 23 pp. (in Russian).

7. Ivanov V.V. et al., Collection of Standard Subroutines for CMWG (Control Machine of Wide Goals), 1966, Kiev: Tehnika, 312 pp. (in Ukrainian).

8. Ivanov V.V. et al., Methods for Algorithmization of Continuous Industrial Processes, M.: Nauka, 1975, 400 pp. (in Russian).

9. Leontief W., Essays in Economics: Theories, Theorizing, Facts, and Politics, New Brunswick and Oxford: Transaction books 1, 1984, 228 pp.

10. Makarov V.L., On Development of Update Economic and Mathematical Techniques, Economics and Mathematical Methods, 22, 3, 1986, 412-425 (in Russian).

11. Silberg G., On the Complex Dynamics of Technical Change and Economic Evolution, J. Sci. & Ind. Res., 5, N 3, 1992, 151-157.

12. Software IDEM for modeling of economic systems development, Glushkov Inst. of Cybernetics, Advertisement, Kiev, 1985, 3 pp. (in Russian).

One link broken, the whole chain is broken

CHAPTER 9

MM OF THE NEO-SPHERE

(by academician Vernadsky)

Mathematical models (MM) of the *neo-sphere* are developed by some combinations both of human activity (HA) models and the biosphere models and also by introduction of two relations: what humanity takes from the biosphere and what gives to it in return.

HA MM with exogenous or endogenous science-technical progress (STP) and the biosphere MM are constructed from the mathematical theory of evolutionary systems (ES) modeling (see Part I).

1. HUMAN ACTIVITY MODELS

1.1. HA models with exogenous STP

The *base*, simplest *MM* of HA as ES

$$m(t) = \int_{a(t)}^{t} \alpha(t,s)y(s)m(s)ds + \Delta m(t);\ P_1(t) = \int_{a(t)}^{t} y(s)m(s)ds,$$

$$c(t) = \int_{b(t)}^{t} \beta(t,s)[1-y(s)]m(s)ds + \Delta c(t);\ P_2(t) = \int_{b(t)}^{t} [1-y(s)]m(s)ds;$$

$$P(t) = P_1(t) + P_2(t),\ M(t) = \int_{0}^{t} m(s)ds,\ G(t) = M(t) - P(t),\ C(t) = \int_{t^\wedge}^{t} c(s)ds;$$

$$f^\wedge(t) \geq f(t) = m(t)+c(t),\ 0 \leq y \leq 1;\ 0 \leq a(t),\ b(t) \leq t;\ t \geq t^\wedge \geq 0. \qquad (1)$$

where m is a rate of creation of new resource numbers in the subsystem A, which provides internal functions of HA; α is an efficiency index for the functioning of A along the channel $y(s)m(s)$--$m(t)$; Δm is an increment of m; $P(t)$ is the total number of resources functioning at the time t; $c(t)$ is a rate of creation of new resource numbers in the subsystem B, which provides external functions of HA; β, and Δc are similar to α, and Δm, but for the subsystem B; $G(t)$ is the total number of out-of-date resources at the time t; $a(t)$, and $b(t)$ are temporal thresholds for elimination (or reservation) of out-of-date resources in the respective subsystems A, B; $f^\wedge(t)$ is a rate of new resource numbers flowing in from the outside ES at the instant t; $t^\wedge$ is a starting point for the model; $[0,t^\wedge]$ is the HA prehistory on which all the functions are given.

Apparently, all the relations in (1) are faithful representations by definition and all the functions in (1) are nonnegative. In general, the functions α and β are dependent on the rest of the functions.

Thus, MM of HA deal with the so-called "gray box" when the structure of dynamic systems is partly revealed.

Under the condition that α, β, y, and P, f or P_i, $i = 1, 2$, are given the system (1) can be used to estimate m, M, a, b, c, C, G, and P_i, $i = 1, 2$, or P and f.

Note that in (1) $P(t)$ can be interpreted as the total number of employees at the instant t, $m(t)$ as the number of new work places (WP), $c(t)$ as the quantity of the new consumer product. The partitioning y is usually found from the solution of a certain optimization problem.

Many results about existence, uniqueness, and structure of the solutions of the equations as well as some optimization problems are given in Part I above.

1.2. HA models with endogenous STP

Let us construct MM of the functions α and β from detailing of the ES structure. Decreasing of α and β of t depends on many factors in the theory of deterioration (aging). As reasonable simple hypotheses we accept

$$\alpha_1(t,s) = \alpha_1(s)\exp[-d_{11}(t-s)],\ \beta_1(t,s) = \beta_1(s)\exp[-d_{21}(t-s)]; \tag{2}$$

or

$$\alpha_2(t,s) = \alpha_2(t-s)\exp(d_{12}t),\ \beta_2(t,s) = \beta_2(t-s)\exp(d_{22}t), \tag{3}$$

where d_{ij} are given constants, $d_{i1} \geq 0$.

Reasoning about the structure of ES as above (see Ch. 2 for detail) results in

$$\alpha_\mu^\nu(t) = \int_{a_1(t)}^t \alpha_\mu(t,s)x(s)m(s)ds,\ \nu = -1,\ 0 \text{ or } 1(\equiv');\ m(t) = \int_{a_2(t)}^t \alpha_\mu(t,s)y(s)m(s)ds,$$

$$M(t) = \int_0^t m(s)ds;\ \beta_\kappa(t) = \int_{a_3(t)}^t \alpha_\mu(t,s)z(s)m(s)ds;\ c(t) = \int_{a_4(t)}^t \beta_\kappa(t,s)[1-x(s)-y(s)-$$

$$z(s)]m(s)ds,\ \mu, \kappa = 1, 2;\ f(t) = m(t) + c(t) + \alpha_\mu^\nu(t) + \beta_\kappa(t);\ P_i(t) = \int_{a_i(t)}^t m_i(s)ds,$$

$$G_i = \int_0^{a_i(t)} m_i(s)ds,\ i = 1, 2, 3, 4;\ m_1 = xm,\ m_2 = ym,\ m_3 = zm,\ m_4 = (1-x-y-z)m;$$

$$P(t) = P_1(t) + P_2(t) + P_3(t) + P_4(t),\ G(t) = M(t) - P(t);\ 0 \leq x, y, z, 1-x-y-z \leq 1;$$

$$0 \leq a_i(t) \leq t,\ i = 1, 2, 3, 4;\ T \geq t \geq t^\wedge \geq 0, \tag{4}$$

where $\alpha_\mu^\nu(t)$ are νth derivatives of $\alpha_\mu(t)$, $\alpha^0_\mu = \alpha_\mu$, $\alpha_\mu^{-1} = \int_0^t \alpha_\mu(s)ds$, $\alpha_\mu(0) > 0$; $x(s)m(s)$--$\alpha_\mu^\nu(t)$ and $z(s)m(s)$--$\beta_\kappa(t)$ are the channels of creation of the new technology quantities in the subsystem S (Science), and the remainder notations previously noted. It is clear that $\alpha_\mu(t)$ is always increasing in the case $\nu = 1$, since $\alpha_\mu'(t) > 0$. For HA as ES P_i, $i = 1, 2, 3, 4$, are usually given and the rest ones are found from the system (4) (and x, y, z are found from a certain optimization problem).

1.3. MM of cost and prices

Let $q(t)$ be *cost function* of the unit product (goods and services) $c(t)$ at the instant t and let $c(t)$, $c_n(t)$, and $c_r(t)$ be quantity of the whole, *social-necessary*, and realized external *product* in the time unit starting from t respectively. Then

$$q^\wedge(t) = q(t)k^\wedge_n(t),\ k^\wedge_n(t) = c_n(t)/c(t), \tag{5}$$

is a *social-necessary cost* of the unit product at the instant t;

$$d(t) = q^\wedge(t) - q(t) = q(t)[k^\wedge_n(t) - 1] \tag{6}$$

is a differential return;

$$r(t) = q^\wedge(t)l^\wedge_r(t) = q(t)l_r(t),\ l^\wedge_r(t) = c(t)/c_r(t),\ l_r(t) = c_n(t)/c_r(t), \tag{7}$$

is a *market price* of the unit product at the instant t.

In turn,

$$q(t) = q(^1)(t)q(^2)(t)q(^3)(t), \tag{8}$$

where $q(^1)$ is the cost of the new labor, $q(^2)$ is the factor cost growth at the expense of a labor earlier acquired, $q(^3)$ is the factor cost growth at the expense of amortization of all the means of a labor under creation of the unit product at the instant t.

Let dT be the segment of the time during which number of social-necessary goods $c_n(t)$ becomes invariable in practice. Then the market income from the sailing of the product obtained during $[t, t+dT]$ will be

$$dI = \int_t^{t+dT} r(t)c_r(t)dt \tag{9}$$

and the *profit* will be

$$dPr = dI - dE,\ dE = \int_t^{t+dT} q(t)c(t)dt, \tag{10}$$

where dE is the *total cost* of the product during $[t, t+dT]$. It is clear that *accumulated profit* $Pr(T)$ and *cost* $E(T)$ can be received by integration of dPr and dE.

The variability of market conjuncture results in changing the structure and composition of variable $c(t)$, which in turn results in the necessity of new WP creation or introduction.

2. BIOSPHERE MODELS

Let $m^*(t)$ be the rate of creation of the new (generalized) *ecological product* (alive or *living substance* [10, Ch.1]) quantity remaining in the biosphere; $y^*(s)m^*(s)$ be a share of $m^*(s)$ for recreation of $m^*(t)$; $g^*(t) = G^{*\prime}(t)$ and $c^*(t)$ be the rate of creation of the new (generalized) *ecological product* (bio-boned *or dead and inorganic substance* respectively) quantity abandoning the biosphere; $[1-y^*(s)]m^*(s)$ be a share of $m^*(s)$ for creation $c^*(t)$; $f^*(t)$ be the rate of inflow into the biosphere of the (generalized) resource amount; α^* and β^* be the indices of efficiency of functioning of the subsystem A and B of the biosphere as ES; $P^*(t)$ be the total quantity of alive substance into the biosphere. Then by analogy with (1) we have the following *MM of the biosphere:*

$$m^*(t) = \int_{a^*(t)}^{t} \alpha^*(t,s)y^*(s)m^*(s)ds,\ M^*(t) = \int_{0}^{t} m^*(s)ds,$$

$$c^*(t) = \int_{a^*(t)}^{t} \beta^*(t,s)[1-y^*(s)]m^*(s)ds\ ,\ C^*(t) = \int_{t^\wedge}^{t} c^*(s)ds,$$

$$P^*(t) = \int_{a^*(t)}^{t} m^*(s)ds,\ G^*(t) = [M^*(t) - P^*(t)],\ f^*(t) = m^*(t) + c^*(t),$$

$$0 \le y^* \le 1, 0 \le a^*(t), b^*(t) \le t,\ t \ge t^\wedge \ge \min\,[a^*(t^\wedge), b^*(t^\wedge)] = 0. \qquad (11)$$

Besides (11) let us introduce the relations

$$f^*(t) = k(f^{*\wedge})m^*(t), f^*(t) = c^*(t-e(t)) + g^*(t-h(t)) + s^*(t), \qquad (12)$$

where coefficient $k(f^{*\wedge})$ depends on the qualitative composition $f^{*\wedge}$ of the resources f^*; $e(t)$ and $h(t)$ are delays of c^* and g^* conversion in f^* by the virtue of the *bio-geo-chemical cycle* of the substances in the biosphere; s^* is a share of the resource f^* incoming into the biosphere from the cosmos.

The first equality in (12) is the additional balance relation arising from the specific ratio of m^*-component entering f^* in different proportions. Below we dwell on this equality in more detail.

Given α^*, β^*, e, h, s^* and also the prehistory of the biosphere, (11) and (12) are the systems of nine equations relative to ten unknowns: m^*, M^*, c^*, C^*, a^*, b^*, P^*, G^*, f^*, and y^*. Giving y^* or solving about it a certain optimization problem, we obtain the determined system of equations.

Difficulties of the estimate of the initial values α^*, β^*, e, h, and s^* are well known (see, for example [1, 3, 6, 7, 8]).

3. NEO-SPHERE MODELS

We assume that the resource of HA f is consisted from a certain parts q_1M^*, q_2C^*, q_3G^*, and q_4S^* of the ecological products of the biosphere and its cosmic resource S^*:

$$f^\wedge = f = q_1M^* + q_2C^* + q_3G^* + q_4S^*; \quad 0 < q_i < 1, i = 1, 2, 3, 4. \tag{13}$$

We also assume that the waste of HA $g = G'$ is the part of the resources incoming into the biosphere:

$$f^*(t) = c^*(t-e(t)) + g^*(t-h(t)) + g(t-u(t)) + (1-q_4)s^*(t), S^{*\prime}(t) = s^*(t), \tag{14}$$

where e, h, and u are lateness of transformation into f^* on the strength of the *bio-geo-chemical cycle* in the neo-sphere.

The *MM of interactions* of HA and the biosphere can be now rewritten in the form (Δm, $\Delta c = 0$)

$$m(t) = \int_{a(t)}^{t} \alpha(t,s)y(s)m(s)ds,\ M(t) = \int_{0}^{t} m(s)ds,\ c(t) = \int_{b(t)}^{t} \beta(t,s)[1-y(s)]m(s)ds,$$

$$P_1(t) = \int_{a(t)}^{t} y(s)m(s)ds,\ P_2(t) = \int_{b(t)}^{t} [1-y(s)]m(s)ds,\ P(t) = P_1(t) + P_2(t),$$

$$G(t) = M(t) - P(t),\ f(t) = m(t) + c(t),\ 0 \le y \le 1, 0 \le a(t), b(t) \le t,\ t \ge t^\wedge \ge 0;$$

$$m^*(t) = \int_{a^*(t)}^{t} \alpha^*(t,s)y^*(s)m^*(s)ds - q_1M^*(t),\ M^*(t) = \int_{0}^{t} m^*(s)ds,$$

$$c^*(t) = \int_{a^*(t)}^{t} \beta^*(t,s)[1-y^*(s)]m^*(s)ds - q_2C^*(t),\ C^*(t) = \int_{t^\wedge}^{t} c^*(s)ds,$$

$$P^*(t) = \int_{a^*(t)}^{t} m^*(s)ds,\ G^*(t) = (1-q_3)[M^*(t) - P^*(t)],\ f^*(t) = m^*(t) + c^*(t) =$$

$$k(f^{*\wedge})m^*(t),\ 0 \le y^* \le 1, 0 \le a^*(t), b^*(t) \le t,\ t \ge t^\wedge \ge \min[a^*(t^\wedge), b^*(t^\wedge)] = 0. \tag{15}$$

Given α, β, y, P, α^*, β^*, y^*, e, h, u, s^*, q_i, $i = 1, 2, 3, 4$ as well the prehistory, (13)-(15) are the systems of 18 equations relative to 17 indeterminate quantities m, M, c, a, b, f, P_1, P_2, G, m^*, M^*, c^*, C^*, a^*, f^*, P^* and G^*.

Combining the systems (4) and (13)-(15), we have the following MM of the neo-sphere:

$$\alpha_\mu^{\nu}(t) = \int_{a1(t)}^{t} \alpha_\mu(t,s)x(s)m(s)ds,\ \nu = -1,\ 0 \text{ or } 1;\ m(t) = \int_{a2(t)}^{t} \alpha_\mu(t,s)y(s)m(s)ds,$$

$$M(t) = \int_{0}^{t} m(s)ds;\ \beta_\kappa(t) = \int_{a3(t)}^{t} \alpha_\mu(t,s)z(s)m(s)ds;\ c(t) = \int_{a4(t)}^{t} \beta_\kappa(t,s)[1-x(s)-y(s)-$$

$z(s)]m(s)ds$, $\mu, \kappa = 1, 2$; $f(t) = m(t)+c(t)+\alpha_\mu{}^{\nu}(t)+\beta_\kappa(t) = q_1M^*+q_2C^*+q_3G^*+q_4S^*$;

$0 < q_i < 1$, $i = 1, 2, 3, 4$; $P_i(t) = \int_{ai(t)}^{t} m_i(s)ds$, $G_i = \int_0^{ai(t)} m_i(s)ds$, $i = 1, 2, 3, 4$;

$m_1 = xm$, $m_2 = ym$, $m_3 = zm$, $m_4 = (1-x-y-z)m$; $P(t) = P_1(t) + P_2(t) + P_3(t) + P_4(t)$,

$G(t) = M(t) - P(t)$; $0 \le x, y, z, 1-x-y-z \le 1$; $0 \le a_i(t) \le t$, $i = 1, 2, 3, 4$; $T \ge t \ge t\hat{} \ge 0$,

$f^*(t) = c^*(t-e(t))+g^*(t-h(t))+g(t-u(t))+(1-q_4)s^*(t) = m^*(t)+c^*(t) = k(f^{*\wedge})m^*(t)$,

$s^*(t) = S^{*\prime}(t)$, $m^*(t) = \int_{a^*(t)}^{t} \alpha^*(t,s)y^*(s)m^*(s)ds - q_1M^*(t)$, $M^*(t) = \int_0^t m^*(s)ds$,

$c^*(t) = \int_{a^*(t)}^{t} \beta^*(t,s)[1-y^*(s)]m^*(s)ds - q_2C^*(t)$, $C^*(t) = \int_{t\hat{}}^{t} c^*(s)ds$,

$P^*(t) = \int_{a^*(t)}^{t} m^*(s)ds$, $G^*(t) = (1-q_3)[M^*(t) - P^*(t)]$, $0 \le y^* \le 1$, $0 \le a^*(t), b^*(t) \le t$,

$$t \ge t\hat{} \ge \min\,[a^*(t\hat{}), b^*(t\hat{})] = 0. \qquad (16)$$

Using a general mathematical approach, it is not difficult to introduce different generalizations of MM (1)-(16): n-subsystem, continuous, stochastic, and discrete *MM of the neo-sphere* (see Ch. 2).

4. INVESTIGATION OF NEO-SPHERE MODELS

The main feature of the models is the concept of development from which follows many important properties of any ES, artificial, natural, and joined ones.

In particular, note that the well-known concept of "not in equilibrium stability" easily follows from that (1)-(4) and (11)-(16) are the open non-singular dynamic systems for which always take place the technical stability and in sequence of significant cases asymptotic stability by Lyapunov. The concept of "maximum of external work" corresponds entirely to the property of adaptive redistribution and regenerative accumulation of resources for any ES, which is connected with optimization of the external production (I1) and grounded in Part I. As to the concept of "complementarity", it is repeatedly expressed in the models by means of the internal and external functions, "useful product" and "waste," etc.

Let us consider the special cases of the systems (13)-(15) in more detail. Let $\alpha = \alpha(t)$, $\beta = \beta(t)$ $(\alpha \ne \beta)$, $\alpha^* = \alpha^*(t)$, $\beta^*(t) = \beta^*$ be given functions and y, y^*, and q_i be given constants. Then (13)-(15) can be rewritten in the form

$$m(t) = \alpha y \int_{a(t)}^{t} m(s)ds,\quad M(t) = \int_0^t m(s)ds,\ c(t) = \beta(1-y)\int_{b(t)}^{t} m(s)ds,$$

$P(t) = y\int_{a(t)}^{t} m(s)ds + (1-y)\int_{b(t)}^{t} m(s)ds$, $G(t) = M(t) - P(t)$, $f(t) = m(t) + c(t) =$

$q_1M^*(t) + q_2C^*(t) + q_3G^*(t) + q_4S^*$, $S^{*\prime}(t) = s^*(t)$, $0 \le y \le 1$, $0 \le a(t), b(t) \le t$,

$$t \ge t^\wedge \ge 0;\ m^*(t) = \alpha^* y^* \int_{a^*(t)}^{t} m^*(s)ds - q_1M^*(t),\ M^*(t) = \int_0^t m^*(s)ds,$$

$$c^*(t) = \beta^*(1-y^*)\int_{a^*(t)}^{t} m^*(s)ds - q_2C^*(t),\ C^*(t) = \int_{t^\wedge}^{t} c^*(s)ds,$$

$$P^*(t) = \int_{a^*(t)}^{t} m^*(s)ds,\ G^*(t) = [M^*(t) - P^*(t)],$$

$$f^*(t) = m^*(t) + c^*(t) = c^*(t-e) + g^*(t-h) + g(t-u) + (1-q_4)s^*(t) = k(f^{*\wedge})m^*(t),$$

$$0 \le y^* \le 1,\ 0 \le a^*(t),\ b^*(t) \le t,\ t \ge t^\wedge \ge \min[a^*(t^\wedge), b^*(t^\wedge)] = 0. \quad (17)$$

It follows from (17) that

$$\alpha y \int_{a(t)}^{t} m(s)ds + \beta(1-y)\int_{b(t)}^{t} m(s)ds = f,\ y\int_{a(t)}^{t} m(s)ds + (1-y)\int_{b(t)}^{t} m(s)ds = P, \quad (18)$$

whence applying $m + c = f$, we find

$$f = \alpha P + (1-\alpha/\beta)c,\ m = \alpha/\beta(\beta P - c),\ g = \alpha/\beta(\beta P - c). \quad (19)$$

Knowing f and m, from (18) one can find a and b. Further, we have

$$f^* = c^*(t-e) + g^*(t-h) + g(t-u) + (1-q_4)s^*, \quad (20)$$

where g is found with the help of (19) and c^*, g^* with the help of the given prehistory. After that we find m^* (and $k(f^{*\wedge})$) from $f^* = k(f^{*\wedge})m^*$, and c^* from $f^* = m^* + c^*$, and thereupon by the usual manner a^*, b^*, M^*, C^*, P^*, and G^*. As the result on the strength of (13) we find one more expression for f.

It is important to note that the relations similar to (17)-(20) can be also obtained in the general case. We have only to put instead of α and β their mean values, which always exist by virtue of the well-known theorem about mean values for integrals.

Comparing (13) and (19) one can see that *the values of P and/or c cannot be arbitrarily given under the condition that the values of M^*, C^*, G^*, and S^* remain limited.* This is the first qualitative corollary of the models in question.

Let us analyze impact of the coefficients q_i, $i = 1, 2, 3, 4$, on f. It is easy to conclude from (15) that

$$m^*(t) < (\alpha^* y^* - q_1)M^*(t) \le (\alpha^* - q_1)M^*(t),\ c^*(t) < (\beta^* - q_2)C^*(t), \quad (21)$$

from which

$$M^*(t) < M^*(t^\wedge)\exp[\int_{t^\wedge}^{t}(\alpha^* - q_1)ds],\ C^*(t) < C^*(t^\wedge)\exp[\int_{t^\wedge}^{t}(\beta^* - q_2)ds]. \qquad (21')$$

Consequently under

$$q_1 > \alpha^*,\ q_2 > \beta^* \qquad (22)$$

the values M^*, C^*, and G^* and hence f will be diminished.

This is the second corollary of the models.

As the third corollary we note that *the value $q_1 > \alpha^* + \varepsilon$, the constant $\varepsilon > 0$, will bring decreasing all M^*, C^*, and G^* values.*

The largest part of *living substance* of the biosphere is so-called *phyto-mass*. So, let M^* be the total generalized phyto-mass quantity, C^* be the total inorganic matter quantity, P^* be the total living phyto-mass quantity, G^* be the total generalized humus mass quantity, and a year be the unit of the time. Then as a rough guide [3, 7]

$$\alpha^* = 10^{-2},\ \beta^* = 10^{-4}. \qquad (23)$$

Let s^* be the solar energy quantity incoming to the neo-sphere per a year, M^* and C^* have the same meaning as earlier, and

$$g^\wedge = g^*(t-h) + g(t-u) \qquad (24)$$

has the meaning of the carbonic gas quantity into atmosphere. It is well known [2] that the new unit of phyto-mass is obtained in the photosynthesis reaction only under the certain portion $s^{*\prime}$, $c^{*\prime}$, and $g^{\wedge\prime}$ of s^*, c^*, and $g^\wedge$ respectively. Therefore if the carbonic gas quantity $g^\wedge$ and the sunny energy s^* are abounding, but we do not have enough phyto-mass quantity M^* (for example, by reason of lack of enough water), *then the additional portions of sunny energy have to be converted to warmth*, which is fourth corollary of the models. It connected with so-called "green house" effect that stems from surplus of solar energy and mostly carbonic gas and the fact that the increasing concentration of mostly carbon dioxide increases the heat trapping between the surface and the level, high in the atmosphere, from which radiation escapes.

On the contrary, if the solar energy is limited then by the virtue of (14)

$$m^*(t) = (1 - q_4)(s^{*\prime} + c^{*\prime} + g^{\wedge\prime})s^*/s^{*\prime}, \qquad (25)$$

from which

$$k(f^{*\wedge}) = [c^*(t-e)+G^*(t-h)+g(t-u)+(1-q_4)s^*\}/[(1-q_4)(s^{*\prime}+C^{*\prime}+g^{\wedge\prime})s^*/s^{*\prime\prime}]. \qquad (26)$$

It follows from (25) and (26) that *under fixed f^* and increasing of q_4 the coefficient $k(f^{*\wedge})$ will be increasing and hence m^* will be decreasing.*

This is the fifth corollary of the models.

It easily concludes from (26) the meaning and the value of the coefficient $k(f^{*\wedge})$.

It should be noted that only a certain part of the solar energy spectrum goes to photosynthesis.

As the sixth corollary we note that *there is the essential impact of g and c (see* (14), (15), (17), (20), (24)) *on quantity and quality of* f^*, *which can be reason of many undesirable anthropologic effects on the neo-sphere even for small g and c.*

On the other hand, *due to human reasoning namely the component g, which turns into* f^* *on the strength of laws of the nature circularity can actually remove the limitation on f indicated in the corollary one.*

This is the seventh corollary that, in particular, corresponds to academician V.I. Vernadsky's conclusion about possible autotroph of humanity.

5. WHAT CAN BE DONE?

To improve the situation in the neo-sphere everyone can do, at least, the following (compare with [8]).

Instead of (8) everyone can introduce

$$q(t) = q(^1)(t)q(^2)(t)q(^3)(t) \ +$$

$$c_1/\min(\alpha^*-q_1,0) + c_2/\min(\beta^*-q_2,0) + c_3/(1-q_3) + c_4/(1-q_4), \qquad (27)$$

where c_i, $i = 1, 2, 3, 4$ are weights decided by experts. Increasing of the *cost function* of the unit of the product of HA in the cases when q_i, $i = 1, 2, 3, 4$ turn out to be closed to their critical for the biosphere values can be the powerful control factor to advantage of the nature.

Since the market has not been mobilized to preserve the environment, the requirement for using of the (27)-type estimation should go through international agreements.

From HA models (2)-(4) with endogenous STP, $\nu = 1$, $\mu = 1$, $\alpha = \beta$, follows (see Ch. 4) that the distribution of given resources for the long-term goal of maximization of HA external product $C(T)$ or HA profit $Pr(T)$ should be

$$x(t) = 1,\ \ y(t) = z(t) = 0,\ t \in [t^\wedge, T^\wedge],\ T^\wedge \asymp T;\ x(t) = z(t) = 0,\ y(t) = 1,\ t \in [T^\wedge, T'],$$

$$T' - T^\wedge \asymp (T-t^\wedge)^{1/2};\ x(t) = y(t) = 0,\ z(t) = 1,\ t \in [T', T],\ T-T' \asymp \text{constant},\ T \to \infty. \qquad (28)$$

Since on the average production of new technology α-type requires much less energy then production of work places m-type and products c-type, it means that the optimal distributions of the given resources among three subsystems of HA: production of new technology S (Science), production of means of production A, and production of consumer product B are energy-preserving ones.

And what is more, let $x, y, z, u, v, x + y + z + u + v = 1$, be the relative shares of internal HA resources distributions among S, A, B, the *subsystem* of HA *recreation* R1 (Health Care and Education) and the *subsystem* of the biosphere *recreation* R2. Then we hope for validity of the following.

Guess. *Under general natural conditions for establishing a certain record of HA (for example,* $C(T)$ = max *over* (x,y,z,u,v) *during "large"* T-$t^\wedge$ *on the large first part of* $[t^\wedge, T]$ *the share v has the highest priority, the next priority has u, then x, after that y, and at last z.*

Since the subsystems R2, R1 and S do not require much energy in comparison with the subsystems A and B, it follows from the Guess that the respective *global change policy* can and should be *energy-preserving* one.

It was emphasized in Ch. 1, the records of efficiency of the natural ES in comparison with the artificial ones are about ten orders higher. So we have immense reserves for *efficiency raising policy*. The strategy to pursue energy efficiency is an example of so-called "tie-in" benefits, which can be obtained anyway, even without any environmental harmful change.

On the whole, even in MM (16) we have at least seven control functions: $x, y, z, q_1, q_2, q_3, q_4$, which can be selected by reasonable manner such to support the neo-sphere flourishing and reach the neo-sphere heyday.

In conclusion, we would like to emphasize that there are plenty of excellent scientific works in the field considered (see, for example [1] and [7, Ch.1]). There are the so-called global-circulation models (GCM's), that were first developed for long-term weather forecasts. But anyway these models and results are not included, as far as we know, MM of the neo-sphere under consideration, which can serve us as important additional means for modeling and solving numerous new difficult originating problems.

REFERENCES

1. Alcamo J.D., Integrated Modeling of Global Climate Change, Kluwer Academic Publishers, 1994, 321 pp.
2. Grodzinsky A.M., Grodzinsky D.M., Shot Reference on Plants Physiology, Kiev: Naukova. dumka, 1973, 592 pp. (in Russian).
3. Ivanov V.V., Mathematical Model of Neo-sphere, M.: VINITI, N 1627-B-89, 1989, 17 pp. (in Russian).
4. Ivanov V.V., Estimation of Industrial Wastes on the Basis of Aerosol Field Scanning, Kiev: Glushkov Inst. of Cybernetics, Reprint 90-42, 1990, 22 pp. (in Russian).
5. Kaznacheev V.P., Michailova L.P., Bio-information Function of Natural Electro-Magnetic Fields, Novosibirsk: Nauka, SO, 1985, 184 pp. (in Russian).
6. Marchuk G.I., Mathematical Modeling in the Problem of Environment, M.: Nauka, 1982, 320 pp. (in Russian).
7. Moiseev N.N., Aleksandrov V.V., Tarko A.M., Men and Biosphere, M.: Nauka, 1985, 275 pp. (in Russian).
8. U.S. Environmental Protection Agency (EPA), Policy Options for Stabilizing Global Climate: Report to Congress-Technical Appendices. EPA Doc: EPA-600/8-90-179. Washington, D.C.: EPA, 1990, 810 pp.

Take care of the pence and the pounds will take care of themselves

CHAPTER 10

MODELING OF FOREIGN CURRENCY CONVERSION PROBLEMS

An attempt of application of the new class of MM to some foreign currency conversion problems (FCCP) is described. The currency exchange rates formulae for the main products of two countries are established. The methods for determination of the prognostic values of currency exchange rates for short and long period of the time are developed. Some currency optimization problems are considered.

1. MM OF ECONOMIC ACTIVITIES WITH EXOGENOUS STP

The MM in question has the form

$$m(t) = \int_{a(t)}^{t} \alpha(t,s)y(s)ds, \quad P(t) = \int_{a(t)}^{t} m(s)ds, \quad c(t) = \int_{a(t)}^{t} \beta(t,s)[1-y(s)]m(s)ds,$$

$$C(t) = \int_{t^\wedge}^{t} c(s)ds, f(t) = m(t) + c(t), \quad M(t) = \int_{0}^{t} m(s)ds = P(t) + G(t),$$

$$0 \le y(s) \le 1, \; a(t) \le t, \; t \ge t^\wedge \ge a(t^\wedge) = 0, \tag{1}$$

where $m(t)$ is the generalized new capacity or the rate of production of the new work places (WP) number at the time instance t; $y(s)m(s)$ is a share of $m(s)$ for restoration of $m(t)$ in the subsystem A; $c(t)$ is the rate of creation of the new consumer product quantity at the instance t; $[1-y(s)]m(s)$ is a share of $m(s)$ for restoration of $c(t)$ in the subsystem B; $\alpha(t,s)$ is the efficiency index for functioning of the subsystem A along the channel ym -- m, i.e., the number of the units of $m(t)$ created in the unit of the time starting from t per one unit of $y(s)m(s)$; $\beta(t,s)$ is the efficiency index for functioning of the subsystem B along the channel $(1-y)m$ -- c which is similar to α; $a(t)$ is the temporal threshold for liquidation of the obsolete WP in the subsystem of A and B; $P(t)$ is the total number of the functioning WP at the instance t; $G(t)$ is the total number of the obsolete WP at the instance t; $[0,t^\wedge]$ is the prehistory of ES for which all the functions are given (their values will be noted by the same symbols but with superscript "^").

The ordinary statement of problems regarding MM (1) is as follows: the exogenous elements α, β, and P to be given, the distribution y to be found from the solutions of some optimization problems and *m, a, c, M, G,* and *C* to be found from the system (1) (usually on the pre-assigned temporal segment $[t, T]$).

If one need to take into account an interaction between AES and NES, then their MM should be considered jointly.

We assume that the resources of the AES f are consisted from a certain parts q_1M^*, q_2C^*, q_3G^*, and q_4S^* of the ecological products of the biosphere and its cosmic resource S^*:

$$f = q_1M^* + q_2C^* + q_3G^* + q_4S^*; \quad 0 < q_i < 1,\ i = 1, 2, 3, 4. \tag{2}$$

We also assume that the waste of the AES $g = G'$ is the part of the resources incoming into the biosphere:

$$f^*(t) = c^*(t-e(t)) + g^*(t-h(t)) + g(t-u(t)) + (1-q_4)s^*(t),\ s^* = S^{*\prime}, \tag{3}$$

where *e, h,* and *u* are delays of transformation into f^* based on the bio-geo-chemical cycle into the biosphere.

The *MM of interactions* of the human activities and the biosphere can be now rewritten in the form

$$m(t) = \int_{a(t)}^{t} \alpha(t,s)y(s)m(s)ds,\ M(t) = \int_{0}^{t} m(s)ds,\ c(t) = \int_{a(t)}^{t} \beta(t,s)[1-y(s)]m(s)ds,$$

$$C(t) = \int_{t^\wedge}^{t} c(s)ds,\ P(t) = \int_{a(t)}^{t} m(s)ds,\ G(t) = M(t) - P(t), f(t) = m(t) + c(t),\ 0 \le y \le 1,$$

$$0 \le a(t) \le t,\ t \ge t^\wedge \ge 0;\ m^*(t) = \int_{a^*(t)}^{t} \alpha^*(t,s)y^*(s)m^*(s)ds - q_1M^*(t),\ M^*(t) =$$

$$\int_{0}^{t} m^*(s)ds,\ c^*(t) = \int_{a^*(t)}^{t} \beta^*(t,s)[1-y^*(s)]m^*(s)ds - q_2C^*(t),\ C^*(t) = \int_{t^\wedge}^{t} c^*(s)ds,$$

$$P^*(t) = \int_{a^*(t)}^{t} m^*(s)ds,\ G^*(t) = [M^*(t) - P^*(t)](1 - q_3),\ f^*(t) = m^*(t) + c^*(t),$$

$$0 \le y^* \le 1, 0 \le a^*(t) \le t,\ t \ge t^\wedge \ge 0. \tag{4}$$

Given α, β, y, α^*, β^*, y^*, e, h, u, s^*, q_i, $i = 1, 2, 3, 4$, as well as the prehistory, (2)-(4) are the systems of 16 equations relative to 16 indeterminate: *m, M, c, C, a, f, P, G,* m^*, M^*, c^*, C^*, a^*, f^*, P^*, and G^*.

It should be noted that under $q_1 \ge \alpha^*$ and/or $q_2 \ge \beta^*$ in the neo-sphere *irreversible process* for the nature is possible. The orientation values of $\alpha^* = 10^{-2}$ and $\beta^* = 10^{-5}$ (see Ch. 9).

Using a general mathematical approach, it is not difficult to introduce different generalizations of MM (1)-(4): *n*-subsystem, continuous, stochastic and discrete MM of ES (see Ch. 2).

2. MM OF ACTIVITIES WITH ENDOGENOUS STP

We will examine (for simplicity, c f Ch. 2)

$$\alpha = \beta = \alpha(s)(1+ks)/(1+kt), \tag{5}$$

where k is the constant defining the rate of deterioration of previously created WP, and introduce MM

$$\alpha^{v}(t) = \int^{t}_{a(t)} \alpha(t,s)x(s)m(s)ds,\ v = 0,\ 1;\ m(t) = \int^{t}_{a(t)} \alpha(t,s)y(s)m(s)ds,$$

$$M(t) = \int^{t}_{0} m(s)ds;\ c(t) = \int^{t}_{a(t)} \alpha(t,s)[1\text{-}x(s)\text{-}y(s)]m(s)ds;\ C(t) = \int^{t}_{t^\wedge} c(s)ds,$$

$$P(t) = \int^{t}_{a(t)} m(s)ds,\ G = \int^{a(t)}_{0} m(s)ds;\ 0 \le x,\ y, 1\text{-}x\text{-}y \le 1;\ 0 \le a(t) \le t;$$

$$T \ge t \ge t^\wedge \ge 0, \tag{6}$$

where $\alpha^{v}(t)$ is the vth derivative of $\alpha(t)$, $\alpha^{0} = \alpha$, $\alpha(0) > 0$; $x(s)m(s)$ is a share of $m(s)$ for restoration of $\alpha^{v}(t)$ in the subsystem S (Science, S+A' = A); all of the rest elements in (6) have the same meaning as in (1).

For MM (6), P and c are usually given (and the prehistory on $[0,t^\wedge]$), x and y are found from the solutions of some optimization problems, and m, a, α, c, M, G, and C are found from the solution of the system (6).

The *MM of interaction* of the human activities and the biosphere can be now rewritten in the form

$$\alpha = \beta = \alpha(s)(1+ks)/(1+kt),\ \alpha^{v}(t) = \int^{t}_{a(t)} \alpha(t,s)x(s)m(s)ds,\ v = 0,\ 1;\ m(t) =$$

$$\int^{t}_{a(t)} \alpha(t,s)y(s)m(s)ds,\ M(t) = \int^{t}_{0} m(s)ds,\ c(t) = \int^{t}_{a(t)} \alpha(t,s)[1\text{-}x(s)\text{-}y(s)]m(s)ds,$$

$$C(t) = \int^{t}_{t^\wedge} c(s)ds,\ P(t) = \int^{t}_{a(t)} m(s)ds,\ G(t) = M(t) - P(t),\ f(t) = \alpha^{v}(t) + m(t) + c(t),$$

$$0 \le x,\ y,\ 1\text{-}x\text{-}y \le 1,\ 0 \le a(t) \le t,\ t \ge t^\wedge \ge 0;$$

$$m^*(t) = \int^{t}_{a^*(t)} \alpha^*(t,s)y^*(s)m^*(s)ds - q_1 M^*(t),\ M^*(t) = \int^{t}_{0} m^*(s)ds,$$

$$c^*(t) = \int^{t}_{a^*(t)} \beta^*(t,s)[1\text{-}y^*(s)]m^*(s)ds - q_2 C^*(t),\ C^*(t) = \int^{t}_{t^\wedge} c^*(s)ds,$$

$$P^*(t) = \int^{t}_{a^*(t)} m^*(s)ds,\ G^*(t) = [M^*(t) - P^*(t)](1\text{-}q_3),\ f^*(t) = m^*(t) + c^*(t),$$

$$0 \le y^* \le 1, 0 \le a^*(t) \le t,\ t \ge t^\wedge \ge 0. \tag{7}$$

Given x, y, k, α^*, β^*, y^*, e, h, u, s^*, q_i, $i = 1, 2, 3, 4$, as well as the prehistory, (2), (3), and (7) are the systems of 18 equations relative to 18 functions: α, β, m, M, c, C, a, f, P, G, m^*, M^*, c^*, C^*, a^*, f^*, P^*, and G^*.

3. MM OF COSTS, MARKET PRICES, AND QUALITIES

Let $s_i(t)$ be the per unit *cost function* of ith product (for example, $i = m, c$) at the instant t and let $i(t)$, $i_n(t)$ and $I_r(t)$ be the respective amounts of all existent, *social-necessary* and realized ith *product* produced in the time units starting from the instant t. Then

$$s\hat{}_i(t) = s_i(t) i_n(t)/i(t) \tag{8}$$

is the *social-necessary* unit *cost* of the unit of the ith product at the instant t;

$$g_i(t) = s\hat{}_i(t) - s_i(t) \tag{9}$$

is the so-called differential gain or return and

$$v_i(t) = s_i(t) i_n(t)/i_r(t) = s\hat{}_i(t) i(t)/i_r(t) \tag{10}$$

is the *market* value or per unit *price* of the *i*th product at the instant t. In turn,

$$s_i(t) = s_{i1}(t) s_{i2}(t) s_{i3}(t) +$$

$$c_1/\min(\alpha^* - q_1, 0) + c_2/\min(\beta^* - q_2, 0) + c_3/(1 - q_3) + c_4/(1 - q_4), \tag{11}$$

where s_{i1} is the cost of new labor, s_{i2} is the factor of the cost accruing from a labor invested earlier, s_{i3} is the factor of the cost attributable to amortization of all means of labor under the creation of the unit of the *i*th product at the instant t, and c_i, $i = 1, 2, 3, 4$, are the coefficients decided by experts of surroundings protection.

The product's quality is equivalent to the social-necessary product growth with the factor $qu_i(t)$, $qu_i(t) \geq 1$ [10], i.e., the real amount of all the *i*th social-necessary product regarding its quality

$$i_{q,n}(t) = qu_i(t) i_n(t). \tag{12}$$

So, the market income from products' selling during the time dT, while the products remain the same,

$$dI = \sum_i \int_t^{t+dT} v_{q,i}(s) i_r(s) ds \ (i = m, c), \quad v_{q,i}(t) = s_i(t) i_{q,n}(t)/i_r(t), \tag{13}$$

and the possible integral (total) *profit*

$$dR = dI - dE = \sum_i \int_t^{t+\Delta T} [v_{q,i}(s)-s_i(s)]i_r(s)ds \ (i = m, c), \tag{14}$$

where dE is the expenditure of the production for the period dT. The *accumulated* values of income $I(T)$, *profit* $R(T)$, and expenditure $E(T)$ are found summing over dT the respective values dI, dR, and dE.

4. ON INVESTIGATION OF MM OF ACTIVITIES

Solution existence and uniqueness theorems for the systems in (1) and in (5), (6) are considered and proved in Part I under rather general conditions. It is clear that all the elements of MM (1), (5), (6) are nonnegative by definition. Under monotone increasing of α and β over s and comparatively insignificant decreasing of α and β over t the solutions m, a, and c for the system (1) will be monotone increasing as well. For the system (6), depending on goals of WP in the subsystem of S (science), it will be α^{ν} on its output for various ν. Under $\nu = 1$ we guarantee increasing of $\alpha(t)$ anyway, for $\alpha'(t) > 0$ by definition.

From the MM of (1)-(14), it is not difficult to do estimates for series of scenarios of economic development of different countries similar to [4] with the same conclusions. Namely: "... the main problem for countries with centralized planning economics during the next decade is the more effective use of their resources." ([4], p. 94). On the setting of MM (1) - (14), it means the necessities of essential reduction of the magnitudes s_{in}, $n = 1, 2, 3$, increasing qu_i, decreasing of deterioration decrement k, and optimization of the distributions x, y about what the question will be later. For the West countries and the USA "... the key global problem is achievement of the best possible balance for international payments until the end of the century." ([4], p.55). On the setting of MM (1)-(14), it means the necessities of essential increasing (for its own country) and/or decreasing (for other countries) of magnitudes $v_i(t)$, $i = m, c$, at the expense, for example, of redistribution of $c(t)$-type contents and alteration of volume, structure, and partners for international trade.

Let us dwell more explicitly on the optimization problem for the consumer product quantity:

$$C(T) = \int_{t^\wedge}^{T} c(t)dt = \max \tag{15}$$

using distributions x, y provided that the relations' (1) or (5) and (6) are observed.

For the first time in [1] under rather broad assumptions, it was proved that for "small" T-t the desired y (maximizing $C(T)$) is minimally possible (by virtue of the restriction $a'(t) \geq 0$), but for "large" T-t the desired y may differ from the minimally possible on the larger initial part of the segment $[t^\wedge, T]$, and only on the smaller final part of $[t\hat{},T]$ the desired y is minimally possible. The notions "small" and "large" depend on the values of the functions α and β, namely, the greater are the functions in question, the nearer to $t\hat{}$ is the boundary between "small" and "large" segments.

This boundary was defined more exactly by numerical modeling of the problem (15) on the computer [3, Ch.4]. As a result, it was determined that for economics of the USSR, the boundary was about $t\hat{} + 20$ years in the 1976, and for economics of the USA, the boundary was about $t\hat{} + 2$ years.

In the sequel, it was found that the above-mentioned regularity is typical for many other optimization problems as for MM (1)-(14). In addition, a rather complicated structure of the desired optimal x, y was determined (see Ch. 4).

As the final result, the important qualitative very general principle was determined:

on a "large" period of the time the record of an external function for any ES can be obtained only under conditions of its sufficiently comfortable state, that is, under the significant fraction of resources sent to the internal needs of ES.

Detailed investigation of the problem (15) in the case of MM (5) and (6) for $v = 1$ has led to the following result. For "large" T-t, the optimal distribution $x(t)$ and $y(t)$ possess the characteristics

$$x(t) = 1,\ t \in [t\hat{},\theta];\ y(t) = 1,\ t \in [\theta,\theta\hat{}];\ x(t) = y(t) = 0,\ t \in [\theta\hat{},T];$$

$$\theta - t\hat{} \asymp T - t\hat{},\ \theta\hat{} - \theta \asymp (T - t\hat{})^{1/2},\ T - \theta\hat{} \asymp 1, \tag{16}$$

where $\asymp$ means the asymptotic equality when T-$t\hat{}$ tends to ∞.

The result qualitatively means that

for the strategic goals under optimal distributions of the given resources among different subsystems of ES, the main priority belongs to the subsystem of S-type created a new more effective technology, on the second place by priority comes the subsystem A creating a new WP and finally comes the subsystem B.

5. STATEMENT OF FCCP

Because of the analysis of the literature over *foreign currency conversion problems (FCCP)* and adjacent problems (see, for example, [1-6, 8, 9]), we conclude that it is necessary to consider the following main problems.

5.1. Relations between prices of two countries

Let us agree later that all the magnitudes having the same name for the first country are designated by the subscript "*1*" and for the second country by "*2*". Let the prices of the unit for the *i*th production at the instance *t* be $v_{qi1}(t)k_{i1}(t)m1$ ($m1$ is a monetary unit of the first country) and $v_{qi2}(t)k_{i2}(t)m2$ ($m2$ is a monetary unit of the second country). Then for the *i*th production at the instance *t*, it is obvious that

$$m1 = v_{qi2}(t)/v_{qi1}(t)k_{i2}(t)/k_{i1}(t)m2. \tag{17}$$

For the simplicity we put below $k_{i2}(t)/k_{i1}(t) = 1$.

Therefore, on average (by products), the relation between $m1$ and $m2$ has the form

$$m1 = \sum_i v_{qi2}(t)/v_{qi1}(t)i_2(t)/\sum_i i_2(t)m2, \tag{18}$$

and

$$m2 = \sum_i v_{qi1}(t)/v_{qi2}(t)i_1(t)/\sum_i i_1(t)m1 \tag{19}$$

and for the production of both countries

$$m1 = \{[\sum_i v_{qi2}(t)/v_{qi1}(t)i_2(t)/\sum_i i_2(t)]\sum_i i_2(t)/[\sum_i i_1(t)+\sum_i i_2(t)] +$$

$$[\sum_i v_{qi1}(t)/v_{qi2}(t)i_1(t)/\sum_i i_1(t)]^{-1}\sum_i i_1(t)/\sum_i i_1(t)+\sum_i i_2(t)\}m2. \tag{20}$$

In particular, for $i = m, c$ we have

$$m1 = [v_{qm2}(t)/v_{qm1}(t)m_2(t)/f_2(t)+v_{qc2}(t)/v_{qc1}(t)c_2(t)/f_2(t)]m2, f_2(t) = m_2(t)+c_2(t), \tag{18'}$$

and

$$m2 = [v_{qm1}(t)/v_{qm2}(t)m_1(t)/f_1(t)+v_{qc1}(t)/v_{qc2}(t)c_1(t)/f_1(t)]m1, f_1(t) = m_1(t)+c_1(t), \tag{19'}$$

and also

$$m1 = \{[v_{qm2}(t)/v_{qm1}(t)m_2(t)/f_2(t) + v_{qc2}(t)/v_{qc1}(t)c_2(t)/f_2(t)]f_2(t)/[f_1(t)+f_2(t)] +$$

$$[v_{qm1}(t)/v_{qm2}(t)m_1(t)/f_1(t) + v_{qc1}(t)/v_{qc2}(t)c_1(t)/f_1(t)]^{-1}f_1(t)/[f_1(t)+f_2(t)]\}m2. \tag{20'}$$

It is presupposed everywhere that m_i, c_i, and f_i are dimensionless variables (see Ch. 2).

On the strength of (8)-(13) and (17')

$$m1/m2 = v_{qi2}(t)/v_{qi1}(t) = s_{i2}(t)\, qu_{i2}(t)i2_n(t)i1_r(t)/s_{i1}(t)qu_{i1}(t)i1_n(t)i2_r(t),$$

$$s_{ij}(t) = s_{ij1}(t)s_{ij2}(t)s_{ij3}(t) + c_{1j}/\min(\alpha^* - q_{1j},0) + c_{2j}/\min(\beta^* - q_{2j},0) +$$

$$c_{3j}/(1 - q_{3j}) + c_{4j}/(1 - q_{4j}). \tag{21}$$

The relation (21) reveals many factors, which determine dependence between prices of two countries.

5.2. The short and long period of time for the FCCP solution

Applying the formulae (17')-(20) and averaging over t on the segment $[t^\wedge, T]$, we have the following relations between $m1$ and $m2$:

$$m1 = \int_{t^\wedge}^{T} v_{qi2}(s)/v_{qi1}(s) ds m2, \tag{22}$$

$$m1 = \int_{t^\wedge}^{T} \sum_i v_{qi2}(s)/v_{qi1}(s) i_2(s)/\sum_i i_2(s) ds m2, \tag{23}$$

$$m2 = \int_{t^\wedge}^{T} \sum_i v_{qi1}(s)/v_{qi2}(s) i_1(s)/\sum_i i_1(s) ds m1, \tag{24}$$

$$m1 = \int_{t^\wedge}^{T} \{[\sum_i v_{qi2}(s)/v_{qi1}(s) i_2(s)/\sum_i i_2(s)] \sum_i i_2(s)/[\sum_i i_1(s)+\sum_i i_2(s)] +$$

$$[\sum_i v_{qi1}(s)/v_{qi2}(s) i_1(s)/\sum_i i_1(s)]^{-1} \sum_i i_1(s)/\sum_i i_1(s)+\sum_i i_2(s)\} ds m2. \tag{25}$$

In particular, for $i = m, c$ we have

$$m1 = \int_{t^\wedge}^{T} [v_{qm2}(s)/v_{qm1}(s) m_2(s)/f_2(s) + v_{qc2}(s)/v_{qc1}(s) c_2(s)/f_2(s)] ds m2, \; f_2(s) = m_2(s) +$$

$$c_2(s), \; m2 = \int_{t^\wedge}^{T} [v_{qm1}(s)/v_{qm2}(s) m_1(s)/f_1(s) + v_{qc1}(s)/v_{qc2}(s) c_1(s)/f_1(s)] ds m1,$$

$$f_1(s) = m_1(s) + c_1(s), \; m1 = \int_{t^\wedge}^{T} \{[v_{qm2}(s)/v_{qm1}(s) m_2(s)/f_2(s) +$$

$$v_{qc2}(s)/v_{qc1}(s) c_2(s)/f_2(s)] f_2(s)/[f_1(s)+f_2(s)] + [v_{qm1}(s)/v_{qm2}(s) m_1(s)/f_1(s) +$$

$$v_{qc1}(s)/v_{qc2}(s) c_1(s)/f_1(s)]^{-1} f_1(s)/[f_1(s)+f_2(s)]\} ds m2. \tag{23')-(25'}$$

Let $T - t^\wedge$ mean the "small" period connected with tactical or short-term goals of the fiscal reform in a country. Then we have the short period of the time for the FCCP solution. If $T - t^\wedge$ means the "large" period connected with strategic or long-term goals of the fiscal reforms, then we have the long period of the time for the FCCP solution.

For all of the mentioned formulae $v_{qi1}(t)/v_{qi2}(t)$ can be determined by the formula (21). To realize the formulae, the prediction of $m_j(t)$ and $c_j(t)$, $j = 1, 2$, for $t \in [t^\wedge, T]$ is necessary, what can be done due to MM of s. 1 - 3 and the results of Ch. 2 and 3.

5.3. Mathematical formulation of FCCP optimization

It is obvious from the formulae (18)-(20) that the desired relation between $m1$ and $m2$ has the form

$$m1 = \lambda(t)m2, \tag{26}$$

where $\lambda(t)$ is a certain function of t. For the formulae (22)-(24), this relation is even simplified:

$$m1 = \lambda m2, \tag{27}$$

where λ is a constant. On the other hand, for the formulae (17) and (22), we come to

$$m1 = \lambda_i(t)m2. \tag{28}$$

In particular, for $i = m, c,$ we obtain

$$m1 = \lambda_m(t)m2,\ m1 = \lambda_c(t)m2. \tag{28'}$$

One of the FCC *optimization problems* for the 1st country can be now

$$C(t) = \int^{t}_{a1(t)} \beta_1(t,s)[1-y_1(s)][m_1(s)+\lambda_m(s)R_m(s)]ds + \lambda_c(t)R_c(t) =$$

$$\max\ (y_1, y_2, \lambda_m, \lambda_c, R_c) \tag{29}$$

provided that the restrictions of MM (1) or (2)-(4) - type, where all the elements of MM have the subscript j, $j = 1, 2,$ and also the relations (28'), (28"), and

$$c_2(t) - \lambda_c(t)R_c(t) > c_2\hat{}(t),\ R_c(t) + R_m(t) = R(t), \tag{30}$$

where $R_i(t)$ is the monetary reserve of the 2nd country in the ith branch, $i = m, c,$ and $c_2\hat{}(t)$ is the given level of consumption observed in the 2nd country. Similarly, some other optimization problems can be formulated regarding possibilities of deeper optimization at the expense of the additional control factors x and q_i, $i = 1, 2, 3, 4$ (see MM (5)-(7)). Besides, the problems of maximization of $C(T)$ over x, y, q_i, $i = 1, 2, 3, 4$ regarding the relations (26) or (27) instead of (28) can be formulated similarly.

6. INVESTIGATION OF FCCP

6.1. Qualitative investigation of FCCP

The main problems of the FCCP qualitative investigation are

(i) $c_j(t)$ and $m_j(t)$ prediction for the "small" period $[t\hat{},T]$ with regard to the tactics of both countries;

(ii) $c_j(t)$ and $m_j(t)$ prediction for the "large" period $[t\hat{},T]$ with regard to the strategies of both countries;

(iii) $i_{jn}(t)$, $s_{ij}(t)$ and $v_{qij}(t)$-values, $i = m, c; j = 1, 2$, estimation for the "small" and "large" period of the time;

(iv) $\lambda(t)$ and λ estimation on the strength of the formulae (17)-(28);

(v) $\lambda(t)$ and λ estimation from the conditions of the FCCP optimization (29), (30).

Let $y\hat{}(s) = y\hat{} =$ constant be the distribution $y(s)$ for the prehistory $[0,t\hat{}]$, the "small" segment $[t\hat{},T]$ be such that the mean values of $\alpha(t,s) = \alpha$ and $\beta(t,s) = \beta$, where α and β are positive constants, and the tactic is the solution of the problem (15) with respect to the restrictions (1). Then under the natural conditions $m\hat{} = \alpha y\hat{} P(t\hat{}) > P'(t\hat{})$ and $P''(t) < \alpha P'(t)$, solving the problem (15), we have $y(t) = 0$,

$$m(t) = m\hat{}e^{-\alpha y\hat{}(t-t\hat{})} + \alpha y\hat{}\int_{t\hat{}}^{t} e^{-\alpha y\hat{}(t-s)}P'(s)ds,\ c(t) = \beta[P(t)-m(t)/\alpha],\ a(t) =$$

$$t\hat{} + \int_{t\hat{}}^{t} m(s)ds/m\hat{} - P(t),\ t\in[t\hat{},z];\ y(t) = P''(t)/[\alpha P'(t)],\ m(t) = P'(t),$$

$$c(t) = \beta[P(t)-P'(t)/\alpha],\ a(t) = t\hat{} + \int_{t\hat{}}^{z} m(s)ds/m\hat{}-P(z),\ t\in[z,T], \tag{31}$$

where z is the root of the equation $m(z) = P'(z)$. In particular, for $P(t) = P\hat{}e^{q(t-t\hat{})}$ ($\alpha y\hat{} > q$) we have

$$z = t\hat{} + 2/(q+\alpha y\hat{})\ln(\alpha y\hat{}/q). \tag{32}$$

It is presupposed here that $T > z$. Otherwise, we have the first part of the formula (32) for $t\in[t\hat{},T]$.

For the "large" segment $[t\hat{},T]$ let P, $\alpha\hat{}$, $m\hat{}$, $x\hat{}$, $y\hat{}$ be constants and the strategy be the solution of the same problem (15) with respect to the restrictions (5) and (6), $k = 0$, $v = 1$. Then on the strength of (16) we have:

$$x(t) = 1,\ y(t) = 0,\ m(t) = \alpha\hat{}y\hat{}m\hat{}(t\hat{}- a(t)),\ P' = 0 = m(t) - m\hat{}a'(t), \alpha'(t) =$$

$$\alpha\hat{}x\hat{}m\hat{}(t\hat{}-a(t))+\int_{t\hat{}}^{t} \alpha(s)m(s)ds,\ \ c(t) = \alpha\hat{}m\hat{}(1-x\hat{}-y\hat{})(t\hat{}- a(t)),\ t\in[t\hat{},\theta], \tag{33}$$

from which it follows that under large T and $t\approx\theta$

$$m(t) = m\hat{}e^{-\alpha\hat{}y\hat{}(t-t\hat{})} \approx 0,\ c(t) \approx 0,\ a(t) = [1- e^{-\alpha\hat{}y\hat{}(t-t\hat{})}]/(\alpha\hat{}y\hat{}) \approx 1/(\alpha\hat{}y\hat{}) = t\hat{},$$

$$\alpha(t) = \alpha\hat{} + \sum_{s=0}^{\infty}\{(t-t\hat{})a_s e^{-s(t-t\hat{})/t\hat{}} + 1/(\alpha\hat{}y\hat{})\sum_{s=1}^{\infty} b_s[1-e^{-s(t-t\hat{})/t\hat{}}]/s\} =$$

$$\alpha\hat{} + x\hat{}/y\hat{}t\hat{}[1- e^{-(t-t\hat{}/t\hat{})}] + (t-t\hat{})\int_{t\hat{}}^{t} \alpha(s)m(s)ds - \int_{t\hat{}}^{t}(s-t\hat{})\alpha(s)m(s)ds \approx(\theta - t\hat{})A,$$

$$A = \int_{t\hat{}}^{\theta} \alpha(s)m(s)ds \approx m\hat{}t\hat{}[\alpha\hat{} + \sum_{s=1}^{\infty} b_s/(s+1) + t\hat{}\sum_{s=1}^{\infty} a_s/(s+1)^2], \tag{34}$$

where a_s and b_s can be determined by the method of the coefficients comparison from

$$\alpha''(t) = (\alpha(t) - \alpha\hat{}x\hat{})m\hat{}e^{-d(t-t\hat{})},\ \alpha'(t\hat{}) = \alpha\hat{}x\hat{}P,\ d = 1/t\hat{}. \tag{35}$$

Indeed, on the strength of (34) and (35) we have

$$\alpha'(t) = a_0+\sum_{s=1}^{\infty} [a_s + b_s - d(t-t\hat{})a_s s]e^{-ds(t-t\hat{})},\ \alpha''(t) = \sum_{s=1}^{\infty} [d^2(t-t\hat{})a_s s^2 - 2da_s s - db_s s]e^{-ds(t-t\hat{})},\ a_s = a_0 r^{s}/(s!)^2,\ s > 0,\ r = m\hat{}/d^2;\ b_s = b_1 r^{s-1}/(s!(s-1)!) -$$

$$2a_0 r^{s}/(s!(s-1)!)(1/2+1/3+...+1/s),\ s > 1, \tag{36}$$

where a_0 and b_1 can be found from the system

$$a_0 + \sum_{s=1}^{\infty}(a_s + b_s) = \alpha\hat{}x\hat{}P,\ 2a_0 r + b_1 + \sum_{s=2}^{\infty}(2a_s+b_s)s = -\alpha\hat{}(1-x\hat{})m\hat{}t\hat{} \tag{37}$$

or the system

$$a_0\{2\sum_{s=2}^{\infty} r^{s}/[s!(s-1)!](1/2+...+1/s) -\sum_{s=1}^{\infty} r^{s}/(s!)^2-1\} - b_1\sum_{s=1}^{\infty} r^{s-1}/[s!(s-1)!] =$$

$$-\alpha\hat{}x\hat{}P,\ a_0[2\sum_{s=2}^{\infty} r^{s+1}/(s!)^2(1/2+...+1/s)-2r] - b_1\sum_{s=0}^{\infty} r^{s}/(s!)^2 = \alpha\hat{}(1-x\hat{})P. \tag{37'}$$

In particular, if $r \ll 1$, then

$$a_0 \approx \alpha\hat{}x\hat{}P,\ b_1 \approx -\alpha\hat{}(1-x\hat{})P. \tag{38}$$

For $t \in [\theta,\theta\hat{}]$: $y(t) = 1,\ x(t) = 0$, and on the basis (36)

$$\alpha'(t) = x\hat{}/y\hat{}e^{-(t-t\hat{})/t\hat{}} + \int_{t\hat{}}^{\theta} \alpha(s)m(s)ds \approx A,\ \alpha(t) \approx A(t-\theta) + \alpha(\theta) \approx A(\theta-t\hat{}),$$

$$m(t) = m\hat{}e^{-(t-t\hat{})/t\hat{}} + \int_{\theta}^{t} \alpha(s)m(s)ds \approx m\hat{}e^{-(\theta-t\hat{})/t\hat{}}e^{A(\theta-t\hat{})(t-\theta)},\ \int_{t\hat{}}^{t^*} m(s)ds = P = m\hat{}t\hat{}$$

$$\approx \int_{t\hat{}}^{\theta} m\hat{}e^{-(s-t\hat{})/t\hat{}}ds + \int_{\theta}^{t^*} m\hat{}e^{-(\theta-t\hat{})[1/t\hat{}+A(t-\theta)]}ds = m\hat{}t\hat{} - m\hat{}t\hat{}e^{-(\theta-t\hat{})/t\hat{}} + m\hat{}e^{-(\theta-t\hat{})/t\hat{}}/[A(\theta$$

$$- t\hat{})][e^{A(\theta-t\hat{})(t^*-\theta)}-1],\ t^* - \theta \sim \ln (A(\theta - t\hat{}))/[A(\theta - t\hat{})],\quad t\in[\theta,t^*],\ a(t^*) = t\hat{}; \tag{39}$$

$$\alpha'(t) = \int_{a(t)}^{\theta} \alpha(s)m(s)ds,\ \alpha(t) = \alpha(t^*) + \int_{t^*}^{t} \alpha'(s)ds \approx \alpha(t^*) \approx A(\theta-t\hat{}),$$

$$m(t) = m(t^*) + \int_{t^*}^{t} \alpha(s)m(s)ds \approx m(t^*)e^{A(\theta-t\hat{})(t-t^*)},\ \int_{t^*}^{t'} m(s)ds \approx P = m\hat{}t\hat{} \approx$$

$$m(t^*) \int_{t^*}^{t'} e^{A(\theta - t^\wedge)(s - t^\wedge)} ds \approx m^\wedge e^{-(\theta - t^\wedge)/t^\wedge + A(\theta - t^\wedge)(t' - t^*)},\ t' - t^* \approx 1/At^\wedge,\ t \in [t^*, t'],\ a(t') = \theta; \quad (40)$$

$$\alpha(t) \approx \alpha(\theta),\ m(t) \approx \alpha(\theta)P \approx A(\theta - t^\wedge)P,\ t - a(t) \approx 1/\alpha(\theta),\ t \in [t', \theta^\wedge]; \quad (41)$$

$$x = y = 0,\ \alpha(t) \approx \alpha(\theta),\ c(t) \approx \alpha(\theta)P \approx A(\theta - t^\wedge)P,\ t \in [\theta^\wedge, T]. \quad (42)$$

Thus, on the strength of (34)-(42) and (2) the summarized optimal strategic value $m(T)$ and $c(T)$ can be estimated by the formula

$$m(T),\ c(T) \approx A(T - t^\wedge)P. \quad (43)$$

The value $v_{qij}(t)$ can be estimated from the relation (21). After this on the strength of (20') and (31) for the end T of the "small" period we have

$$\lambda(T) = [v_{qm2}(T)/v_{qm1}(T)P_2'(T)/f_2(T) + v_{qc2}(T)/v_{qc1}(T)\beta_2[P_2(T) - P_2'(T)/\alpha_2]/f_2(T)]f_2(T)/[f_1(T)$$

$$+ f_2(T)] + \{v_{qm1}(T)/v_{qm2}(T)P_1'(T)/f_1(T) + v_{qc1}(T)/v_{qc2}(T)\beta_1[P_1(T) -$$

$$P_1'(T)/\alpha_1]/f_1(T)\}^{-1} f_1(T)/[f_1(T) + f_2(T)]\},\ f_j(T) = P_j'(T) + \beta_j[P_j(T) - P_j'(T)/\alpha_j, \quad (44)$$

and on the strength of (44) for the end of "large" period

$$\lambda(T) = [v_{qm2}(T)/v_{qm1}(T) + v_{qc2}(T)/v_{qc1}(T)]A_2P_2/[2(A_1P_1 + A_2P_2)] +$$

$$[v_{qm1}(T)/v_{qm2}(T) + v_{qc1}(T)/v_{qc2}(T)]^{-1} 2A_1P_1/[A_1P_1 + A_2P_2]. \quad (45)$$

In a similar way, on the strength of (25'), (31) and (45) we can obtain the average values of $\lambda(t)$ on "small" and "large" period of the time for the solution of FCCP.

Let us estimate $\lambda(T)$ going from the condition (29) and (30) in the simplified case of the large T when

$$R_m(T) = 0,\ R_c(T) = R(T),\ c_1(T) \approx A_1TP_1 + \lambda R(T) =$$

$$\max (\lambda | c_2(T) - \lambda R(T) > c_2^\wedge(T)). \quad (46)$$

Apparently, the desired values are

$$\lambda = [c_2(T) - c_2^\wedge(T)]/R(T),\ c_1(T) \approx A_1TP_1 + c_2(T) - c_2^\wedge(T). \quad (47)$$

6.2. Methods of calculation for FCCP

According to the results of the s. 5 and the point 6.1. the FCCP methods of calculation consist:

For the "small" period

in the $\lambda(t)$- and/or λ-estimate with the help of the formulae (1)-(14), (17)-(28), (18')-(20'), (22') - (25'), (28'), (31), (32), (44) or from the optimization problem (29) and (30) - type solution.

For the "large" period

in the $\lambda(t)$- and/or λ- estimate with the help of the formulae (1)-(14), (17)-(28), (18')-(20'), (22')-(25'), (28'), (33)-(43), and (45) or from the solution of optimization problem (29) and (30)-types, in particular, from the result (46), (47).

6.3. Example of calculation

Let us agree that a ratio u/v of the compared approximate values u and v to be equal to 1 if the values have the same exponent (at the binary number system), $3/4 < u/v \leq 3/2$, to 2 if the value in numerator exceeds the value in denominator on 1 in order, $3/2 < u/v \leq 3$, and so on. Then according to the facts in [1 - 7] for the former USSR (1st country) and the USA (2nd country) in $t = 1987$ we have

$$s_{i1}(t)/s_{i2}(t) = 2,\ \ i = m,\ c;\ qu_{m1}(t)/qu_{m2}(t) = 2^{-1};\ qu_{c1}(t)/qu_{c2}(t) = 2^{-2};$$

$$m1_n(t)/m2_n(t) = 1;\ c1_n(t)/c2_n(t) = 1;\ \ m1_r(t)/m2_r(t) = 2^{-2};\ c1_r(t)/c2_r(t) = 2^{-2}; \qquad (48)$$

and therefore for $i = m,\ c$

$$1\,\text{ruble}/1\,\text{dollar} = s_{m2}(t)qu_{m2}(t)m2_n(t)m1_r(t)/[s_{m1}(t)qu_{m1}(t)m1_n(t)m2_r(t)] = 2^{-2},$$

$$1\,\text{ruble}/1\,\text{dollar} = s_{c2}(t)qu_{c2}(t)c2_n(t)c1_r(t)/[s_{c1}(t)qu_{c1}(t)c1_n(t)c2_r(t)] = 2^{-1}. \qquad (49)$$

In addition, $f_1(t) = \$\ 1.2\times10^{12}$, $f_2(t) = \$\ 3.6\times10^{12}$, and

$$m_1(t)/f_1(t) = c_1(t)/f_1(t) = 2^{-1},\ m_2(t)/f_2(t) = 0.2,\ c_2(t)/f_2(t) = 0.8. \qquad (50)$$

Therefore, on average (by product) in $t = 1987$

$$1\ \text{ruble} = \lambda(t)\ \text{dollars} = \{[v_{qm2}(t)/v_{qm1}(t)m_2(t)/f_2(t) +$$

$v_{qc2}(t)/v_{qc1}(t)c_2(t)/f_2(t)]f_2(t)/[f_1(t) + f_2(t)] + [v_{qm1}(t)/v_{qm2}(t)m_1(t)/f_1(t) +$

$v_{qc1}(t)/v_{qc2}(t)c_1(t)/f_1(t)]^{-1}f_1(t)/[f_1(t)+f_2(t)]\}$ dollars =

$$\{[2^{-2}0.2 + 2^{-1}0.8]3.6/4.8 + [2 + 1]^{-1}1.2/4.8\} \text{ dollars} \approx 2^{-1} \text{ dollars}. \quad (51)$$

7. CONCLUSION

A new class of the dynamic macro-models has been adopted to the FCCP analysis for the first time. It has allowed prognostic modeling and estimates of the nominal exchange rates [8, 9] and some other indices for the main branches of economy of both countries during their both the short- and the long-term development.

Besides the well-known scenario-type prediction [4], the method of optimization-type prediction regarding the maximization of consumer goods and services during short-term development of the 1st country, which corresponds now to the tactic of undeveloped country, and the strategic optimization of production of both countries has been developed here.

All the above-mentioned results are the first attempt of application of a new mathematical approach to the modeling of FCCP, which requires further perfection and development. As it was emphasized in the introduction of this book, the goal of each example of application is not revising of many results accumulated in the given field before, but demonstration of new possibilities and some certain new results following from the attempt of application of MM under consideration.

REFERENCES

1. Basic Directions for Radical Perfecting of the Price-Formation Methodology , M.: In-t ekonomiki AN SSSR, 1987, 258 pp. (in Russian).
2. Davis M.H.A. et al, Mathematical Finance, New York: Springer, 1995, 148 pp.
3. FAO Yearbook, Vol. 42, Rome: 1989, 350 pp.
4. General Economic Perspective up to 2000 y., UN, ESE/ES. AD/32, 1989, 292 pp. (in Russian).
5. Handbook of International Trade and Development Statistics, UN, 1989, 605 pp.
6. Handbook of Pricing, M.: Ekonomika, 1985, 399 pp., (in Russian).
7. Industry and Development: Global Report 1989/90, Vienna: UNIDO, Document unit F-355N, 1989, 444 pp.
8. Krueger A.O., Exchange-Rate Determination, Cambridge Univ. Press, 1985, 218 pp.
9. Lial M.L., Miller C.D., Hungerford T.W., Mathematics with Applications in the Management, Natural and Social Sciences, Harper Collins Publ., 1991, XVI + 828 + A-74 pp.
10. L'vov D.S., Foundations of Machines Economic Design, M.: Economika, 1977, 256 pp. (in Russian).

Who shall decide when doctors disagree?

CHAPTER 11

NEW TECHNIQUE FOR SIMULATION OF ORGANISM SUBSYSTEMS

It is shown that the earlier considered technique for development mathematical modeling can be applied and grounded for simulation of various organism subsystems (OS) with applications to medicine and health care. In this connection, the mathematical models obtained can be useful for both common and nontraditional modes of a disease treatment. General theory is displayed on the examples of the subsystem of blood sugar and primary-pancreatic diabetes, cancers modeling, and the novel approach to low dose exposure effects' extrapolation of OS modeling.

1. INTRODUCTION

With the base of the techniques under consideration one can be acquainted in Part I. Now, there are about 150 works of different authors devoted to the further development of the techniques with applications mainly in economics and biology. However, until now there is not a systematic presentation of these mathematical modeling techniques for various subsystems of an organism with applications to medicine and health care. Meanwhile, as we will show below, the new technique allows us to describe much more thoroughly the main structure and functions of an organism both in the norm and pathology. In addition, the description is essentially different from the traditional one [2, 8, 9] in the form of ordinary differential equations, which requires us to revise the whole statement and realization of technology of necessary biomedical experiments. We believe new difficulties, which arise here, can be overlapped by new useful results.

2. BASE MM OF OS

Regarding the structure and physiological functions of OS, its minimal, *base MM* (see Ch. 2) can be represented in the form

$$m(t) = \int_{a(t)}^{t} \alpha(t,u)y(u)m(u)du, \quad c(t) = \int_{b(t)}^{t} \beta(t,u)[1-y(u)]m(u)du,$$

$$P(t) = \int_{a(t)}^{t} y(u)m(u)du + \int_{b(t)}^{t} [1-y(u)]m(u)du, \quad G(t) = M(t) - P(t),$$

$$M(t) = \int_0^t m(u)du, f(t) = m(t) + c(t),$$

$$0 \le y \le 1, 0 \le a(t), b(t) \le t, T \ge t \ge t\wedge \ge 0, \tag{1}$$

where *m* is a new *resource* (per unit of the time) of OS providing its internal functions (recreation of *m*); *c* is some new *resource* providing external function of OS; *y* is a distribution function; α and β are efficiency *indices* of functioning of OS along the channels *ym* -- *m* and [1-*y*]*m* -- *c* respectively; *a*(*t*) and *b*(*t*) are temporal bounds: on the segments [*a*(*t*^),*a*(*t*)] and [*b*(*t*^),*b*(*t*)] the respective resources *ym* and [1-*y*]*m* are considered as obsolete; *P*(*t*) is the total amount of *functioning resource* of OS; *G*(*t*) is the total amount of the *obsolete resource* of OS; *t*^ is the starting time for modeling; [0,*t*^] is the prehistory period on which all the elements of the model are considered as given; the "horizon" *T* is usually given too.

In different cases we can give different interpretations for the *resource m*. It can be a quantity of new living bio-mass, tissues, cells, active or reactive centers, organelles, averaged new compositions of chemicals in biology similar to averaged new WP in economics, a number of macromolecules, ionic particles, etc. The magnitudes *P* and *G* should be interpreted correspondingly. As to the function *c*-type, the *resources* above are some new external products and/or heat, sweat, etc. that exude from an organism or its subsystems.

Apparently, all the relations' (1) are exact by definition and all the elements of the model are nonnegative. A diminution of the output values is regulated not by a sign but, for example, by the rate of *a*(*t*) and/or *b*(*t*) growth in (1). The functions α and β can depend on the other elements in the general case.

The total number of different elements in (1) is 12 of which 2, namely *y* and *f*, can be considered as the control elements of the model. The total number of the equations in (1) is 6. If the exogenous functions *c*, α, and β (and *T*, *y*, and *f*) are given, we will have a determined system of the equations.

Knowing *T*, *c*, *y*, and *f*, we can find $m = f - c$, *M*, and *a*; knowing *c* and *m*, we can find *b*; knowing *a* and *b*, we can find *P* and *G*. Of course, there can be other variants of given and desired elements of the model.

It should be emphasized that MM (1) deals with the so-called "gray box" when the structure of a dynamic system is partly revealed.

3. STATEMENT OF PROBLEMS

3. 1. Problems of the type "If. . ., then . . .":

1.1. If the control elements *y*, *f*, and functions *c*, α, and β are given in a predetermined manner, then what kind of solution of the system (1) will the functions *m*, *M*, *a*, *b*, *P*, and *G* be.

1.2. If α, β, *y*, *P*, and *G* are given in a predetermined manner, then what kind of solution will a, b, m, c, f, and M be.

1.3. If α, β, c, P, and y are given, then what kind of solution will m, a, b, f, G, and M be.

3. 2. Optimization problems:

2.1. It is required to find y such that the functional

$$I1 = \int_{t^\wedge}^{T} c(t)dt = \max \tag{2}$$

(maximum of external activities of OS) under observing the relations (1).

2.2. It is required to find y such that

$$I2 = \int_{t^\wedge}^{T} P(t)dt = \min \tag{3}$$

(minimum of bio-mass of OS or, for example, a cancer or HIV population) provided that the relation's (1) are observed.

2.3. It is required to find y for which

$$I3 = T - t^\wedge = \min \tag{4}$$

(optimal control by speed for restoration of the norm of OS functioning) provided that the relations (1) and the conditions

$$C(T) = \int_{t^\wedge}^{T} c(t)dt \in C^*,\ P(T) \in P^*, \tag{5}$$

where the sets C^* and P^* are given, are observed.

2.4. It is required to find y for which

$$I3 = T - t^\wedge = \max \tag{6}$$

(maximum of an active OS life) provided that the relations (1) and the restriction

$$c(t) \geq c^0(t), \tag{7}$$

where $c^0(t)$ is given, are observed.

3. 3. Identification problems:

3.1. Let

$$\alpha(t,u) = \alpha(u)\exp(-d_1 t),\ \beta(t,u) = \beta(u)\exp(-d_2 t),\ d_1,\ d_2 > 0, \tag{8}$$

and let $c(t)$,

$$P_1(t) = \int_{a(t)}^{t} y(u)m(u)du,\ G_1(t) = \int_{a(0)}^{a(t)} y(u)m(u)du,\ M_1(t) = P_1(t) + G_1(t),$$

$$P_2(t) = P(t) - P_1(t),\ G_2(t) = G(t) - G_1(t),\ M_2(t) = P_2(t) + G_2(t), \tag{9}$$

and d_1, d_2 are given on the prehistory (d_1 and d_2 can be obtained at the verification stage of the prediction). It is required to find $\alpha(u)$ and $\beta(u)$. Knowing P_i, G_i, $i = 1, 2$, we can find $m(t)$:

$$m(t) = [P(t) + G(t)]', \tag{10}$$

$a(t)$, $b(t)$, and $y(t)$:

$$a(t) = M_1^{-1}[P_1(t) - M_1(t)],\ b(t) = M_2^{-1}[P_2(t) - M_2(t)],\ y(t) = M_1'(t)/m(t), \tag{11}$$

where M_i^{-1}, $i = 1, 2$, are the reciprocal functions for M_i. After that, knowing in addition c, we can find $\alpha(u)$ and $\beta(u)$:

$$\alpha(u) = [(m(u)\exp(d_1 t))' + \alpha(a(u))y(a(u))m(a(u))a'(u)]/[y(u)m(u)],$$

$$\beta(u) = [(c(u)\exp(d_2 t))' + \beta(b(u))(1-y(u))m(b(u))b'(u)]/[1-y(u))m(u)]. \tag{12}$$

We have two different cases here:

$$t\hat{} = a(t\hat{}) = b(t\hat{}) = \alpha(0) = \beta(0) = 0 \tag{13}$$

and

$$t\hat{} > 0. \tag{14}$$

In the first case, $\alpha(u)$ and $\beta(u)$ can be found with the help of (12) immediately. In the second case, we are in a need of the approximate restoration of data back to the instant 0 of OS origin.

It should be noted that for realization of the formulae (10)- (12), we have to use the technique of the solution of the *ill-posed problem* of numerical differentiation (see Ch.5, 6).

3.2. Let

$$\alpha(t,u) = \alpha(t - u),\ \beta(t,u) = \beta(t - u), \tag{15}$$

and let P, G, y, and c be given on the prehistory. It is required to find $\alpha(t-u)$ and $\beta(t-u)$. For this, we should solve the Volterra-type equations of the first kind:

$$m(t) = \int^{t-a(t)}_{0} \alpha(u)y(t-u)m(t-u)du, \; c(t) = \int^{t-b(t)}_{0} \beta(u)[1-y(t-u)]m(t-u)du. \qquad (16)$$

3.3. In the general case, it is required to find the functions α and β from the conditions:

$$\min\|m - \int^{t}_{a(t)} \alpha ymdu\| \; (\alpha \in V), \; \min\|c - \int^{t}_{b(t)} \beta[1-y]mdu\| \; (\beta \in W), \qquad (17)$$

where V and W are reasonable given sets, $\|\bullet\|$ is a sign of the measure of $\bullet$.

3.4. Functions α and β depend on environmental parameters: pressure, concentrations of oxygen, carbon dioxide and water vapor, temperature, the intensities of lighting, sound, radiation, the control element y, f, etc. It is required to find the relations α and β depending on these elements and parameters.

3.5. Knowing α and β in the prehistory, it is required to find these functions in the determined period in the future depending on environmental parameters forthcoming.

4. MORE GENERAL MM

It is not difficult to introduce different extensions of the two-subsystem model (1). In particular, introducing the equation

$$c\hat{}(t) = \int^{t}_{b(t)} \alpha\hat{}(t,u)x(u)m(u)du, \qquad (18)$$

where $\alpha\hat{}$ is the index of efficiency for functioning of OS along the channel xm -- $c\hat{}$, $c\hat{}$ is a new resource providing the external function of this OS, and interchanging $1-y$ on $1-x-y$ and $f = m + c$ on $f = m + c + c\hat{}$, we have a three-subsystem model. Introducing for OS the model of (1)-type beside the model of an organism, we will have a four-subsystem model. Introducing for every OS the model of (1)-type, we will have a $2N$-subsystem model, where N is the number of subsystems of an organism. Introducing for every OS, in addition, the model of (18)-type, we will have a $3N$-subsystem model, and so on.

As an example, the more general *MM of OS* can be the following:

$$m_i(t) = \int^{t}_{ai(t)} \alpha_i[t,s,m_i(s),C^1(s),\ldots,C^n(s)]y_i(s)m_i(s)ds,$$

$$c^i_j(t) = \int^t_{ai(t)} \beta^i_j[t,s,m_i(s),C^1(s),...,C^n(s)]z^i_j(s)m_i(s)ds, j = 1,..., ri, C^i = (c^i_1,..., c^i_{ri}),$$

$$P_i(t) = \int^t_{ai(t)} m_i(s)ds,\ G_i(t) = \int^{ai(t)}_0 m_i(s)ds,\ 0 \le y_i, z_i \le 1, y_i + \sum z^i_j\ (j = 1,..., ri) = 1,$$

$$ai(t) \le t, t \ge t\^{} > 0,\ f_i(t) = m_i(t) + \sum c^i_j(t)\ (j = 1,..., ri),\ i = 1,..., n, \tag{19}$$

where n is the number of OS components (organs or active centers); t is the time; $m_i(t)$ is the new generalized resource recreated in the unit of time by ith active center; $y_i(t)m_i(t)$ is a share of $m_i(t)$ going to the creation of $m_i(t)$; $c^i_j(t)$ is the new generalized resource created in the unit of time by ith active center; $z^i_j(t)m_i(t)$ is a share of $m_i(t)$ going to the creation of $c^i_j(t)$; f_i is the external generalized resource (m_i and c^i_j have the unit of measurement f_i); α_i is the index or capacity of biosynthesis (analog of the specific rate) for functioning along the channel y_im_i -- m_i: quantities of the units of $m_i(t)$ created in the unit of the time per the unit of $y_i(t)m_i(t)$; β^i_j is the index of efficiency of functioning along the channel $z^i_jm_i$ -- c^i_j, similar to α_i; $a_i(t)$ is the temporal bound for liquidation of the obsolete resources; $P_i(t)$ is the total quantity of functioning resources of m_i-type; $G_i(t)$ is the total quantity of the obsolete resources; $t\^{}$ is the start for modeling; $[0,t\^{}]$ is the prehistory of OS functioning on which all the elements of MM (19) are given by definition. Since the rate of recreation of the new resource cannot be negative, all the values entering (19) are nonnegative. But this does not mean that their behavior is a monotone. The increase of $a_i(t)$ can reduce to the decrease of other values. On the whole, (19) is the system of nonlinear equations of Volterra-type with the prehistory and series of the relations as inequalities.

If we agree that nth component of the subsystem in question is the remained part of the general organism, it will be always $n > 1$ excepting, perhaps, the case when OS is the whole organism.

5. EXISTENCE AND UNIQUENESS THEOREM

For the numerous theoretical results of investigation of MM of the (19)-type and the respective above-mentioned problems see Part I, Ch. 3 and 4. We dwell here on only one result of the theory to illustrate an important advantage of MM in question.

Theorem. *Let the segment* $[t\^{},T]$ *be such that the distributions* $y_i(t)$ *and* $z^i_j(t)$, *the indices of efficiencies* α_i *and* β^i_j, *and the functions* P_i (*or* G_i) *are given on it. Besides, there exist constants* y^-, $\alpha^\pm$, β^+, *and* $P^\pm$ *such that* $y_i(s) \ge y^- > 0$, α_i, β^i_j, *and* P_i *are piecewise continuously-differentiable over all the arguments,* $0 < \alpha^- \le \alpha_i \le \alpha^+$, $0 \le \beta^i_j \le \beta^+$, *and* $0 < P^- \le P_i \le P^+$. *Then the system equations* (19) *relative to* m_i, a_i, c^i_j,

$j = 1,\ldots, ri$, $i = 1,\ldots, n$, has the unique piecewise continuously differentiable solution on $[t^\wedge,T]$. In addition, $m_i(t) > 0$, $t \in [t^\wedge,T]$, if $m_i^\wedge(t) > 0$, $t \in [0,t^\wedge]$, and the solution will be uniformly bounded under $T \to \infty$ if α_i, β^i_j, and P_i are uniformly bounded.

Proof. From the conditions of the theorem and the first three equalities in (19) it easily follows that

$$y^-\alpha^-P^- < y^-\alpha^-P_i(t) < m_i(t) \leq \alpha^+P_i(t) \leq \alpha^+P^+,$$

$$c^i_j(t) \leq \beta^+P_i(t) \leq \beta^+P^+, \tag{20}$$

i.e., $m_i(t)$ and $c^i_j(t)$ are bounded and $m_i(t)$ are positive (if they exist). Let $t \in [t^\wedge,t_1]$, where $ai(t) \leq t^\wedge$, $i = 1,\ldots, n$, so that

$$P_i(t) = \int^{t^\wedge}_{ai(t)} m_i^\wedge(s)ds + M_i(t),\ M_i(t) = \int^{t}_{t^\wedge} m_i(s)ds,\ i = 1,\ldots, n, \tag{21}$$

where $m_i^\wedge(s)$ is given value $m_i(s)$ on the prehistory. Existence of $ai(t)$ follows from the condition of Theorem (c f with the condition and the proof of Theorem 2, Ch. 2). Therefore

$$ai(t) = M_i^{\wedge -1}[M_i^\wedge(t^\wedge) + M_i(t) - P_i(t)],\ i = 1,\ldots, n, \tag{22}$$

where $M_i^\wedge$ is the anti-derivative for $m_i^\wedge$ and $M_i^{\wedge -1}$ is the inverse function for $M_i^\wedge$, which always exists and is piecewise continuously differentiable on the strength of positiveness of $m_i^\wedge(s)$. Substituting $ai(t)$ as the right part of (22) in (19), we have as the result Volterra nonlinear system $n + \sum^n_{i=1} ri$ equations about unknowns $m_i(t)$ and $c^i_j(t)$. Solvability of the system and uniqueness of the solutions on the segment $[t^\wedge,t_1]$ follow from the well-known classical theorem, for due to (20), the classic Lipschitz condition is valid for all the variables m_i, c^i_j of the system.

Later the segment $[t_1,t_2]$ on which $a_i(t) \leq t_1$ plays the role $[t^\wedge,t_1]$, and so on. It should be shown that thus the whole given segment $[t^\wedge,T]$ will be passed. It is sufficient to show that $t - ai(t) \geq d > 0$, because here we can put $t_{k+1} = t_k+d = t^\wedge+(k+1)d$ and therefore any given finite segment will be passed by a finite number of steps. On the strength of the theorem's conditions we have

$$P_i(t) \leq \max[m_i](t - ai(t)) \leq \alpha^+P^+(t - ai(t)), \tag{23}$$

from which

$$t - ai(t) \geq P^{-}/(\alpha^{+}P^{+}) = d > 0, \qquad (24)$$

which is needed. Piecewise continuous differentiability m_i and c^i_j follows from the relation's (19) themselves. □

It follows from the proof of the theorem that MM (19) has the important property as follows. Independently from that how the indices of efficiency depend on their arguments (only they should be bounded, positive and piecewise continuously differentiable) the system equation (19) relative to m_i, a_i, c^i_j is always solvable on the $[t^{\wedge}, T]$ and has the unique solution. The result is remarkable due to the Lipschitz conditions for the kernels of the initial system are not supposed to be given in advance, but are obtained because of the internal properties and structure of the system.

In practice, instead of $P_i(t)$ one can take "heap" or "waste" or "manure" (*obsolete resource*) $G_i(t)$, $i = 1,..., n$. Then instead of (22) we have

$$ai(t) = M_i^{\wedge -1}[M_i^{\wedge}(0) + G_i(t)], \; i = 1,..., n. \qquad (25)$$

Thus, one of the essential demands that produces MM (19) to the technology of biomedical experiments is the necessity of careful measurement of the whole *functioning resource* $P_i(t)$ for the *i*th reaction or active center of OS under consideration or corresponding "heap" $G_i(t)$. Another demand is the necessity of measurement of the concentrations for the *new substances* that are always nonnegative.

6. CONCEPTIONS OF NORM AND PATHOLOGY

All the values in (19) have rather profound sense from the point of view of the structure and functions of the subsystem under consideration. So, if the vector of these values is denoted by $R = (r_1,..., r_q)$, where $q = 7n+3\sum^{n}_{i=1} ri$ (q can be greater, if some certain entered integers r_s are combinations of the initial values), then it is natural to suppose that one of the *conceptions* of the *norm* consists in the validity of the relations

$$r^{-}_{s} \leq r_s \leq r^{+}_{s}, \; s = 1,..., q, \qquad (26)$$

where r^{-}_{s} and r^{+}_{s} are admissible bounds for variations of the introduced values from point of view of the norm. On the other hand, if at least one of the relations in (26) is violated, then it is natural to speak about *conception* of the *pathology* or the pathological state of the subsystem.

The pathology can be detailed by discrimination of two classes of (26) violations: A) those of the resource values, namely m_i, c^i_j, P_i, G_i and f_i, and B) those of the structural values, namely *ai*, α_i, y_i, β^i_j, and z^i_j. In turn, the indices of the class A) can be decomposed into two groups: A') the indices connected with the internal functions of resource recreation, namely m_i, P_i and G_i, and A") the indices connected with the external functions of active centers, namely c^i_j, and f_i.

Of course, in the clinical practice there are various other indices of the norm and the pathology. Those indices and their bonds as well as the problems of introducing generalized indices and reducing drug toxicity are considered in detail in [2, 6, 7, and 10].

7. NORM RESTORATION PROBLEMS

The most simple and expanded restoration is that the determined values of the group A") are injected into the organism directly or on the contrary are removed from the organism and/or the respective medicines are injected into the organism so that to achieve validity of the necessary bounds (4) relative to this group. However, under the condition that the indices of violations of the group A") are accompanied by those of the group A') and class B) such a simple method is usually the temporal measure and does not deduce restoration.

Let us consider at first the case of index violation of the group A', for example, m_i. The distributive function $y_i(t)$ and $z^i_j(t)$ are the main control magnitudes for MM (19). It depends on their value how the whole internal resource $m_i(s)$ of the subsystem is distributed among its various parts.

For increasing (decreasing) of the values of $m_i(t)$, provided that the values of the other magnitudes are not to be broken, it is usually necessary to increase (decrease) the value $y_i(s)$, $s < t$.

As it appears (see Part I), for increasing (decreasing) of the values $c^i_j(t)$ for the majority of values t from $[t^\wedge, T]$ under sufficiently large T-$t^\wedge$, the respective values z^i_j do not usually have to be increased (decreased), but on the contrary decreased (increased). It follows from outstripping of $m_i(s)$ increasing (decreasing) on the $[t^\wedge, T]$ such that $z^i_j(s)m_i(s)$ for the majority of s from $[t^\wedge, T]$ is found greater (smaller) in spite of decreasing (increasing) of z^i_j. Thus, the crucial condition for restoration (under weakened functions of the class A) is the increase of a share of all resources of the subsystems to its internal requirements. It is possible to hope that any modes of action, among them nontraditional and non-medicinal ones, connected with the various physical loading on a certain part of OS and contributing to increasing of resource distribution in advantage to the internal sphere of the sick subsystem will usually bring essential positive effect (under weakened function of the class A).

The most serious and profound violations are connected to the structural shift of the subsystem, i.e., with the violation of (26) for the indices of the class B. Here only the interference into the genome apparatus can likely restore the norm. Although in principle, in this case there can be also developed the theory similar to (1) - (19).

Indeed, for creation, support and variation of the structures of α_i and $\beta^i j$ types, a certain share of the whole resources should be also used. To complete MM (19) in the case, for example, of α_i, we can assume

$$\alpha_i(t,s) = \alpha_i(s)\exp[d_i(s-t)],\ d_i \geq 0,\ s \leq t,\ \alpha_i(t) = \int^{t}_{ai(t)} \alpha_i(t,s)x_i(s)m_i(s)ds,$$

$$0 \leq x_i(s) \leq 1,\ x_i(s)+y_i(s)+\sum z^i_j = 1\ (j = 1,\dots, ri),\ i = 1,\dots, n,\ t \geq t^\wedge, \tag{27}$$

where d_i are decrements of deterioration characterizing the rate of decreasing $\alpha_i(t,s)$ by t; $x_i(s)m_i(s)$ are a share of $m_i(s)$ incoming for the creation of the new technologies $\alpha_i(t)$; the rest magnitudes are the same as in (19).

Investigation of MM similar to (27) (see Ch. 4) shows that the priorities of x_i are usually higher than those of y_i, which, in turn, as was underlined earlier, are higher than those of z^i_j. It means that

the crucial condition for restoration (under weakened functions of the classes A and B) is the increase of the share of all the resources to the needs of a comfortable state of OS structure, first, then its internal sphere, and after that its external sphere.

8. ON OPTIMIZATION PROBLEMS

Let us consider one of the natural *optimization problems*, namely that of optimal control by high speed, i.e., the problem of determination of those distributions x_i, y_i, and z^i_j that reduce the pathological values to the norm in the minimal possible time so that they will remain inside the norm limits later on. Let for definiteness be $r_1 < r^-_1$ and be required to find

$$T = T^* = \min\,(x_i, y_i, z^i_j|\ r^-_i \leq r_i \leq r^+_i,\ t > T^*) \tag{28}$$

As it is shown in Ch. 4, the general regularity for the desired distribution x^*_i, y^*_i, and z^{*i}_j behavior is the same as noted above: an essential share of resources must go to maintain the structure and internal sphere of the subsystem. However, the requirement of optimization can result in the rather complicated character for the desired distributions. They can have a rather large, theoretically the infinite number of switching times from their minimal to maximal possible values. The instances of these steps are unknown in advance and have to be found in the solution process.

Many other optimization problems for OS can be considered in a similar way as above (see Ch. 4).

9. EXAMPLES OF APPLICATIONS

9.1. A blood sugar subsystem and diabetes

Let us take for granted the description of a blood sugar subsystem given in [3, 6, and 12]. In that case n = 7.

1) Hypophysis (H) ($i = 1$):
$m_1(t)$ is a new H bio-mass differential in the time unit 1 (in mg/sec) at the time instant t; $c^1_j(t)$, $j = 1,\ldots, 5$, are the quantities respectively of the pancreatic, adrenotrophic, corticotrophic, thyrotrophic, and hypophysiotrophic hormone created by H (and Hypothalamus) in the time unit (in mg/sec); $a1(t)$ is the temporal bound for liquidation of the obsolete H bio-mass; $P_1(t)$ is the functioning H bio-mass; $G_1(t)$ is "manure", i.e., the total quantity of the obsolete H bio-mass.

2) Pancreas (P) ($i = 2$):
$m_2(t)$ is a new P bio-mass differential in the time unit at the instant t; $c^2_1(t)$ is the quantity of the pancreatic insulin extracted by P in the time unit; $c^2_2(t)$ is the quantity of the glucagon in the time unit; $a2(t)$ is the temporal bound for liquidation of the obsolete P bio-mass; $P_2(t)$ is the functioning P bio-mass at the instant t.

3) Adrenal (Ad) ($i = 3$):
$m_3(t)$ is a new Ad bio-mass differential in the time unit; $c^3_1(t)$ is the quantity of adrenaline extracted by Ad in the time unit; $c^3_2(t)$ is the quantity of glucocorticoid extracted by Ad in the time unit; $P_3(t)$ is the functioning bio-mass of Ad at the instant t.

4) Thyroid gland (TG) ($i = 4$):
$m_4(t)$ is a new TG bio-mass differential in the time unit; $c^4_1(t)$ is the quantity of tyrosine (in mg/sec) created by TG in the time unit; $P_4(t)$ is the functioning TG bio-mass.

5) Liver (L) ($i = 5$):
$m_5(t)$ is a new L bio-mass differential in the time unit; $c^5{}_1(t)$ is the quantity of glycogen accumulated by L in the time unit; $c^5{}_2(t)$ is the quantity of glucose extracted by L in the time unit; $P_5(t)$ is functioning L bio-mass.

6) Kidneys (K) ($i = 6$):
$m_6(t)$ is a new K bio-mass differential in the time unit; $c^6{}_1(t)$ is the quantity of glucose eliminated by K in the time unit; $P_6(t)$ is the functioning K bio-mass.

7) The remained part of an organism (RPO) including Muscle tissue (the main "consumer" of glucose), Alimentary canal (the main "supplier" of glucose), Cardiovascular subsystem (the main "mixer" of glucose), and Immune subsystem (the main "protector" or in case pathology one of the main "wrecker") ($i = 7$):
$m_7(t)$ is the new PRO bio-mass differential in the time unit; $c^7{}_1(t)$ is the quantity of glucose consumed by RPO in the time unit; $c^7{}_2(t)$ is the quantity of glucose provided by Alimentary canal into blood in the time unit; $c^7{}_3(t)$ is the present-day concentration of glucose formed in blood in the time unit; $c^7{}_4(t)$ is the quantity of immune competent cells formed by Immune subsystem in the time unit to the AG-determinant of any matter such that it can eliminate some components of the blood sugar subsystem.

In addition, we have at least the following relations (see [3, 6]):

$$\alpha_1 = \alpha_1(t,s,c^2{}_1,c^3{}_1,c^3{}_2,c^7{}_4),\ \beta^1{}_j = \beta^1{}_j(t,s,c^2{}_1,c^3{}_1,c^3{}_2,c^7{}_3,c^7{}_4),\ j = 1,\dots,\ 5;$$

$$\alpha_2 = \alpha_2(t,s,c^1{}_1,c^7{}_4),\ \beta^2{}_j = \beta^2{}_j(t,s,c^1{}_1,c^7{}_4),\ j = 1,\ 2;\ \alpha_3 = \alpha_3(t,s,c^1{}_2,c^1{}_3,c^7{}_4),$$

$$\beta^3{}_j = \beta^3{}_j(t,s,c^1{}_2,c^1{}_3,c^7{}_4),\ j = 1,\ 2;\ \alpha_4 = \alpha_4(t,s,c^1{}_4,c^7{}_4),\ \beta^4{}_1 = \beta^4{}_1(t,s,c^1{}_4,c^7{}_4);$$

$$\alpha_5 = \alpha_5(t,s,c^1{}_5,c^2{}_1,c^2{}_2,c^3{}_1,c^3{}_2,c^4{}_1,c^7{}_3,c^7{}_4),\ \beta^5{}_j = \beta^5{}_j(t,s,c^1{}_5,c^2{}_1,c^2{}_2,c^3{}_1,c^3{}_2,c^4{}_1,c^7{}_3,c^7{}_4),$$

$$j = 1,\ 2;\ \alpha_6 = \alpha_6(t,s,c^7{}_3,c^7{}_4),\ \beta^6{}_1 = \beta^6{}_1(t,s,c^7{}_3,c^7{}_4);\ \alpha_7 =$$

$$\alpha_7(t,s,c^5{}_1,c^5{}_2,c^6{}_1,c^7{}_1,c^7{}_2,c^7{}_3,c^7{}_4),\ \beta^7{}_j = \beta^7{}_j(t,s,c^5{}_1,c^5{}_2,c^6{}_1,c^7{}_1,c^7{}_2,c^7{}_3,c^7{}_4),\ j = 1,\ 2,\ 3. \quad (29)$$

In primary-pancreatic *diabetes* [6, 7, 12], P β-cells generating insulin are stricken. The result is that there is an essential shortage of insulin ($c^2{}_1$) in the organism and/or functions strengthening of P β-cells generating hormone glycogen ($c^2{}_2$), which raises sugar level in blood. In the norm the functioning β-cells have the weight 0.64-0.75 g on average, but for the patients of diabetes the same cells have the weight in the limits 0.22 - 0.3.

Thus, regarding the notations, the value $\int^t_{a2(t)} m_2(s) z^2{}_1(s) ds$ should be transferred from the range 0.22-0.3 g to the range 0.64 - 0.75 g. The mind has the desire to increase a relative share z^2_1 of P resources providing formation of insulin. But as we noted above this can be followed by greater exhaustion of β-cells.

From the qualitative investigation of *MM of diabetes* it follows that for correction of the situation almost all the P resource must go to the internal needs of P, i.e., it is necessary to increase y_2 (at the expense of z^2_1 and/or z^2_2 due to $y_2+z^2_1+z^2_2= 1$).

In this connection, the temporal strengthening of the insulin deficit can be compensated at the expense of insulin injection outside, for example, intro-muscular one. Thus, the injection of external insulin must be not the basic, but an auxiliary temporal curative reception.

Unfortunately, the such data connecting to the diabetes that they would show the essential structural shift in P., i.e., pathological behavior of indices efficiency α_2, β^2_1, and β^2_2, are absent in practice.

9. 2. Base model of cancers

On the strength of the theory above the base, minimal *MM* of a *cancer* has the form

$$m(t) = \int_{a(t)}^{t} \alpha(t,s)y(s)m(s)ds, \; M(t) = \int_{0}^{t} m(s)ds, \; c(t) = \int_{a(t)}^{t} \beta(t,s)[1-y(s)]m(s)ds,$$

$$P(t) = \int_{a(t)}^{t} m(s)ds, \; G(t) = M(t)-P(t), \; x(t)f(t) = m(t) + c(t), \; n(t) = \int_{b(t)}^{t} \gamma(t,s)z(s)n(s)ds,$$

$$N(t) = \int_{0}^{t} n(s)ds, \; d(t) = \int_{b(t)}^{t} \delta(t,s)[1-z(s)]n(s)ds, \; Q(t) = \int_{b(t)}^{t} n(s)ds, \; H(t) = N(t)$$

$$- Q(t), \; [1-x(t)]f(t) = n(t) + d(t), \; 0 \le x, y, z \le 1, \; 0 \le a(t), b(t) \le t, \; t \ge t\hat{} \ge 0, \qquad (30)$$

where m is a new resource of a healthy part of an organism (HPO) providing its internal functions (of recreation of m); c is some new resource providing external functions of HPO; y is a distribution function for HPO; α and β are efficiency indices of functioning of HPO along the channel m--ym and the channel c--$(1-y)m$ respectively; $a(t)$ is a temporal bound: outside $a(t)$, on the $[0, a(t)]$, resource m is considered as obsolete; P is the total amount of functioning resource (mass) of HPO; G is the total amount of the obsolete resource of HPO; f is the external resource; n, d, z, γ, δ, b, Q, and H are the similar elements, but for a cancer (C); x is a relative share of f between HPO and C; $t\hat{}$ is the starting time for modeling; $[0, t\hat{}]$ is the prehistory on which all the elements of the model are considered as given.

Thus, the total number of different elements in (30) is 20 of which 4, namely x, y,

z, and f can be considered as the control elements of the model. The total number of the equations in (30) is 12. If the exogenous character functions α, β, γ, and δ are given, we will have the determined system of the equations.

The problem of C elimination can be stated with the help of the model (30) as follows:

$$P'(t) = 0 \;\&\; Q'(t) < 0, \tag{31}$$

that is, under HPO mass constant we have C mass decreasing. On the strength of (30) and (31) we have

$$P'(t) = m(t) - m(a)a'(t) = 0,\ G'(t) = m(a)a'(t),$$

$$Q'(t) = n(t) - n(b)b'(t) < 0,\ H'(t) = n(b)b'(t), \tag{32}$$

from which

$$m(t) = G'(t),\ n(t) < H'(t). \tag{33}$$

We have, in particular, if $c = d = 0$, then

$$G'(t) < x(t)/[1-x(t)]H'(t), \tag{34}$$

and if $x = 1/2$, then

$$G'(t) < H'(t). \tag{35}$$

The relations (33)-(35) have obviously the determined physiological interpretation and can serve as experimental testing of C elimination. Note that we always can know $G' + H'$ as the total increment of a whole organism.

Let us consider the problem of C elimination in more detail depending on the control functions x, y, z, and f.

We consider the case when α, β, γ, and δ depend on s (or t-s) only. This assumption may be correct regarding a short period of the time for development of C. On the strength of (30), it is not difficult to find

$$P'(t) = m(t) - [(\alpha ym)(t) - m'(t)]/(\alpha y)(a) = m(t) - [(\beta(1-y)m)(t) - c'(t)]/(\beta(1-y))(a) \tag{36}$$

and

$$Q'(t) = n(t) - [(\gamma zn)(t) - n'(t)]/(\gamma z)(b) = n(t) - [(\delta(1\text{-}z)n)(t) - d'(t)]/(\delta(1\text{-}z))(b). \quad (37)$$

On the strength of (31), from (36) and (37) it follows

$$m(t) = [(\alpha ym)(t) - m'(t)]/(\alpha y)(a) = [(\beta(1\text{-}y)m)(t) - c'(t)]/(\beta(1\text{-}y))(a) \quad (38)$$

and

$$n(t) < [(\gamma zn)(t) - n'(t)]/(\gamma z)(b). \quad (39)$$

If $m' = c' = 0$, then from (38) it follows

$$y(t)/y(a) = \alpha(a)/\alpha(t),\ [1\text{-}y(t)]/[1\text{-}y(a)] = \beta(a)/\beta(t). \quad (40)$$

Let, for example,

$$\alpha(a) > \alpha(t)\ \&\ \beta(a) < \beta(t), \quad (41)$$

which seems to be natural. Then from (40) it follows

$$y(t) > y(a), \quad (42)$$

that is, a greater share of HPO resource should go to realization of its internal functions.

If $n' = d' = 0$, then from (39) and the similar relation for δ and $1\text{-}z$ it follows

$$\gamma(t)z(t) > \gamma(b)z(b),\ \delta(t)(1\text{-}z(t)) > \delta(b)(1\text{-}z(b)). \quad (43)$$

If , for example,

$$\gamma(b) < \gamma(t)\ \&\ \delta(b) = \delta(t), \quad (44)$$

then

$$\gamma(b)z(b)/\gamma(t) < z(t) < z(b), \quad (45)$$

that is, a certain smaller share of C resource should go to fulfillment of its internal functions. Similar estimates can be also done in the case of on $t - s$ dependence.

Let us consider now the comparative conducts of xf and $(1\text{-}x)f$ depending on values $\alpha y + \beta(1\text{-}y)$ and $\gamma z + \delta(1\text{-}z)$. It is easy to see that

$$xf = \int{}^{t}_{a} [\alpha y + \beta(1\text{-}y)]mds, \; (1\text{-}x)f = \int{}^{t}_{b} [\gamma z + \delta(1\text{-}z)]nds. \qquad (46)$$

So, if f, m, and n are positive constants and

$$\gamma z + \delta(1\text{-}z) \gg \alpha y + \beta(1\text{-}y), \; a = b, \qquad (47)$$

then

$$xf \ll (1\text{-}x)f \qquad (48)$$

starting from a certain instant of the time. It means that we will have $P \ll Q$, i.e., C suppresses HPO.

We can prevent it, in principle, with the help of control factors z. Indeed, under sufficiently small z

$$n(t) = \int{}^{t}_{b} \gamma z n(s)ds \qquad (49)$$

will become sufficiently small too that, in turn, due to (46) will result in sufficiently small $(1\text{-}x)f$. Unfortunately, we do not know, and will not learn [4], C's peculiarities in order to select in practice such a small z as above as well as x, and/or f such that $(1\text{-}x)f \ll xf$.

We dwell still on the control factor z in more detail. From $Q' < 0$ it follows that

$$n(t) = \int{}^{t}_{b} \gamma znds < n(b)b' = b'\int{}^{b(t)}_{b(b)} \gamma znds =$$

$$b'\int{}^{t}_{b} \gamma(b(t),b(s))z(b(s))n(b(s))b'(s)ds. \qquad (50)$$

The condition (50) means qualitatively that new $z(s)$ should be essentially less than previous ones on the average. As it has shown in Ch. 4, a response of any ES on an external short time, but significant increasing impulse (stress) (d) is just an essential decreasing impulse (z).

Thus, one of the desired therapies is likely increasing intensive impulse effects (stresses) localized on C. The frequency ratio of the effects is proportional to the values γ and δ, and can be determined more exactly on the basis of MM (30) if γ and δ are given. An example of such therapy can be an exposure of C, but not the way which kills the C cells and adjacent cells of HPO: it should probably spare the cells' life exposure, but be sufficiently strong to be a stress.

Of course, there are many other approaches to the C mathematical modeling and

C therapy. A survey of the traditional approaches based on differential equations and certain new results one can find in [1].

9. 3. An approach to low dose exposure effects extrapolation modeling

The importance of this problem was emphasized by Dr. H. Salem [10]. Toxic-kinetic *low dose exposure* effects' extrapolation modeling allows us to avoid cruel experiments with animals. For the simplicity, we consider the minimal, base MM of ES having the form

$$m(t) = \int_{a(t)}^{t} \alpha(t,s)y(s)m(s)ds,\ M(t) = \int_{0}^{t} m(s)ds,\ c(t) = \int_{a(t)}^{t} \beta(t,s)[1-y(s)]m(s)ds,$$

$$0 \le y(s) \le 1,\ P(t) = \int_{a(t)}^{t} m(s)ds,\ G(t) = M(t) - P(t),\ f(t) = m(t) + c(t),\ t \ge t^{\wedge} \ge 0, \qquad (51)$$

where the sense of the magnitudes is the same as above. In the general case, (51) is a system of the nonlinear integro-functional relations which consists of 6 equalities and 6 inequalities connecting 12 magnitudes, namely: m, c, α, β, a, y, P, M, G, t, $t^{\wedge}$, and f, which all are non-negative.

For MM (51), usually α, β, and P or α, β, and f are to be given and the other elements to be found (y can be usually found from a certain extreme principle).

We consider MM (51) to be the model of an organism. Then m is the rate of creation of the new bio-mass quantity, P is the functioning bio-mass of the organism, G is the obsolete bio-mass, c is the external work of the organism in the unit of the time, and the remaining elements of the model have approximately the same meaning as in the general case.

Let us introduce the condition of the organism's *homeostasis* in the form

$$P'(t) = m(t) - m(a(t))a'(t) = 0. \qquad (52)$$

and denote the results of the low dose exposure df by δm, δc, δa, δy and the high dose exposure Df by Δm, Δc, Δa, Δy.

Our problem: knowing the estimates of the *low dose exposure*, it is required to estimate missing elements of the MM (51), (52) with a view to estimate the results of the *high dose exposure*.

It is easy to see that knowing δm, one can find $\delta c = df - \delta m$ and δa from (52), after which one can find $\alpha\delta y$ and $\beta[1-\delta y]$ from (51) (under different reasonable assumptions about the form of these functions). Knowing, in addition, δy, one can find the desired, usually, missing elements α and β. Knowing the $m^\wedge$ and $y^\wedge$ on the prehistory of the organism functioning, one can find in a similar way $\alpha^\wedge$ and $\beta^\wedge$ that can serve as desired estimates for the modeling forward.

Of course, in both cases one is in a need of development and application of the new technique concerning to estimation of nontraditional values m and y type.

Knowing α and β, one can find Δm, Δc, Δa, Δy from the system equation in (51), (52), where $f = Df$ to be given. If it is appeared that Δa and Δy violate the restrictions in (51), then Df is the high dose such that the homeostasis (52) cannot be valid.

We have supposed that α (and β) do not depend on the value f and a control with the help of the distribution y is enough for preserving of the condition of homeostasis (52). In reality the control by y cannot be enough. What is more, y can be fixed from another extreme principle.

A way out of the situation is the subsequent *factorization* of α (see Ch.2):

$$\alpha(t,s) = u(s)v(s)/v(t)\alpha^*(t-s), \tag{53}$$

where the controls by u and by v likely relate to reactions of the endocrine system and the nervous system of the organism respectively, α^* probably corresponds to the organism's *genome*.

Thus, α^* perhaps does not depend on f and the control u and/or v can be found in a similar way to the above.

In conclusion of this s., we would like to emphasize that exploiting the new technique under consideration, one can hope for possibility of determination of all the elements of the models without using animals. For examples of such models see the next Ch. (12). Besides, not infrequently knowing only qualitative properties of initial data, we can exploit the models for receiving significant qualitative results if we apply only qualitative methods of investigation.

10. ON THE PROBLEMS OF INDIVIDUALIZATION OF MM

The high value of MM of the (1), (19), (30), (51)-types is in, particularly, that only rather general qualitative information about initial data, usually about the functions α and β-type, is sufficient for investigation of these models by qualitative methods for the future period of the time and, by the same token, for anticipation of possible undesirable consequences. Of course, the numerical values of initial data are necessary in the cases of bringing the results to the numbers. These numerical initial data can be obtained with the help of the models themselves under the known prehistory of functioning of the system considered. However, here we should consider in detail new requirements to the biomedical technology for realization of the respective procedures and experiments about what we have already spoken above. All these are related to the so-called identification problems, solving of which are usually obtained at the expense of application of the mathematical methods based on the regularities of mass events.

Thus, we have the fact of a certain controversy here: identification is needed for the concrete object, but the method requires mass events. Therefore, recently the problem of identification for one given organism is called the *problem* of *individualization* or personification, and the new respective approach was created. We dwell on some results from [11], one of the important works in this new direction.

The main parameter of personification - Q - characterizes the relative flow liquids (lymph, plasma of blood) velocity in the organism considered in comparison with the "base" or "basal" organism. In our notations

$$Q^2 = Q^2(t) = \sum^n_{i=1} \int^{t+t2}_t m_i(s)ds / [\sum^n_{i=1} \int^{t+t2}_t m^*_i(s)ds], \tag{54}$$

where $t2$ is the typical time ($t2 \approx 1$ minute) for which sufficiently complete intermixing of separate micro-motions into liquid mediums of the organism takes place ($t1$ is the typical time of a separate particle pulsation, $t1 \approx 1$ second), $\sum^n_{i=1} m^*_i(s)$ is the internal generalized resource circulating into the subsystem (organ) of the "basal" organism or into the "basal" organism on the whole. It is supposed that for the given and "basal" organism the properties of the liquid mediums as well as disposition and sizes of organs do not have essential differences. The parameter Q^2 is called in [11] as (relative) "alive" temperature of the organism. The scale of Q has been selected such that $Q = 1$ coordinate with the "norm", the case $Q = 0$ means complete stopping of "alive" motion. It has been shown in [11] that Q is decreased with the aging: from 25 to 75 age Q varies from 1 to 0.82. Q is considered as constant during the typical time $t3$ of keeping by the organism of a certain physiological state, $t3 \approx 1$ hour.

Let the main characteristics of MM for the "basal" organism be α^*_i and β^{*i}_j. Then the respective averaged characteristics of MM for the given organism is

$$\alpha_i = Q^2\alpha^*_i,\ \beta^i_j = Q^2\beta^{*i}_j,\ j = 1,\dots, n,\ i = 1,\dots, n. \tag{55}$$

The ground of the relations of (55)-type is given in [11] for linear MM in the form of ordinary differential equations.

For the time that is much more than $t3$ instead of Q^2 we can use, in particular,

$$Q_\delta^2 = Q^2 + \zeta_Q(t/\delta),\ \delta = t3/t4, \tag{56}$$

where $t4$ is the typical time for changing of the phase variables, $t4 \approx 1$ day; ζ_Q is stationary stochastic time series with a finite variance and zero mean value. Substituting (56) in MM, we have the perturbed system of the equations the solutions of which are converging to the solutions of the originate systems when $\delta \to 0$.

11. ON ALGORITHMS AND SOFTWARE

One mass problem under realization of a new apparatus of mathematical modeling, based on MM (1), (19), and (30), is the problem of a numerical solution of the system of Volterra-type integral equations. For the economic numerical solution of this class of the equation with the pre-assigned accuracy and with regard to the errors of three types (at the expense of the initial data inaccuracy, approximation methods and round-off errors under realization of the approximation methods on the computer) see Ch. 5. It has been shown in Ch. 6. that the solution of optimization problems can be also reduced to the solution of the systems of integral equations of Volterra-type.

Very sophisticated software, including the apparatus for the application of the new technique under consideration is contained in [7, 8, 13, Ch. 7]. Peculiarity of the program complex [7] consists in that it contains subroutines resolving principal standard problems of numerical mathematics with subroutines estimating the errors of three types and also necessary numbers of operations and memory of the computer. The program complex [8, 13] is based on [7] and focused to the solution of the problems of identification, prediction and optimization for MM under consideration.

An essential part of the programs [13, Ch. 7] is devoted to the solution of different problems in biology, in particular, to the problems of modeling of OS.

REFERENCES

1. Adam J.A., Bellomo N., A Survey of Models for Tumor-Immune System Dynamics, Birkhäuser, 1997, 344 pp.
2. Asachenkov A., Marchuk G., Mohler R., Zuev S., Disease Dynamics, Birkhäuser, 1994, 316 pp.
3. Cobelli C., Federspil G., Pocini G., Salvan A., and Scandellari C., An Integrated Mathematical Model of the Dynamics of Blood Glucose and Its Normal Control, Mathematical bio-sciences, 58, 1982, 27-60.
4. Glushkov V.M., Cancer Problem from a Position of Systems General Theory, Kiev: Inst. of Cybernetics, Reprint 79-26, 1979, 20 pp. (in Russian).
5. Glushkov V.M., Ivanov V.V., Yanenko V.M., Methodological Problems of Mathematical Methods Application to Biology, Kiev: Inst. of Cybernetics, Reprint 79-60, 1979, 65 pp. (in Russian).
6. Guide on Physiology. Physiology of Endocrine System, L.: Nauka, 1979, 679 pp. (in Russian).
7. Handbook for Therapeutist, M.: Meditsina, 1979, 655 pp. (in Russian).
8. Kaplan D., Glass L., Understanding Nonlinear Dynamics, Springer, 1995, 400 pp.
9. Murray J.D., Mathematical Biology, New York: Springer, 1993, 767 pp.
10. New Technologies and Concepts for Reducing Drug Toxicity, CRC Press, 1993, 154 pp.
11. Pogozhev I.B., Application of Mathematical Models of Diseases in Clinic Practice, M.: Nauka, 1988, 192 pp. (in Russian).
12. Toffolo G., Bergman R.N., Finegood D.T., Bowdon C.R. and Cobelli C., Quantitative Estimate of Beta-Cells Sensitivity to Glucose in the Intact Organism: a Minimal Model of Insulin Kinetics in the Dog, Diabetes, 29, 1980, 979-990.

Little knowledge is a dangerous thing

CHAPTER 12

MODELING OF THE IMMUNE NETWORK

Most assumptions, biological in nature that follow from the known experimental data on idio-type anti-idio-typical interactions of various sub-populations of cells and antibodies of the *immune network* (IN), and from the foundation of the mathematical models (MM) of IN are considered. Such assumptions as the refractoriness of cells towards the action of antigen (Ag) during their proliferation stimulated by Ag and their passage out of proliferation according to a certain law after which the cells can be stimulated again lead to MM different from well-known ones.

The *immune network MM* are important here not only by themselves but also as verifying phase of validity and quality of the models of ES developed in this book (see PART I and Ch. 11). We show below that the immunology models following from the biological data and observations are entirely corresponding to MM of ES. And what is more, MM of ES allow us to deepen understanding of IN dynamics.

1. INTRODUCTION

This Ch. is aimed at development of the models of functioning of *IN* and *immune system* (IS) [1-11, 13, 14] regarding available experimental data on the idio-type anti-idio-typical (Id-AId) interactions among distinctive sub-populations of lymphocytes [2, 6, 10] as well as at the application of these models to the immune response regulation [1-14]. The Ch. discusses tolerance and the rapid immune response. The role of the macrophages that examine foreign particles and then present their findings to other cells is omitted here because their number is usually not limiting.

The postulates or assumptions underlying the models were mainly developed by the Moscow immunologists Drs. L. N. Fontalin and N. G. Nesterenko [2, 6, 10]. These assumptions lead to provide rather sophisticated difference and continuous integro-functional MM different from the well-known ones [13]. In particular, they essentially different from the models of the so-called +,- - theory of IN [3] and the second-generation theory of IN [14]. Among the models reviewed by G. Rowe in [13], the models by A. S. Perelson [11, 13], arisen from certain extreme principles of IS functioning, remind some important features of the models under consideration.

The comparison of these MM with the new class of models of ES one can find in s. 5. Some results of investigations of the models and an attempt of application to the simulation of HIV and immune-therapy of AIDS are also given here and in the Ch.13.

2. BASIC ASSUMPTIONS

The following postulates or assumptions underlie the models.

1) Receptors of *lymphocytes* (*Lc*) and *immune-globulin* (*Ig*) that recognize the *antigen* (*Ag*) have centers of fixation of Ag, the para-tops, and individual antigenic determinants, the idio-tops, located in the variable region of the Lc receptors and Ig.

2) Various levels of IN are distinguished. The first level consists of idio-type-positive (Id) *T Lc* (T helpers - Th, T suppressors - Ts, T effectors of hypersensitivity and T killers - Tef), *B Lc* - B cells and their products (idio-type-positive immune-globulin - Ig). The second level consists of anti-idio-type (AId) Lc (Th, Ts, Tef, B) and Ig. The third level consists of anti-anti-idio-type (AAId) Lc (Th, Ts, Tef, B) and Ig, etc.

3) Behavior of IN essentially depends on the presence of the epi-tops of Ag (immune-gene, tolerant-gene) or on their absence. The situation is discussed when the elements of the first level react with the *molecular-disperse Ag* of the protein nature. For the higher order levels the *antigens* are the cells and Ig of the previous level.

4) A supplement of the population of Lc due to the differentiation of stem cells as well as a natural loss in the population of lymphoid cells caused by aging are ignored.

5) For Th of every level, the molecular-disperse Ag (for the cells of the second, third and succeeding levels such Ag are the Ig of the first, second and the higher levels, respectively) can induce three states according to its concentration: suppression by small doses, stimulation by average doses and suppression by large doses. For Ts, Tef, and B the molecular-disperse Ag results in two states: stimulation by average doses and suppression by large doses. The cells of every level, being the Ag for the cells of the higher level, induce only stimulation.

6) Among the B cells of every level we distinguish, first, the B_2 cells (90% of all B cells) which are lunching into proliferation and differentiation only by Th Lc and, secondly, the B_1 cells (10% of B cells) proliferating and differentiating without Th.

7) A part of Th Lc that are in the state of suppression due to small doses of Ag is determined depending on the concentration of Ag ($10^{11.5}$ - 10^{12} molecules) by the truncated normal law of the decimal logarithm of Ag concentration with the mean $x_1 = 11.75$ and mean square deviation $y_1 = 0.0833$ (dispersion is y_1^2).

8) The part of the cells that are stimulated by mean doses of Ag and suppressed by large doses can be determined also according to the concentration of Ag using the truncated normal law for its decimal logarithm. The concentrations of Ag inducing the stimulation of Th, Ts, and Tef are in the interval 10^{12} - 10^{15} molecules ($x_2 = 13.5$, $y_2 = 0.500$), for B_1 the range is 10^{13} - 10^{15} molecules ($x_3 = 14$, $y_3 = 0.333$) and for B_2 the range is $10^{12.5}$- 10^{15} molecules ($x_4 = 13.5$, $y_4 = 0.417$).

9) The concentrations of Ag inducing for all the cells suppression with large doses lay within the range 10^{15} - 10^{16} molecules ($x_5 = 15.5$, $y_5 = 0.167$).

10) The part of Id, AId, AAId, etc. cells eliminated by AId-, AAId-, AAAId-Ig, etc.

is determined according to the concentration of Ig $10^{12.5}$-$10^{14.5}$ by the truncated normal law of the decimal logarithm of Ig concentration with $x_6 = 13.5$ and $y_6 = 0.333$.

11) The Lc of adult mice can recognize about 10^6 different epi-topes of Ag. Therefore, before the introduction of Ag, on each level (regarding the relative part of the cells of each type) the number of Th, Ts, Tef is set to be equal to 2×10^2, the number of B_1 equals to 0.4×10^2, the number of B_2 equals to 3.6×10^2, and the number of molecules Ig is equal to 10^{11}.

Here and later the concentration of Ag and cells is indicated in calculation per 1 mm^3 of blood.

12) The period of half-decay of Ig (and of the protein Ag) is assumed to be 50 -- 100 hours and the time of the mitotic cycle of cells is assumed to be 8 --10 hours.

13) The stimulation of Th, Ts, Tef causes their proliferation in 12 hours and the proliferation of B in 18 hours. B cells produce Ig in 30 hours after stimulation.

14) B_2 cells stimulated by Ag become ready to accept a signal of stimulated Th during 18 hours. If contact B_2 cell with Th cell takes place, B_2 cell will start to proliferate in 12 hours, otherwise B_2 cell will go to the rest (r) phase.

15) The part of X target-cells that are in one of the states: stimulation, suppression or elimination, due to the interaction with the respective Y control (or effector)-cells is determined by the truncated normal law for the decimal logarithm of Y with the mean $x_7 = 5$ and deviation $y_7 = 0.667$ and anyway do not exceed $k^*\min(X, Y)$, where k^* is the average productivity of Y-cells. Thus, we accept the approach for the cell-cell interactions similar to the Ag (Ig)-cell interactions.

16) The following properties of Lc are employed:

a) the start of the proliferation of Lc is impossible without their stimulation;

b) the proliferating Lc are characterized by refractoriness with respect to the stimulus, i.e. they proliferate by themselves;

c) the Lc that began to proliferate stopped their proliferation spontaneously by the truncated normal law for the number of hours ($x_8 = 114$, $y_8 = 10$);

d) the Lc that stopped to proliferate can be stimulated again and then, in 6 hours, they are the mature (m) cells, which cannot proliferate;

e) the period of half-decay of the mature cells is assumed to be 48 -- 96 hours.

17) The interaction of cells within each level and between these levels is caused by Ag and Id-AId bridges that realize the functional potentials of Lc of the appropriate sub-populations. Therefore, IdTh help IdB_2 and $AIdB_2$; IdTs suppress IdTh, IdTef, IdB, AIdTh, AIdTef, and AIdB; IdTef eliminate IdTh, IdTs, IdTef, IdB, AIdTh, AIdTs, AIdTef, and AIdB; IdB produce IdIg; IdIg eliminate Ag; AIdTh help IdB_2, $AIdB_2$ and $AAIdB_2$; AIdTs suppress IdTh, IdTef, IdB, AIdTh, AIdTef, AIdB, AAIdTh, AAIdTef, and AAIdB; AIdTef eliminate IdTh, IdTs, IdTef, IdB, AIdTh, AIdTs, AIdTef, AIdB, AAIdTh, AAIdTs, AAIdTef, and AAIdB; AIdB produce AIdIg; AIdIg eliminate IdIg and kills Id cells. Behavior of the elements of the 3-rd, 4-th, and higher levels (if they exist) is similar to the behavior of the elements of the second level.

18) Ts suppress the cell-targets within their level at the instant of their stimulation (through the antigen bridge). Ts suppress the cell-targets of the lower and higher levels (due to Id-AId interactions) except the plasma B cells due to the latter has no Ig receptors. Because of the interactions between Ts and the cell-targets the former are inactivated and the latter are suppressed throughout the reactions being considered.

19) Tef eliminate the cell-targets except the plasma B cell.

20) Every control-cell can affect only one target-cell during three hours. The general productivity of Ts, Tef, and Th can make up: for the proliferating cells from 2 to 4 targets and for the mature cells from 6 to 10 targets.

21) The mean rate of Ig production by one proliferating B cell is 10^7 molecules per hour. The mature (plasma) B cell gives 10^8 molecules per hour.

22) One molecule Ig eliminates on the average one molecule of Ag in the large concentrations of Ag and Ig being considered.

23) Within all the interactions Ig have priority over cells.

Detailed analysis of MM of IN [6] reduces to the several corrections and additions:

24) Ag are subdivided into T-dependent and T-independent ones. T-dependent Ag stimulate all the sub-populations of T cells and B_2 cells. T-independent Ag stimulate Ts and B_1 cells. Ig of any levels are related to T-dependent Ag.

25) Two classes of Ig are introduced: IgM and IgG. Half-live of IgM is 50 hours and that of IgG (75% of all Ig) is 100 hours. They are similar in all the other respects.

26) Under primary response IgG are derived only by the plasma B_2 cells; all the other B cells (proliferating B_2 and proliferating and plasma B_1) produce IgM. Under secondary response IgG are derived by B_2 cells and IgM by B_1 cells.

27) There are the priorities in action of different T cells on the same target: Tef $\succ$ Ts $\succ$ Th, i.e., first Tef acts, then Ts and at last Th, and also $Ti^m \succ Ti^p$, where *m* means mature cells, *p* means proliferating cells. Tef and Ts of the next, higher level have the priorities in action in comparison with Tef and Ts of the antecedent level; on the contrary, Th of the higher level acts after Th of the antecedent level.

28) All the cells except the proliferating and mature Tef, Ts, Th, and plasma cells enter the quantity of the cells of the given level that stimulate the cells of the next level.

29) The resting Lc (or the memory cells) are not reactive to the action of any cells of the same level.

30) The following *MM* is accepted as a basis of *cell-cell interactions*:

$$z = kxy/Y = k^{\wedge}xy,\ k = k(Y) = 1/3F_7(lg\ Y),\ k^{\wedge} = k/Y, \tag{1}$$

where $3k^{\wedge}$ is probability of two cells of a control-type and target-type contact; Y is the total number of control cells (regarding priorities in action) at the time instant t for x; x is any part of the total number of target cells $X = X(Y)$ at the time instant t for Y; y is any part of the total number of control cells Y; z is the result of x-cells and y-cells interactions during an hour. Interactions, while the time is Δt, result $z\Delta t$ instead of z.

Thus, we accept "acting mass" law, where $k^\wedge$ depends on t, but not on x, y. Under large concentrations of Y when $F_7(lg\ Y)$ is 1, we have expression (1) usually used.

From (1), it follows that under fixed t the results of the cell-cell interactions are additive over all x or y for which X, and Y are fixed values.

31) All the cells must be subjected to the laws of *affinity* to Ag (Ig) (see, for example, [9]). In particular, the capability of each cell to react to a certain dose of Ag is kept by its off-springs. And also, we have to employ the following relations

$$F_i(v')F_i(v) = F_i(v')\cup F_i(v) = \max\,[F_i(v'), F_i(v)];\ F_i(v')(1-F_i(v)) = F_i(v')\text{-}\min\,[F_i(v'),F_i(v)];$$

$$(1-F_i(v'))(1-F_i(v)) = 1 - F_i(v')(1 - F_i(v)). \tag{2}$$

where F_i, $i = 1, 2,..., 6,$ are the functions defined above and below

3. DIFFERENCE EQUATIONS OF THE MODEL

From the assumptions considered we obtain rather sophisticated and tedious MM of IN. We accept below the explicit difference scheme with the step 1-hour.

Let M_{5n+1}, M_{5n+2}, M_{5n+3}, M_{5n+4}, and M_{5n+5} be the numbers of Th, Ts, Tef, B, and Ig respectively at the instant t at the level $n+1$, where $n = 0,1,2,...;$ $M_{5n+4,1}$ and $M_{5n+4,2}$ be the numbers of the cells B_1 and B_2 - type respectively. Let M_0 be the number of Ag molecules at the instant t and M_{i0} be the numbers of the cells and molecules at the instant $t = 0$ of the introduction of Ag. Let

$$y_0 = \lg M_0,\ \ D0 = M_0,\ Di = M_{5i} + 10^7 M_{5i-1}{}^{p,l} + 10^8 M_{5i-1}{}^{pl}\ \ i = 1, 2,... \tag{3}$$

where $M_{5i-1}{}^{p,l}$ is the number of the cells from $M_{5i-1}{}^{p}$ that produce Ig (see the assumptions 13), 21)), and

$$\Phi_s(v) = 1/(2\pi y_s^2)^{1/2}\int_{-\infty}^{v}\exp[-(u-x_s)^2/(2y_s^2)]du;\ F_s(v) = \Phi_s(v).\ v\in[x_s-3y_s, x_s+3y_s];$$

$$F_s(v) = 0,\ v < x_s-3y_s;\ F_s(v) = 1,\ v > x_s+3y_s,\ s = 1, 2,... \tag{4}$$

and let $M_i^{\,r}$, $M_i^{\,s}$, $M_i^{\,s'}$, $M_i^{\,p}$, $M_i^{\,ss}$, $M_i^{\,ss'}$, and $M_i^{\,m}$ be the numbers of the resting (memory), stimulated by Ag, stimulated by Ag and Th-cells (for B_2-cells), proliferating, second time stimulated by Ag, second time stimulated by Ag and Th-cells (for B_2-cells), and mature (plasma for B) cells respectively from M_i, where

$$M_i = M_i^{\,r} + M_i^{\,s} + M_i^{\,s'} + M_i^{\,p} + M_i^{\,ss} + M_i^{\,ss'} + M_i^{\,m} \tag{5}$$

If $Di\text{-}1 - Di \geq 0$, $i = 1, 2,...,$ then (see the assumptions 12), 22))

$$M_0(t+1) = [D0 - D1]0.99,\ M_5(t+1) = 0,\ M_{10}(t+1) = [D2 - D3]0.99,$$

$$M_{15}(t+1) = 0,\ \text{etc.} \tag{6}$$

If $D0 - D1 < 0$, $D1 - D0 - D2 \geq 0$, $Di\text{-}1 - Di \geq 0$, $i > 2$, then

$$M_0(t+1) = 0,\ M_5(t+1) = [D1 - D0 - D2]0.99,\ M_{10}(t+1) = 0,$$

$$M_{15}(t+1) = [D3 - D4]0.99,\ \text{etc.} \tag{7}$$

If $D0 - D1 < 0$, $D1 - D0 - D2 < 0$, $D2 - D1 + D0 - D3 \geq 0$, then

$$M_0(t+1) = M_5(t+1) = 0,\ M_{10}(t+1) = [D2 - D1 + D0 - D3]0.99,$$

$$M_{15}(t+1) = 0,\ M_{20}(t+1) = [D4 - D5]0.99,\ \text{etc.} \tag{8}$$

and so on.

The interactions between target cells and control cells are summarized in the following table:

Control cells M_i^j, $j = m, p$	Target cells M_i^j
$i = 1$	$i = 4{,}2,\ 9{,}2,\ j = s,\ s{,}s$
$i = 2$	$i = 1, 3, j \neq r;\ i = 4, j \neq r,m;\ i = 6, 8;\ i = 9, j \neq m$
$i = 3$	$i = 1, 2, 3, j \neq r;\ i = 4, j \neq r,m;\ i = 6, 7, 8;\ i = 9, j \neq m$
$i = 6$	$i = 4{,}2,\ 9{,}2,\ 14{,}2, j = s,\ s{,}s$
$i = 7$	$i = 1, 3;\ i = 4, j \neq m;\ i = 6, 8, j \neq r;\ i = 9, j \neq r,m;\ i = 11, 13;\ i = 14, j \neq m$
$i = 8$	$i = 1, 2, 3;\ i = 4, j \neq m;\ i = 6, 7, 8,\ j \neq r;\ i = 9, j \neq r,m;\ i = 11, 12, 13;\ i = 14, j \neq m$
Σ_1	$i = 6, 7, 8, 9;\ j = r$

Target cells M_i^j	Control cells M_i^j, $j = m, p$
$i = 1, 3, 4, j = r$	$i = 7, 8$
$i = 1, 3, j \neq r;\ i = 4, j \neq r, m;\ i = 4{,}2, j \neq s,\ s{,}s$	$i = 2, 3, 7, 8$
$i = 2, j = r$	$i = 8$
$i = 2, j \neq r$	$i = 3, 8$
$i = 4{,}2, j = s,\ s{,}s$	$i = 1, 6, 2, 3, 7, 8$
$i = 4, j = m$	none
$i = 6, 8, 9, j = r$	$i = 2, 3, 12, 13$
$i = 6, 8, j \neq r;\ i = 9, j \neq r, m;\ i = 9{,}2, j \neq s,\ s{,}s$	$i = 2, 3, 7, 8, 12, 13$
$i = 7, j = r$	$i = 3, 13$
$i = 7, j \neq r$	$i = 3, 8, 13$
$i = 9{,}2, j = s,\ s{,}s$	$i = 1, 6, 11, 2, 3, 7, 8, 12, 13$
$i = 9, j = m$	none
$i = 6, 7, 8, 9;\ j = r$	Σ_1

(9)

Here (see the assumption 28))

$$\Sigma_1 = \Sigma^4_{i=1}(M_i - M_i^p - M_i^m) + M_4^p, \tag{10}$$

where M_i is given by (5). For the third and the remaining levels the interactions are the same as for the second level with addition 5, 10, etc. to the respective values of subscripts (in the table (9) and in (10)).

According to the assumption 16), 17), 20), 30), the notations in (1), and the table (9), we have the following cases

$$k_i^j = k(M_i^j) = 1/3F_7[\lg (M_i^j)],\ k\hat{}_i^j = k_i^j/M_i^j,\ i = 1, 2, 3, 6, 7, 8, ..., j = m, p;$$

$$k_i = 1/3F_7[\lg (\Sigma_i)],\ k\hat{}_i = k_i/\Sigma_i,\ i = 1, 6,.... \tag{11}$$

We denote

$$[1-k_{i1,...,ij}{}^{m,p}] = [1-k_{i1}{}^m][1-k_{i1}{}^p]...[1-k_{ij}{}^m][1-k_{ij}{}^p],\ i1,..., ij = 1, 2, 3, 6, 7, 8,...,$$

$$U_1(t) = U_1 = U_3(t) = U_3 = U = [1-F_6(y_5)][1-k_{8,7,3,2}{}^{m,p}],$$

$U_2(t) = U_2 = [1-F_6(y_5)][1-k_{8,3}{}^{m,p}]$, $U\hat{}_1(t) = U\hat{}_1 = U\hat{}_3(t) = U\hat{}_3 = U\hat{} = [1-F_6(y_5)][1-k_{8,7}{}^{m,p}]$, $U\hat{}_2(t) = U\hat{}_2 = [1-F_6(y_5)][1-k_8{}^{m,p}]$, $V(t) = V = [1-F_6(y_{10})][1-k_{13,12,8,7,3,2}{}^{m,p}]$,

$$W(t) = W = k_1{}^m + k_1{}^p(1-k_1{}^m) + k_6{}^m(1-k_1{}^{m,p}) + k_6{}^p(1-k_6{}^m)(1-k_1{}^{m,p}), \tag{12}$$

Therefore, for the first level in the case $15 > y_0(0) > 12$ (for $11.5 \le y_0(0) \le 12$ and $y_0(0) \ge 15$ the model differs accordingly by the initial value M_{10} decreased $[1-F_1(y_0(0))]$ times and the initial values M_{i0}, $i = 1, 2, 3, 4$, decreased $[1-F_5(y_0(0))]$ times) on the strength of the assumptions 1)-5), 8), 10)-13), 15)-20), 23), 27), 29)-31), and also the table (9), we find

$M_i^s = M_{i0}F_2(y_0(0))$, $0<t<12$, $M_i^s = 0$, $11<t<361$; $M_i^p = 0$, $0<t<12$, $M_i^p(12) = N_i^p(11) = M_{i0}F_2(y_0(0))$, $M_i^p(t+1) = N_i^p[1-F_8(t-12)]$, $N_i^p = N_i^p(t-1)e^{0.08}U_i(1-c_i^p)$, $c_1^p = 1/3k\hat{}_1{}^p(1-k_1{}^m)[U\Sigma M_i^j(i=4,2; j=s, s,s)+V\Sigma M_i^j(i=9,2;\ j=s, s,s)]$, $c_2^p = 1/3k\hat{}_2{}^p/(1-k_2{}^p)[UM_i^j(i=1,3, j\ne r;\ i=4, j\ne r,m)+V\Sigma M_i^j(i=6,8;\ i=9, j\ne m)]\}$, $c_3^p = 1/3k\hat{}_3{}^p/(1-k_3{}^p)[U\Sigma M_i^j(i=1,2,3, j\ne r;\ i=4, j\ne r,m)+V\Sigma M_i^j(i=6, 7, 8;\ i=9, j\ne m)\}$, $11<t<157$,

$M_i^p = 0$, $157<t<361$; $M_i^{s,s} = 0$, $0<t<97$, $M_i^{s,s}(t+1) = [M_i^{s,s}+dM_i^{s,s}-dM_i^{s,s}(t-6)U_i(t-6)...U_i(t-1)]U_i$, $95<t<360$, $dM_i^{s,s} = N_i^p dF_8(t-12)F_2(y_0)$, $dF_8(t-12) = F_8(t-12)-F_8(t-13)$, $95<t<158$, $dM_i^{s,s} = 0$, $0<t<96$, $157<t<360$; $M_i^m = 0$, $0<t<102$, $M_i^m(t+1) = M_i^m+dM_i^{s,s}(t-6)U_i(t-6)...U_i(t-1) - M_i^m c_i^m$, $c_1^m = 1/8k\hat{}_1{}^m[U\sum M_i^j\,(i=4,2; j=s,\ s,s)+ V\sum M_i^j\,(i=9,2; j=s,\ s,s)]\}Ue^{-0.01}$, $c_2^m = 1/8k\hat{}_2{}^m/(1-k_2^{m,p})[U\sum M_i^j\,(i=1,3, j\neq r;\ i=4, j\neq r,m)+V\sum M_i^j(\ i=6,8;\ i=9, j\neq m)]\}U_2e^{-0.01}$, $c_3^m = 1/8k\hat{}_3{}^m/(1-k_3^{m,p})[U\sum M_i^j\,(i=1,2,3, j\neq r;\ i=4, j\neq r,m)+V\sum M_i^j\,(i=6,7,8;\ i=9, j\neq m)]\}Ue^{-0.01}$, $100<t<164$, $M_i^m(t+1) = M_i^m(1-c_i^m)U_ie^{-0.01}$, $163<t<360$; $M_i^r(1) = M_{i0}[1-F_2(y_0(0)]$, $M_i^r(t+1) = (M_i^r+dM_i^r)U\hat{}_i$, $0<t<360$, $dM_i^r = N_i^p dF_8(t-12)[1-F_2(y_0)]$, $95 < t < 158$, $dM_i^r = 0$, $0<t<96$,

$$157<t<360,\ i = 1, 2, 3. \tag{13}$$

From the assumptions 6), 8), 10)-20), 23), 27), 29)-31), and also the table (9) it follows

$M_{4,1}^s = M_{4,10}F_3(y_0(0))$, $0<t<18$, $M_{4,1}^s = 0$, $17<t<361$; $M_{4,1}^p = 0$, $0<t<18$, $M_{4,1}^p(18) = N_{4,1}^p(17) = M_{4,10}F_3(y_0(0))$, $M_{4,1}^p(t+1) = N_{4,1}^p[1-F_8(t-18)]$,

$N_{4,1}^p = N_{4,1}^p(t-1)e^{0.08}U$, $17<t<163$, $M_{4,1}^p = 0$, $163<t<361$;

$M_{4,1}^{s,s} = 0$, $0<t<103$, $M_{4,1}^{s,s}(t+1) = [M_{4,1}^{s,s}+dM_{4,1}^{s,s}-dM_{4,1}^{s,s}(t-6)U(t-6)...U(t-1)]U$,

$dM_{4,1}^{s,s} = N_{4,1}^p dF_8(t-18)F_3(y_0)$, $101<t<360$; $M_{4,1}^m = 0$, $0<t<109$,

$M_{4,1}^m(t+1) = [M_{4,1}^m+dM_{4,1}^{s,s}(t-6)U(t-6)...U(t-1)]e^{-0.01}$, $107<t<171$,

$M_{4,1}^m(t+1) = M_{4,1}^m e^{-0.01}$, $169<t<360$; $M_{4,1}^r(1) = M_{4,10}[1-F_3(y_0(0)]$,

$M_{4,1}^r(t+1) = (M_{4,1}^r+dM_{4,1}^r)U\hat{}$, $dM_{4,1}^r = N_{4,1}^p dF_8(t-18)[1-F_3(y_0)]$, $0<t<360$.

$M_{4,2}^s = M_{4,20}F_4(y_0(0))$, $0<t<19$, $M_{4,2}^s = M_{4,20}F_4(y_0(18))$, $18<t<37$, $M_{4,2}^s(36) = M_{4,20}F_4(y_0(36))$, $M_{4,2}^s(t+1) = M_{4,2}^sU(1-W)$, $36<t<55$, $M_{4,2}^s(t+1) = M_{4,2}^sU(1-$

$W)F_4(y_0(54))$, $54<t<73,\ldots$, $M_{4,2}^{s}(t+1) = M_{4,2}^{s}U(1-W)F_4(y_0(342))$, $342<t<361$; $M_{4,2}^{s'} = 0$, $0<t<37$, $M_{4,2}^{s'}(t+1) = M_{4,2}^{s'}U+dM_{4,2}^{s'}- dM_{4,2}^{s'}(t-12)U(t-12)\ldots U$, $dM_{4,2}^{s'} = M_{4,2}^{s}UW$, $35<t<360$; $M_{4,2}^{p} =0$, $0<t<49$, $M_{4,2}^{p}(u+12,u) = N_{4,2}^{p}(u+11,u) = dM_{4,2}^{s'}(u)U(u+1)\ldots U(u+12)$, $36<u<349$, $M_{4,2}^{p}(t+1,u) = N_{4,2}^{p}(t,u)[1-F_8(t-u-12)]$, $N_{4,2}^{p}(t,u) = N_{4,2}^{p}(t-1,u)e^{0.08}U$, $u+11<t<u+145$, $M_{4,2}^{p}(t,u) = 0$, $u+144<t<361$, $M_{4,2}^{p} = \sum M_{4,2}^{p}(t,u)$ $(36<u<t-11)$, $48<t<361$; $M_{4,2}^{s,s} = 0$, $0<t<133$, $M_{4,2}^{s,s}(t+1,u) = M_{4,2}^{s,s}(t,u)U+dM_{4,2}^{s,s}-dM[1-F_4(y_0)U(1-W)]$, $dM= \{\wedge(dM_{4,2}^{s,s})+\wedge(\wedge(dM_{4,2}^{s,s})F_4(y_0)U(1-W))+\wedge(\wedge(\wedge(dM_{4,2}^{s,s})F_4(y_0)U(1-W))F_4(y_0)U(1-W))+\ldots\}$,

$\wedge(h) = h(t-18)U(t-17)[1-W(t-17)]\ldots U(t-1)[1-W(t-1)]$, $M_{4,2}^{s,s}(u+95,u) = 0$, $dM_{4,2}^{s,s} = N_{4,2}^{p}(t,u)dF_8(t-u-12)F_4(y_0)U(1-W)$, $dM_{4,2}^{s,s} = 0$, $u+157<t<360$, $M_{4,2}^{s,s} = \sum M_{4,2}^{s,s}(t,u)$ $(36<u<t-95)$, $132<t<361$; $M_{4,2}^{s,s'} = 0$, $0<t<133$, $M_{4,2}^{s,s'}(t+1,u) = M_{4,2}^{s,s'}(t,u)U+dM_{4,2}^{s,s'}-dM_{4,2}^{s,s'}(t-6,u)U(t-5)\ldots U$, $dM_{4,2}^{s,s'} = dM_{4,2}^{s,s}W/(1-W)+dMF_4(y_0)UW$, $M_{4,2}^{s,s'}(u+95,u) = 0$, $M_{4,2}^{s,s'} = \sum M_{4,2}^{s,s'}(t,u)$ $(36<u<t-95)$, $132<t<361$; $M_{4,2}^{m} = 0$, $0<t<139$, $M_{4,2}^{m}(t+1,u) = [M_{4,2}^{m}(t,u)+dM_{4,2}^{s,s'}(t-6,u)U(t-5)\ldots U(t-1)]e^{-0.01}$, $95+u<t<360$, $M_{4,2}^{m}(u+95,u) = 0$, $M_{4,2}^{m} = \sum M_{4,2}^{m}(t,u)$ $(36<u<t-95)$, $138<t<361$; $M^{\wedge}_{4,2}{}^{r} = M_{4,20}[1-F_4(y_0(0)]$, $0<t<18$, $M^{\wedge}_{4,2}{}^{r} = M_{4,20}[1-F_4(y_0(18))]$, $17<t<36$, $M^{\wedge}_{4,2}{}^{r} = M_{4,20}[1-F_4(y_0(36))]U^{\wedge}$, $35<t<54$, $M^{\wedge}_{4,2}{}^{r}(t+1) = M^{\wedge}_{4,2}{}^{r}(t)[1-F_4(y_0(54)]U^{\wedge}$, $53<t<72,\ldots$, $M^{\wedge}_{4,2}{}^{r}(t+1) = M^{\wedge}_{4,2}{}^{r}(t)[1-F_4(y_0(342)]U^{\wedge}$, $341<t<360$, $M_{4,2}^{r}(u,u) = 0$, $0<u<361$, $M_{4,2}^{r}(t+1,u) = M_{4,2}^{r}(t,u)U^{\wedge}+dM_{4,2}^{r}+dM[1-F_4(y_0)]U^{\wedge}$, $dM_{4,2}^{r} = N_{4,2}^{p}dF_8(t-u-12)[1-F_4(y_0)]U^{\wedge}$, $dM_{4,2}^{r} = 0$, $157+u<t<360$, $M_{4,2}^{r}(u,u) = 0$, $M_{4,2}^{r} = M^{\wedge}_{4,2}{}^{r} + \sum M_{4,2}^{r}(t,u)$ $(0<u\leq t)$, $0<t<361$. (13')

Here +... means continuing the respective operation until t-18k<19. And also if we agree on the condition that all the functions considered become to equal 0 for negative values of arguments, then we can simplify analytical forms of many relations.

The equations of the model for the second level are similar to the equations (12), (13), and (13'), but their writing is essentially more tedious.

Let

$y_i = y_i(t) = \lg M_i$, $z = z(t) = \lg [M_1 + M_2 + M_3 + M_4 - \sum M_i^j (i=1,2,3,4; j=p,m) + M_4^p]$, $V_6(t) = V_6 = V_8(t) = V_8 = V = [1-F_5(y_5)][1-F_6(y_{10})][1-k_{13,12,8,7,3,2}^{m,p}]$, $V_7(t) = V_7 = [1-F_5(y_5)][1-F_6(y_{10})][1-k_{13,8,3}^{m,p}]$, $V\hat{}_6(t) = V\hat{}_6 = V\hat{}_8(t) = V\hat{}_8 = V\hat{} = [1-F_5(y_5)][1-F_6(y_{10})][1-k_{13,12,3,2}^{m,p}]$, $V\hat{}_7(t) = V\hat{}_7 = [1-F_5(y_5)][1-F_6(y_{10})][1-k_{13,3}^{m,p}]$, $X(t) = X = k_1^m + k_1^p(1-k_1^m) + k_6^m(1-k_1^{m,p}) + k_6^p(1-k_6^m)(1-k_1^{m,p}) + k_{11}^m(1-k_{6,1}^{m,p}) + k_{11}^p(1-k_{11}^m)(1-k_{6,1}^{m,p})$. (14)

Then on the strength of the assumptions 2), 3), 5), 8)-13), 15)-20), 23), 27), 29)-31)

$M_i^s(1) = 0$, $M_i^s(t+1) = M_i^s V_i + dM_i^s - dM_i^s(t-12)[1-F_1(y_5(t-11))]V_i(t-11)\ldots[1-F_1(y_5(t-1))]V_i(t-1)$, $dM_i^s = dM_{i,Ig} + dM_{i,cells}$, $dM_{i,Ig} = M_{i0}\{\prod[1-F_1(y_5(u))]F_2(y_5(u))V_i(u)\ (0<u<t+1) - \prod[1-F_1(y_5(u))]F_2(y_5(u))V_i(u)V_i\ (0<u<t)\}$, $dM_{i,cells} = M_{i0}\{\prod[1-F_1(y_5(u))][1-F_2(y_5(u))]V_i(u)(0<u<t+1) - \prod[1-F_1(y_5(u))][1-F_2(y_5(u))V_i(u)V_i](0<u<t)\}F_7(z)$, $0<t<360$; $M_i^p(t,u) = 0$, $0< t-u<12$, $M_i^p(t+1,u) = N_i^p(t,u)[1-F_8(t-u-12)]$, $N_i^p(t,u) = N_i^p(t-1,u)e^{0.08}V_i(1-c_i^p)$, $M_i^p(u+12,u) = N_i^p(u+11,u) = dM_i^s(u)V_i(u+1)\ldots V_i(u+11)$, $c_6^p = 1/3k\hat{}_6^p/(1-k_6^m)\{1/(1-k_1^{m,p})[U\sum M_i^j (i=4,2; j=s, s,s) + V\sum M_i^j (i=9,2; j=s, s,s)] + V_{+5}\sum M_i^j (i=14,2; j=s, s,s)\}$, $c_7^p = 1/3k\hat{}_7^p\{1/[(1-k_{2,3}^{m,p})(1-k_7^p)][U\sum M_i^j (i=1,3; i=4, j\neq m) + V\sum M_i^j (i=6,8, j\neq r; i=9 j\neq r,m)] + 1/(1-k_7^p)V_{+5}\sum M_i^j (i=11,13; i=14\ j\neq m)\}$, $c_8^p = 1/3k\hat{}_8^p(1/[(1-k_{2,3}^{m,p})(1-k_{7,8}^p)][U\sum M_i^j (i=1,2,3; i=4, j\neq m) + V\sum M_i^j (i=6,7,8, j\neq r; i=9, j\neq r,m)] + 1/(1-k_{7,8}^p)V_{+5}\sum M_i^j (i=11,12,13; i=14, j\neq m))\}$, $11+u<t<360$, $M_i^p =$

$\sum M_i^p(t,u)$ $(u<t)$; $M_i^{s,s}(t,u) = 0$, $0< t\text{-}u<97$, $M_i^{s,s}(t+1,u) = M_i^{s,s}(t,u)V_i+dM_i^{s,s}- dM_i^{s,s}(t\text{-}6)[1\text{-}F_1(y_5(t\text{-}5))]V_i(t\text{-}5)...[1\text{-}F_1(y_5(t\text{-}1))]V_i(t\text{-}1)$, $dM_i^{s,s} = N_i^p(t,u)e^{0.08}dF_8(t\text{-}u\text{-}12)[1\text{-}F_1(y_5)]\{F_2(y_5)+[1\text{-}F_2(y_5)]F_7(z)\}V_i$, $95+u<t<360$, $M_i^{s,s} = \sum M_i^{s,s}(t,u)$ $(u< t)$, $0<t<361$; $M_i^m(t,u) = 0$, $0< t\text{-}u<103$, $M_i^m(t+1,u) = (M_i^m(t,u)+dM_i^{s,s}(t\text{-}6)[1\text{-}F_1(y_5(t\text{-}5))]V_i(t\text{-}5)...[1\text{-}F_1(y_5(t\text{-}1))]V_i(t\text{-}1) \text{-}M_i^m(t,u)c_i^m)V_ie^{-0.01}$, $c_6^m = k\hat{}_6{}^m/(1\text{-}k_6{}^{m,p})\{1/(1\text{-}k_1{}^{m,p})[U\sum M_i^j$ $(i=4,2; j=s, s,s)+V\sum M_i^j$ $(i=9,2; j=s, s,s)]+V_{+5}\sum M_i^j$ $(i=14,2; j=s, s,s)\}$, $c_7^m = 1/8M_7{}^m(t,u)k\hat{}_7{}^m\{1/[(1\text{-}k_{2,3}{}^{m,p})(1\text{-}k_7{}^p][U\sum M_i^j$ $(i=1,3;$ $i= 4, j\neq m)+V\sum M_i^j(i=6,8, j\neq r;$ $i=9, j\neq r,m)]+1/(1\text{-}k_7{}^{m,p})V_{+5}\sum M_i^j(i=11,13;$ $i=14, j\neq m)\}$, $c_8^m = 1/8M_8{}^m(t,u)k\hat{}_8{}^m\{1/[(1\text{-}k_{2,3}{}^{m,p})(1\text{-}k_{7,8}{}^{m,p}][U\sum M_i^j(i=1,2,3;$ $i= 4, j\neq m)+V\sum M_i^j(i=6,7,8, j\neq r;$ $i=9, j\neq r,m)]+1/(1\text{-}k_{7,8}{}^{m,p})V_{+5}\sum M_i^j(i=11,12,13;$ $i=14, j\neq m)\}$, $101+u< t<360$, $M_i^m = \sum M_i^m(t,u)$ $(u< t)$, $0<t<361$; $M\hat{}_i{}^r = M_{i,0}\prod[1\text{-}F_1(y_5(u))][1\text{-}F_2(y_5(u)][1\text{-}F_7(z)]V\hat{}_i(u)$ $(0<u<t)$, $0<t<361$, $\hat{}M_i{}^r(t+1,u) = \hat{}M_i{}^r(t,u)V\hat{}_i+dM_i{}^r$, $\hat{}M_i{}^r(u,u) = 0$, $dM_i^r = N_i^p(t,u)e^{0.08}dF_8(t\text{-}u\text{-}12)[1\text{-}F_1(y_5)][1\text{-}F_2(y_5)][1\text{-}F_7(z)]V\hat{}_i$, $0<t<360$, $\hat{}M_i{}^r = \sum \hat{}M_i{}^r(t,u)$ $(u< t)$, $0<t<361$,

$$M_i^r = M\hat{}_i{}^r+\hat{}M_i{}^r,\ 0< t<361,\ i = 6, 7, 8. \tag{15}$$

From the assumptions 2), 3), 6), 8), 10)-20), 23), 27), 29)-31), it follows that

$M_{9,1}{}^s(1) = 0$, $M_{9,1}{}^s(t+1) = M_{9,1}{}^sV+ dM_{9,1}{}^s\text{-} dM_{9,1}{}^s(t\text{-}17)V(t\text{-}16)...V(t\text{-}1)$,

$dM_{9,1}{}^s = dM_{9,1,Ig}+ dM_{9,1,cells}$, $dM_{9,1,Ig} = M_{9,10}\{\prod F_3(y_5(u))V(u)$ $(0<u<t+1)\text{-}\prod F_3(y_5(u))V(u)V$ $(0<u<t)\}$, $dM_{9,1,cells}=M_{9,10}\{\prod[1\text{-}F_3(y_5(u))]V(u)$ $(0<u<t+1)\text{-}\prod[1\text{-}F_3(y_5(u))]V(u)V$ $(0<u<t)\}F_7(z)$, $0<t<360$;

$M_{9,1}{}^p(t,u) = 0$, $0< t\text{-}u<18$, $M_{9,1}{}^p(t+1,u) = N_{9,1}{}^p(t,u)[1\text{-}F_8(t\text{-}u\text{-}18)]$, $N_{9,1}{}^p(t,u) = N_{9,1}{}^p(t\text{-}1,u)e^{0.08}V$, $M_{9,1}{}^p(u+18,u) = N_{9,1}{}^p(u+17,u) = dM_{9,1}{}^s(u)V(u+1)...V(u+17)$,

$17+u<t<360$, $M_{9,1}{}^{p} = \sum M_{9,1}{}^{,p}(t,u)\ (u<t)$, $0<t<361$; $M_{9,1}{}^{s,s}(t,u)=0$,

$0<t-u<103$, $M_{9,1}{}^{s,s}(t+1,u)=M_{9,1}{}^{s,s}(t,u)V+dM_{9,1}{}^{s,s}-M_{9,1}{}^{s,s}(t-6)V(t-5)\ldots V(t-1)$,

$dM_{9,1}{}^{s,s} = N_{9,1}{}^{p}(t,u)e^{0.08}dF_8(t-u-18)\{F_3(y_5)+[1-F_3(y_5)]F_7(z)\}V$, $101+u<t<360$,

$M_{9,1}{}^{s,s} = \sum M_{9,1}{}^{,s,s}(t,u)\ (u<t)$, $0<t<361$;

$M_{9,1}{}^{m}(t,u)=0$, $0<t-u<109$, $M_{9,1}{}^{m}(t+1,u) = [M_{9,1}{}^{m}(t,u)+dM_{9,1}{}^{s,s}(t-6)V(t-5)\ldots V(t-1)]e^{-0.01}$,

$107+u<t<360$, $M_{9,1}{}^{m} = \sum M_{9,1}{}^{,m}(t,u)\ (u<t)$, $0<t<361$;

$M\hat{}_{9,1}{}^{r} = M_{9,10}\prod[1-F_3(y_5(u))][1-F_7(z)]V\hat{}(u)\ (0<u<t)$, $0<t<361$,

$\hat{}M_{9,1}{}^{r}(t,u) = 0$, $0<t-u<97$, $\hat{}M_{9,1}{}^{r}(t+1,u) = \hat{}M_{9,1}{}^{,r}(t,u)+dM_{9,1}{}^{r}$,

$dM_{9,1}{}^{r} = N_{9,1}{}^{p}(t,u)e^{0.08}dF_8(t-u-18)[1-F_3(y_5)][1-F_7(z)]V\hat{}$, $95+u<t<360$,

$\hat{}M_{9,1}{}^{,r} = \sum \hat{}M_{9,1}{}^{,r}(t,u)\ (u<t)$, $0<t<361$, $M_{9,1}{}^{r} = M\hat{}_{9,1}{}^{r}+\hat{}M_{9,1}{}^{r}$, $0<t<361$.

$M_{9,2}{}^{s}(0)=0$, $M_{9,2}{}^{s}(t+1) = M_{9,2}{}^{s}V(1-X)+dM_{9,2}{}^{s} - g(dM_{9,2}{}^{s})\{1-[F_4(y_5)+[1-F_4(y_5)]F_7(z)]V(1-X)\}$, $g(dM_{9,2}{}^{s}) = \hat{}(dM_{9,2}{}^{s})+\hat{}(\hat{}(dM_{9,2}{}^{s})[F_4(y_5)+[1-F_4(y_5)]F_7(z)](1-X))+\hat{}(\hat{}(\hat{}(dM_{9,2}{}^{s})[F_4(y_5)+[1-F_4(y_5)]F_7(z)](1-X))[F_4(y_5)+[1-F_4(y_5)]F_7(z)](1-X))+\ldots\}$,

$\hat{}(f) = f(t-18)V(t-17)[1-X(t-17)]\ldots V(t-1)[1-X(t-1)]$, $dM_{9,2}{}^{s} = (dM_{9,2,Ig}+dM_{9,2,cells})(1-X)$,

$dM_{9,2,Ig} = M_{9,20}\{\prod F_4(y_5(u))V(u)\ (0<u<t+1) - \prod F_4(y_5(u))V(u)V\ (0<u<t)\}$, $dM_{9,2,cells} = M_{9,20}\{\prod[1-F_4(y_5(u))]V(u)\ (0<u<t+1) - \prod[1-F_4(y_5(u))]V(u)V\ (0<u<t)\}F_7(z)$, $0<t<360$;

$M_{9,2}{}^{s'}(1)=0$, $M_{9,2}{}^{s'}(t+1) = M_{9,2}{}^{s'}V+dM_{9,2}{}^{s'} - M_{9,2}{}^{s'}(t-11)V(t-10)\ldots V(t-1)$,

$dM_{9,2}{}^{s'} = (dM_{9,2,Ig}+dM_{9,2,cells})X+g(dM_{9,2}{}^{s})[F_4(y_5)+[1-F_4(y_5)]F_7(z)]VX$, $0<t<360$;

$M_{9,2}{}^{p}(t,u)=0$, $0<t-u<12$, $M_{9,2}{}^{p}(t+1,u) = N_{9,2}{}^{p}(t,u)[1-F_8(t-u-12)]$, $N_{9,2}{}^{p}(t,u) = N_{9,2}{}^{p}(t-1,u)e^{0.08}V$, $M_{9,2}{}^{p}(u+12,u) = N_{9,2}{}^{p}(u+11,u) = dM_6{}^{s'}(u)V(u+1)\ldots V(u+11)$,

$11+u<t<360$, $M_{9,2}{}^{p} = \sum M_{9,2}{}^{,p}(t,u)\ (u<t)$, $0<t<361$;

$M_{9,2}{}^{s,s}(t,u)=0$, $0<t-u<97$, $M_{9,2}{}^{s,s}(t+1,u) = M_{9,2}{}^{s,s}(t,u)V(1-X)+dM_{9,2}{}^{s,s}-g(dM_{9,2}{}^{s,s})\{1-[F_4(y_5)+[1-F_4(y_5)]F_7(z)]V(1-X)\}$, $dM_{9,2}{}^{s,s} = N_{9,2}{}^{P}(t,u)e^{0.08}dF_8(t-u-12)\{F_4(y_5)+[1-F_4(y_5)]F_7(z)\}V(1-X)$, $95+u<t<360$, $M_{9,2}{}^{s,s} =\sum M_{9,2}{}^{s,s}(t,u)$ $(u<t)$, $0<t<361$;

$M_{9,2}{}^{s,s'}(t,u)=0$, $0<t-u<97$, $M_{9,2}{}^{s,s'}(t+1,u) = M_{9,2}{}^{s,s'}(t,u)V+dM_{9,2}{}^{s,s'}-dM_{9,2}{}^{s,s'}(t-6)V(t-5)\dots V(t-1)$, $dM_{9,2}{}^{s,s'} = \{N_{9,2}{}^{P}(t,u)e^{0.08}dF_8(t-u-12)+g(dM_{9,2}{}^{s,s})\}[F_4(y_5)+[1-F_4(y_5)]F_7(z)]VX$, $95+u<t<360$, $M_{9,2}{}^{s,s'} =\sum M_{9,2}{}^{s,s'}(t,u)$ $(u<t)$, $0<t<361$; $M_{9,2}{}^{m}(t,u)=0$, $0<t-u<103$, $M_{9,2}{}^{m}(t+1,u) = [M_{9,2}{}^{m}(t,u)+dM_{9,2}{}^{s,s'}(t-6)V(t-5)\dots V(t-1)]e^{-0.01}$, $101+u<t<360$, $M_{9,2}{}^{m} =\sum M_{9,2}{}^{m}(t,u)$ $(u<t)$, $0<t<361$; $M\hat{}_{9,2}{}^{r}(t+1) = \{M_{9,20}\prod[1-F_4(y_5(u)][1-F_7(z(u))]V\hat{}(u)$ $(0<u<t)+g(dM_{9,2}{}^{s})\}[1-F_4(y_5][1-F_7(z)]$,

$0<t<360$, $\hat{}M_{9,2}{}^{r}(t,u) = 0$, $0<t-u<97$, $\hat{}M_{9,2}{}^{r}(t+1,u) = \hat{}M_{9,2}{}^{r}(t,u)+dM_{9,2}{}^{r}$, $dM_{9,2}{}^{r} = \{N_{9,2}{}^{P}(t,u)e^{0.08}dF_8(t-u-12)+g(dM_{9,2}{}^{s,s})\}[1-F_4(y_5)][1-F_7(z)]V\hat{}$, $95+u<t<360$, $\hat{}M_{9,2}{}^{r} =\sum \hat{}M_{9,2}{}^{r}(t,u)$ $(u<t)$, $0<t<361$, $M_{9,2}{}^{r} = M\hat{}_{9,2}{}^{r}+\hat{}M_{9,2}{}^{r}$, $0<t<361$. (15')

The explanations for +... are the same as for the model of the first level. The sign V_{+5} means the same as V with all the subscripts increased by 5.

The equations of the model for the third level can be obtained by means of addition of 5 to all the subscripts of the previous level. The following levels can be obtained in the same way.

Note that the model of the first level can be formally obtained by means of subtraction of 5 from the subscripts of the second level and giving 0 to all the values with the negative subscripts. It may be useful for an arbitrary rule of Ag insertion.

Note more that in the case of Ig increasing many relations in (15) can be simplified. For example, in this case we have

$$dM_{9,1,Ig} = M_{9,10}\{\prod F_3(y_5(u))V(u)\ (0<u<t+1)-\prod F_3(y_5(u))V(u)\ (0<u<t)V\} =$$

$$M_{9,10}[F_3(y_5(t)) - F_3(y_5(t-1))]\prod V(u)\ (0<u<t+1). \quad (16)$$

If Ig increase until $t < t\hat{}+1$ and decrease for $t > t\hat{}$, then we have (16) until $t < t\hat{}+1$ and

$$dM_{9,1,1g} = M_{9,10}[F_3(y_5(t^\wedge)) - F_3(y_5(t^\wedge-1))]\prod V(u)\,(0<u<t+1),\ t>t^\wedge. \quad (17)$$

Note at last that we have not considered the models corresponding to the assumptions 24)-26). These models can be constructed in a similar way.

4. BASE CONTINUOUS EQUATIONS

The models (12)-(13') and (14)-(17) are difference-type (the explicit scheme). It is not difficult to introduce the respective continuous analogy of these models following the ordinary procedure of passage to the limit. Indeed, let

$$u = \sum k_i^j\ (i=2,3,7,8;\ j=m,p),\ u^\wedge = \sum k_{8,7}^{m,p}\ (i=7,8;\ j=m,p),$$

$$f(t,s) = \lim \prod[1 - F_6(y_5(k\Delta t)]\ (k\Delta t=s,\dots,t),\ \Delta t \to 0. \quad (18)$$

Then instead of, for example, the $M_1^{\,p}$ in (13) we have

$$M_1^{\,p} = M_1^{\,p}(12)+\int_{12}^{t} m_1^{\,p}(s)ds = M_1^{\,p}(12)f(t,12)\exp\left(\int_{12}^{t}[0.08-u(s)-c_1^{\,p}(s)]ds\right)[1-$$

$$F_8(t-12)],\ M_1^{\,p}(12) = M_{10}F_2(y_0(0)),\ t>12. \quad (19)$$

It follows under $\Delta t \to 0$ from the relation

$$M_1^{\,p}(s\Delta t)=M_1^{\,p}(12)\{\prod[1- F_6(y_5(k\Delta t)]\ (k\Delta t=12,12+\Delta t\dots,s\Delta t)\}\exp(0.08(s\Delta t-12)+$$

$$\sum\{\ln U_\Delta(k\Delta t)+\ln[1-c_1^{\,p}(k\Delta t)\Delta t]\}\ (k\Delta t=12,12+\Delta t,\dots,s\Delta t))[1- F_8(s\Delta t-12)],$$

$$U_\Delta = [1-k_{8,7,3,2}^{\,m,p}\Delta t],\quad s\Delta t = t. \quad (20)$$

In a similar way, we have

$$M_1^{\,m}(t+\Delta t) = \{M_1^{\,m}+dM_1^{\,s,s}(t-6)\{\prod[1- F_6(y_5(k\Delta t)]\ (k\Delta t=t-6-\Delta t,\dots,t)\}\exp[\ln U_\Delta(t-6-$$

$$\Delta t)+\ln U_\Delta(t-6)+\dots+\ln U_\Delta(t-\Delta t)] - M_1^{\,m}c_1^{\,m}\Delta t\}U_\Delta e^{-0.01\Delta t},$$

$$dM_1^{\,s,s} = N_1^{\,p}dF_8(t-12)F_2(y_0)U,\ dF_8(t-12) = F_8(t-12)-F_8(t-\Delta t-12);$$

$$M_1^{\,r}(t+\Delta t) = \{M_1^{\,r}+N_1^{\,p}dF_8(t-12)[1-F_2(y_0)][1-F_6(y_5)]\Delta t\}U^\wedge_\Delta,\ U^\wedge_\Delta = [1-k_{8,7}^{\,m,p}\Delta t], \quad (21)$$

from where

$m_1{}^m(t) = -(u+0.01+c_1{}^m)M_1{}^m+N_1{}^p(t\text{-}6)F'_8(t\text{-}18)F_2(y_0(t\text{-}6))f(t,t\text{-}6)\exp[-\int{}^t{}_{t\text{-}6}u(s)ds]$,

$M_1{}^m(18) = 0,\ t\geq 18;\ m_1{}^r(t) = -u\hat{}M_1{}^r+N_1{}^pF'_8(t\text{-}12)[1\text{-}F_2(y_0)],\ t\geq 12,$

$$M_1{}^r(12) = M_{10}[1\text{-}F_2(y_0(0))]. \qquad (22)$$

And instead of the relations for $M_i{}^j$, $i=4,1,\ 4,2;\ j=p,m,r$, in (13') we have

$M_{4,1}{}^p = M_{4,1}{}^p(18)+\int{}^t{}_{18}m_{4,1}{}^p(s)ds = M_{4,1}{}^p(18)f(t,18)\exp(\int{}^t{}_{18}[0.08 - u(s)]ds)[1\text{-}$

$F_8(t\text{-}18)],\ M_{4,1}{}^p(18) = M_{4,10}F_3(y_0(0)),\ t>18;\ m_{4,1}{}^m(t) = -0.01M_{4,1}{}^m+N_{4,1}{}^p(t\text{-}6)F'_8(t\text{-}$

$24)f(t,18)F_3(y_0(t\text{-}6)\exp[-\int{}^t{}_{t\text{-}6}u(s)ds],\ M_{4,1}{}^m(24) = 0,\ t \geq 24;$

$m_{4,1}{}^r(t) = -u\hat{}M_{4,1}{}^r+N_{4,1}{}^pF'_8(t\text{-}18)[1\text{-}F_3(y_0)],\ M_{4,1}{}^r(18) = M_{4,10}[1\text{-}F_3(y_0(0))],\ t\geq 18.$

$M_{4,2}{}^p(r+12,r) = M_{4,2}{}^{s'}(r)f(r+12,r)\exp[-\int{}^{r+12}{}_r u(s)ds],\ r>0,$

$M_{4,2}{}^p(t,r) = M_{4,2}{}^p(r+12,r)f(t,r+12)\exp(\int{}^t{}_{r+12}[0.08 - u(s)]ds)[1\text{-}F_8(t\text{-}u\text{-}12)],\ t>r+12,$

$m_{4,2}{}^{s'}(t) = -u[1\text{-}F_6(y_5)]M_{4,2}{}^{s'}+M_{4,2}{}^sUW\text{-}M_{4,2}{}^{s'}(t\text{-}12)f(t,t\text{-}12)\exp[-\int{}^t{}_{t\text{-}12}u(s)ds],$

$m_{4,2}{}^s(t) = M_{4,20}f(t,0)\exp\{-\int{}^t{}_0[u(s)+W(s)]ds\}F_4(y_0(18k)),\ 18k<t\leq 18(k+1),\ k=0,1,...,$

$M_{4,2}{}^p(t) = \int{}^{t\text{-}12}{}_0 M_{4,2}{}^p(t,r)dr,\ t>12;$

$m_{4,2}{}^m(t,r) = -0.01M_{4,2}{}^m(t,r)+dM_{4,2}{}^{s,s'}(t\text{-}6)f(t,t\text{-}6)\exp[-\int{}^t{}_{t\text{-}6}u(s)ds],\ dM_{4,2}{}^{s,s'} =$

$[N_{4,2}{}^p(t,r)F'_8(t\text{-}r\text{-}12)+dM]F_4(y_0)W,\ t>r+12,\ dM=\{\hat{}(dM_{4,2}{}^{s,s})+\hat{}(\hat{}(dM_{4,2}{}^{s,s})F_4(y_0)U(1\text{-}$

$W))+\hat{}(\hat{}(\hat{}(dM_{4,2}{}^{s,s})F_4(y_0)U(1\text{-}W))F_4(y_0)U(1\text{-}W))+...\},\ \hat{}(h) =h(t\text{-}18)f(t,t\text{-}18)\exp(-\int{}^t{}_{t\text{-}18}$

$[u(s)+W(s)]ds]),\ m_{4,2}{}^m(t) = \int{}^t{}_0 m_{4,2}{}^m(t,r)dr,\ t>0;m_{4,2}{}^r(t,\tau) = -u\hat{}[1\text{-}F_6(y_5)]$

$M_{4,2}{}^r(t,\tau)+dM_{4,2}{}^r,\ dM_{4,2}{}^r = [N_{4,2}{}^p(t,\tau)F'_8(t\text{-}\tau\text{-}12)+dM][1\text{-}F_4(y_0)],\ M\hat{}_{4,2}{}^r(t) = M_{4,20}$

$f(t,0)\exp\{[-\int{}^t{}_0[W(s)+u\hat{}(s)]ds\}[1\text{-}F_4(y_0(18k))],\ 18k<t\leq 18(k+1),\ k=0,1,...,$

$$m_{4,2}{}^r(t) = [M\hat{}_{4,2}{}^r(t)]'+\int{}^t{}_0 m_{4,2}{}^r(t,\tau)d\tau,\ t>0. \qquad (23)$$

In the case of the relations for the second level of IN, let us put

$$v = \sum k_i^j\ (i=2,3,7,8,12,13;\ j=m,p),\ v^\wedge = \sum k_i^j\ (i=2,3,12,13;\ j=m,p),$$

$$f_1(t,s) = \lim \prod\{[1-F_1(y_5(k\Delta t))][1-F_5(y_5(k\Delta t)][1-F_6(y_{10}(k\Delta t)]\}\ (k\Delta t=s,...,t),\ \Delta t \to 0,$$

$$f_{1,2}(t,s) = \lim \prod\{[1-F_1(y_5(k\Delta t))][1-F_2(y_5(k\Delta t))][1-F_5(y_5(k\Delta t)][1-F_6(y_{10}(k\Delta t)]\}$$

$$(k\Delta t=s,...,t),\ f^\wedge_{1,2}(t,s) = \lim \prod\{[1-F_1(y_5(k\Delta t))]F_2(y_5(k\Delta t))[1-F_5(y_5(k\Delta t)][1-$$

$$F_6(y_{10}(k\Delta t)]\},(k\Delta t=s,...,t), \tag{24}$$

where F_i is one of the functions in (2) (see also (16) and (17)). Then instead of, for example, M_6^j, $j=p,m,r$, in (15), we have

$$M_6^P(t,u) = M_6^P(u+12,u)f_1(t,u+12)\exp(\int_{u+12}^{t} [0.08 - v(s) - c_6^P]ds)[1-F_8(t-u-12)],$$

$$M_6^P(u+12,u) = dM_6^s(u)f_1(u+12,u)\exp[-\int_u^{u+12} v(s)]ds],\ u>0,$$

$$dM_6^s(u)= M_{60}\ \{f^\wedge_{1,2}(u,0)[1-F_1(y_5(u))]F_2(y_5(u))+f_{1,2}(u,0)([1-F_1(y_5(u))][1-F_2(y_5(u))]-$$

$$1)F_7(z(u))\}\exp[-\int_0^u v(s)ds],\ M_6^P(t) = \int_0^{t-12} M_6^P(t,u)du,\ t>12;$$

$$m_6^m(t,u) = -(v+0.01+c_6^m)M_6^m(t,u)+dM_6^{s,s}(t-6,u)f_1(t,t-6)\exp[-\int_{t-6}^{t} v(s)ds],$$

$$dM_6^{s,s}(t,u)= N_6^P(t,u)F'_8(t-u-18)\{F_2(y_5)+[1-F_2(y_5)]F_7(z)\},$$

$$M_6^m(u,u) = 0,\ t>u>0,\ m_6^m(t) = \int_0^t m_6^m(t,u)du,\ t>0;$$

$$M^\wedge{}_6^r = M_{6,0}f_{1,2}(t,0)\exp(-\int_0^t [F_7(z(u))+v^\wedge(u))du),\ t>0,\ m_6^r(t,u) = -v^\wedge M_6^r(t,u)+dM_6^r,$$

$$M_6^r(u,u) = 0,\ dM_6^r = N_6^P(t,u)F'_8(t-u-12)[1-F_2(y_5)][1-F_7(z)],\ t>u>0,$$

$$M_6^r = M^\wedge{}_6^r+\int_0^t M_6^r(t,u)du,\ t > 0. \tag{25}$$

5. ON THE CLASSICAL AND THE NEW APPROACH

Let us show that the proposed models belong to the class of MM in the general theory of ES (see, for example, (1, Ch.2) and (19, Ch.11)). Indeed, the base equations of the class are the equation of restoration and guarantee of external functions,

$$m_i(t) = \int_0^t \alpha_i(t,s)\lambda_i(t,s)y_i(s)m_i(s)ds,\ P_i(t) = \int_0^t \lambda_i(t,s)m_i(s)ds,\ G_i(t) = \int_0^t [1-$$

$$\lambda_i(t,s)]m_i(s)ds,\ \lambda_i(t,s) = \mu_i(t,s),\ c_i(t) = \int_0^t \beta_i(t,s)\mu_i(t,s)[1-y_i(s)]m_i(s)ds,$$

$$f_i(t) = m_i(t)+c_i(t), \tag{26}$$

and the equation of realization of the external functions,

$$c_i^j(t) = \int^t_0 \beta_i^j(t,s)\mu_i^j(t,s)z_i^j(s)m_i(s)ds,$$

$$j=1,\dots,n_i,\ 0 \le y_i,\ z_i^j \le 1,\ y_i + \sum z_i^j\ (j=1,\dots,n_i) = 1,\ t \ge 0,\ i=1,\dots,n. \tag{27}$$

In our case for every level of IN we have 5 equations of (26)-type relative to m_i^P, $i=5n+1,\ 5n+2,\ 5n+3,\ 5n+4,1,\ 5n+4,2$, and 16+6+1 = 23 equations of (27)-type relative to m_i^j, $j=s,\ s',\ s,s,\ s,s',\ m,\ r$, and m_{n+5} (the number of Ig per unit of the time).

Comparing (26) and (19), we have

$P_1 = \lambda_1(t)[M_1^P(12)+\int^t_{12} m_1(s)ds] = \lambda_1(t)M_1 = M_1^P = M_1^P(12)\exp\{\int^t_{12}[0.08-u(s)-c_1^P(s)]ds\}[1-F_8(t-12)]$, $M_1 = P_1+G_1$, $G_1 = [1-\lambda_1(t)]\int^t_0 m_1(s)ds = \int^t_{12} M_1^P(s)[u(s)+c_1^P(s)]ds$; $m_1(t) = M_1^P/M_1 \int^t_{12}\alpha_1(t,s)y_1(s)m_1(s)ds = (P_1+G_1)' = M_1^P[0.08-d(t)]$, $d(t) = d = F'_8(t-12)/[1-F_8(t-12)]$, $m_1(t) = 0,\ t \ge t^*,\ d(t^*) = 0.08,\ t^* \approx 126,\ \alpha_1(t,s) = [1+25\delta(s-12)]\alpha(t,s)$; $c_1(t) = M_1^P/M_1\int^t_{12}\beta_1(t,s)[1-y_1(s)]m_1(s)ds = M_1^P d,\ c_1(t) = 0,\ t<96,\ t>156,\ \beta_1(t,s) = [1+2M_1(96)d(96)/\{[1-y_1(96)]m_1(96)\delta(s-96)]\}\beta(t,s)$;

$$f_1(t) = m_1(t)+c_1(t) = M_1^P 0.08,\quad f_1(t) = 0,\ t>156, \tag{28}$$

where δ is as usually Dirac δ-function. It is naturally supposed (see Ch. 2, s. 6.2.) that

$$\alpha(t,s) = 0.08,\quad \beta(t,s) = z(t)/z(s), \tag{29}$$

Then

$y_1(t) = 1,\ M_1(12)\alpha(t,12)+\int^t_{12}\alpha(t,s)m_1(s)ds = M_1\,0.08,\ 12<s<t<96,\ y_1(t) = 0,$

$[M_1(12)+\int^{96}_{12} m_1(s)ds]x(t) = M_1\,[0.08-d(t)],\ x(t) = M_1/M_1(96)[0.08-d(t)],\ 96 \le t <t^*$; $c_1(t) = 0,\ 12<t<96,\ M_1(96)d(96)+\int^t_{96} m_1(s)/z(s)ds = m_1(t)/z(t)M_1(t)/m_1(t)d(t) =$

$M_1(t)/z(t)d(t) = M_1(96)d(96)exp\{\int^t_{96} m_1(s)/[M_1(s)d(s)]ds\},$

$$z(t) = M_I(t)d(t)/[M_I(96)d(96)]\exp\{-\int^t_{96} m_I(s)/[M_I(s)d(s)]ds\},\ 96 \le t \le 156. \quad (30)$$

Note that in (28)-(30) $m_I(t)$ is the increment of new bio-mass of the cells arisen from the proliferating cells, $f_I(t)$ is the resource that is necessary for new bio-mass and the external product output, and

$$\int^t_{12} M_I^P(s)u(s)ds,\ \int^t_{12} M_I^P(s)F'_8(s-12)/[1-F_8(s-12)]ds,\ \int^t_{12} M_I^P(s)c_I^P(s)ds \quad (31)$$

are respectively the suppressed cells, stopped proliferating cells, and turned out to be stimulated B_2-cells of the first and second level (with the help of M_I^P cells).

It is not difficult to see that in our case the description by the new approach is likely more complicated than by the classical one, but in return the new approach gives us more information and proves useful for accounting new, likely more profound, assumptions. Probable examples are different assumptions about properties of bio-mass of the cells depending on its age. If immunologists confirm that older target-cells are open to control-cells more than younger ones, then $\lambda_I(t,s)$ can make sense instead of $\lambda_I(t)$, in particular,

$$\lambda_I(t,s) = 0,\ 0<s<a(t),\ \lambda_I(t,s) = 1,\ a(t)<s<t, \quad (32)$$

can make sense. With (32) we would have the typical expression

$$M_I^P = \int^t_{a(t)} m_I(s)ds,\ G_I = \int^{a(t)}_0 m_I(s)ds,\ f_I(t) = m_I(t)+c_I(t),$$

$$m_I(t) = \int^t_{a(t)} \alpha_I(t-s)y_I(s)m_I(s)ds = (M_I^P+G_I)',\ c_I(t) = \int^t_{a(t)} \beta_I(t-s)[1-y_I(s)]m_I(s)ds, \quad (33)$$

which cannot be obtained easily by the classical approach. Note that knowing M_I^P, G_I, and $f_I(t)$ in (33) and finding y from an experiment and/or a certain extreme principle, we can find the others.

Comparing (27) and (22), we have

$$c_I^{\,1}(t) = \int^t_{12} \beta_I^{\,1}(t,s)\mu_I^{\,1}(t,s)m_I(s)z_I^{\,1}(s)ds = m_I^{\,s,s}(t) = c_I - c_I(t-6)\exp[-\int^t_{t-6} u(s)ds],$$

$$c_I^{\,2}(t) = \int^t_{18} \beta_I^{\,2}(t,s)\mu_I^{\,2}(t,s)m_I(s)z_I^{\,2}(s)ds = m_I^{\,m}(t) = -[u+0.01+c_I^{\,m}]M_I^{\,m}+$$

$$c_I(t-6)F_2(y_0(t-6))f(t,t-6)\exp[-\int^t_{t-6} u(s)ds],\ M_I^{\,m}(18) = 0,\ t>18,$$

$$c_I^{\,3}(t) = m_I^{\,r}(t) = \int^t_{12} \beta_I^{\,3}(t,s)\mu_I^{\,3}(t,s)m_I(s)z_I^{\,3}(s)ds = -u\hat{\ }M_I^{\,r}+c_I[1-F_2(y_0)],$$

$$M_I^{\,r}(12) = M_{I0}[1-F_2(y_0(0))],\ t>12,\ z_I^{\,1} = z_I^{\,2} = z_I^{\,3} = 1. \quad (34)$$

Solving the equations in the relations (34) with respect to $m_I^{\,m}(t)$ and $m_I^{\,r}(t)$, we can find the kernels $\beta_l^{\,2}(t,s)\mu_l^{\,2}(t,s)$ and $\beta_l^{\,3}(t,s)\mu_l^{\,3}(t,s)$.

Here and in many other cases, we need to use the well-known formulae of the type

$$x'(t)+g(t)x(t) = f(t),\ x(t) = x(t^\wedge)\exp[-\int_{t^\wedge}^{t} g(v)dv]+\int_{t^\wedge}^{t} f(s)\exp[-\int_{s}^{t} g(v)dv]ds,$$

$$x'(t) = g(t)[-x(t^\wedge)\exp[-\int_{t^\wedge}^{t} g(v)dv]+\int_{t^\wedge}^{t} [f(t)/g(t)-f(s)/g(s)]\exp[-\int_{s}^{t} g(v)dv]g(s)ds. \quad (35)$$

It should be noted that

$$m_l^{\,g} = c_l - m_l^{\,s,s} - m_l^{\,r} - m_l^{\,m} + M_l^{\,P}(u+c_l^{\,P}) \quad (36)$$

is the increment of the general "obsolete" product under output of $m_l^{\,P} + m_l^{\,s,s} + m_l^{\,r} + m_l^{\,m}$, but the part of it, which is $M_l^{\,P} c_l^{\,P} + M_l^{\,m} c_l^{\,m}$, turns out to be stimulated B_2 cells of the first and second level.

6. SOME RESULTS OF INVESTIGATION

It is not difficult to see that the equations (10)-(15) have the unique nonnegative boundary solutions. Boundaries of these solutions follow from boundaries of $M_i^{\,r}(t)$ and $M_i^{\,P}(t)$ which, in turn, follow from the postulates in 16).

The following experimental facts are known (which are the results of multilevel interaction) about the first level cells:

1. The number of antibody generating B cells (AGC) is maximal on fourth-sixth day after the primary or on fourth-fifth day after the secondary immunization.

2. The number of $B_2^{\,P}$ cells is maximal on 4-6th day and the number of plasma cells is maximal on 6-8th day after the primary immunization. After the secondary immunization the total number of AGC is maximal on 4-6th day.

3. The total number of AGC of the given specificity is maximal on the fourth-sixth day.

4. The number of memory cells of the given specificity is less than 10^4.

5. The number of memory B cells must be less than the initial number of stimulated B-cells (so-called exhausting differentiation) over the range that is less or equal to 0.3 of the decimal logarithm of the Ag doses (on the right edge of the dose curve, before the tolerance concentration).

6. The "plateau" of the curvature for the maximal number of AGC depending on the dose of Ag should be no wider than 0.6 of the Ag dose logarithm.

7. Free Ig should be under the condition that the logarithm of number Ag is not less than one logarithm of the stimulating Ag doses.

8. Contents of free Ig at the top of the primary immune response should not be below the initial Ig dose if only in some intervals (no less than 0.6 of logarithm) of the dose range.

9. Contents of free Ig at the top of the secondary immune response should exceed the summary Ag dose by no less than 0.5-1 order, if only over some intervals of the dose range.

10. The total number of the observed levels of IN does not exceed 3-4.

Guess 1. *All the experimental facts* 1.-10. *are contained in the properties of the MM of IN* (*under some certain subintervals of IN parameters falling within the possible intervals*). *In addition, the solutions of the MM equations have, as a rule* (*due to the assumption* 17, *namely* Tef *eliminate themselves*), *fluctuating nature* (*effect of "cannibalism"*) *and only their averaged values correspond to the usually expected ones.*

Guess 2. *The solutions of the models have the structural stability* [3], *which is provided by a dominating idio-type on the each level* (*under the strong asymmetry of the matrix of IN interactions*). *Here, perhaps, the evolutionary principle of economizing of the interactions' number and structure under guarantee of the required functions manifests itself.*

The attempt [6] of qualitative investigation of this kind of models confirms that some of the properties 1.-10. follow from the models. However, the proof of the guesses to a complete extent is still ahead of us. This investigation needs to be sponsored.

The attempt of evaluation of the respective software for the models was done by programmers V.A. Dul'sky and T.A. Dyn'ko in Glushkov Institute of Cybernetics, Kiev, the Ukraine. However, it was also not finished.

Some of the simplest numerical results of modeling on a computer confirming the observations of IS can be taken from [6] (the Russian version).

It should be emphasized here that this is by no means completed work. Besides the above mentioned open problems there are many others, in particular, the problems of the control with the help of changing of the initial data of the type M_{i0}. The latter are possible due to the biotechnology of the production of *hybridomas* that are the fusion of cancerous Lc and antibody-producing cells [5].

Guess 3. *Each possible steady behavior of IN and, in general, of the immune system can be achieved as a rule by a tremendous number of controls. It is one of the manifestations of immunology mobile principle* [12].

7. ON APPLICATIONS

We dwell only on the problem of possible applications of the models to understanding of HIV immune dynamics.

Let us suppose that Ag (Ag-determinants of HIV) suppress most of M_l (T helpers of the first level) and that Ig against of Ag are only $M_{4,2}{}^{m}$ (plasma B_2 cells of the first level) production. Of course, $M_{4,2}{}^{s}$ and, the more so, $M_{4,2}{}^{s,s}$ can not be stimulated by M_1-cells, but in principle they can be stimulated by ample $M_6{}^{m,p}$ (mature and proliferating T-helpers of the second level). In turn, for $M_6{}^{m,p}$ we need the normal M_{60} that can be stimulated by ample $M_{4,1}{}^{p}$ (proliferating B_1 cells) for which, in turn, we need ample $M_{4,10}$.

Thus, the problem is to determine $M_{4,10}$, $M_{4,20}$, and M_{60} as well as the instants of their insertions in order to have immune response to Ag in question.

If Ag are in the number M which is greater than 10^{16}, then we have the case of high-zone tolerance to Ag, that is, no immune response possible. We suppose for definiteness that $F_5(\ln M) = 0.99$. Then for the steady state of IN we have approximately 10--4 rest cells and 10^3 mature cells of each kind except B_2 plasma cells. Regarding the possible increased source of new T cells from the thymus [5, 8] we have, in addition, possibly more than 10^3 proliferating cells of each kind except B_2 cells. $M_{4,1}{}^{p}$ can stimulate rest cells of the second level what can result in more than 10^3 proliferating, mature and rest cells of each kind. About this, maybe $M_6{}^{m,p}$ stimulate $M_{4,2}{}^{s}$ and $M_{4,2}{}^{s,s}$ what result in a certain immune response to Ag and, in turn, can explain rather long period of Ag stability (while the thymus can give the increased source of new T cells).

To summon the effective immune response (independently from the thymus), we are in a need of restoration of the normal initial data ($M_{4,10} = 40$ and $M_{60} = 200$) and the normal stimulation of $M_{4,2}{}^{s}$ and $M_{4,2}{}^{s,s}$ for what we should insert additionally $M_{4,10} = 40/0.01$, and about 5 days later the normal M_{60} (if it is not so), and about more 5 days later $M_{4,20} = 10^6$ in order. Such a large number is used because we are in need of the following $M_{4,2}{}^{m}$:

$$M_{4,2}{}^{m} \approx 10^6 \approx 0.01 M_{4,2}{}^{p},\ M_{4,2}{}^{p} \approx 0.01 M_{4,20}\, 10^4,\ 10^6 \int_0^\infty e^{-0.01s} ds \text{ cells } 10^8 \text{ Ig/cell} = 10^{16} \text{ Ig}.$$

Of course, all these numbers are most likely heuristic consideration. To make the estimates more precise, we need to use the respective MM, software, and computer.

HIV immune dynamics in more details one can see in the next Ch. (13).

8. CONCLUSION

On the whole, we discuss the IN models constructed for the first time in a complete form according to 31 quite detailed postulates or assumptions. It is well known that 7 simple postulates of Euclidean geometry have generated numerous important results both in the theory and practice. Therefore, we believe that above-mentioned 31 immunology postulates will also generate numerous significant results. However, the respective investigation is still ahead of us. It exceeds the limits of this book.

The IN models are also important here due to the main goals of the book, especially to prove the validity of ES models for not only artificial, but also natural systems.

What has one learned by using this application of the general theory in PART I (see also Ch. 11) to IN modeling ?

First of all, we have showed in s. 5. that all the equations of the IN models belong to the class of the models of the general theory. It has been very important to receive explicit expressions for the functions of α, β-types, which are usually not entirely determined things.

Then, one can see that the distribution function y-type takes in IN the value 1 or 0 what corresponds to the optimal strategy of control of the general theory (see Ch. 4). From this it follows that the control of y-type has been likely selected in natural systems. Therefore, it is important to study more profound *factorization* of initial dynamic characteristics of natural systems in comparison with artificial systems. This factorization has been given in Ch. 2, s. 6.2 (see also Ch. 11, s. 9.3).

Besides, we have showed in s. 5. that in contrast to artificial systems the "obsolete products" in IN turn out to be useful for realization of its external functions. It should be one of the important goals of any bionic system.

At the same time, on the strength of the general laws of conservation of energy the functioning of IN cannot be without some waste. A certain possibility to regard this fact and, by the same token, deepen the models of IN follows from the general theory of development modeling, and has also been shown in s. 5..

On the whole, we can believe much more now that MM of ES are very useful for mathematical modeling of a wide spectrum of objects including biological ones.

REFERENCES

1. Bell G.I., Mathematical Model of Clone Selection and Antibody Production, I-III, J. Theor. Biol., 29, 1970, 191-232; 33, 1971, 339-378, 379-398.
2. Fontalin L.N., Pevnitskiy L.A., Immunology Tolerance, M.: Meditsina, 1978, 313 pp. (in Russian).
3. Hoffman G.W., On Network Theory and H-2 Restriction, Contemporary topics in immunology, 11, 1980, 186-226.
4. Hoffman G.W., The Application of Stability Criteria in Evaluating Network Regulation Models, in: Regulation of Immune Response Dynamics, CRC Press, 1, 1982, 137-162.
5. Immunology Methods, Vol. III, Academic Press, 1985, 583 pp.
6. Ivanov V.V. et al., Modeling of Idio-type Anti-idio-typical Interactions of Immune Network with Regard to Distinctive Sub-populations Among Lymphocytes. In: Mathematical modeling, North-Holland, 1983, 141-149 (see also the revised article of the same title in the book: Mathematical modeling, M.: Mir, 1986 (in Russian))
7. Jerne N.K., Toward a Network Theory of the Immune System, Ann. Immunol., Inst. Pasteur, 1256, 1974, 373-389.
8. Mikhalevich V.S, Ivanov V.V. et al., Integro-functional Model of Blood Regulation System, Kibernetika, N 3, 1986, 69-77, 86 (in Russian).
9. Mohler R.R., Bruni C. and Gandolfi A., A Systems Approach to Immunology, IEEE Trans. Automat. Contr., AC-25: 1980, 964-990.
10. Nesterenko V.G., Auto-logic Idio-type Anti-idio-typical Interactions and Immune Response Regulation, Immunologiya, 2, 1982, 5-15 (in Russian).
11. Perelson A.S., Immune Network Theory, Immunol. Rev., 110, 1989, 5-36.
12. Petrov R.V., Stepanenko R.N., Auto-regulative Processes in Immune System and Principle of Immunology Mobile, Kibernetika, N 3, 1986, 78-86 (in Russian).
13. Rowe G., Theoretical Models in Biology. The origin of Life, the Immune System, and the Brain, Oxford Univ. Press, 1994, 420 pp.
14. Vareba F.J. and Continto A., Second-Generation Immune Networks, Immunology Today, 12, N5, 1991, 159-166.

Desperate diseases must have desperate cures

CHAPTER 13

MM OF HIV, HIV POPULATION, AND AIDS

The first part of the chapter is devoted to a formalized description of the immune status of an organism. The central part is mainly devoted to a formalized description of HIV based on some published information on the properties of HIV, and HIV population based on the MM of immune network of an organism. The joint mathematical models of a patient with HIV infection, i.e., AIDS is considered in short. The final part is devoted to the preliminary consideration of the applied problem consisting in selection of immune and other correctors that help the organism to neutralize HIV and to induce the immune status to the norm.

1. INTRODUCTION

Let us assume the following abbreviations of the expressions: immune system of an organism (IS), *immune status of an organism* (ISO), immune network of an organism (IN), immune competent cells (ICC), antigens (Ag), immune-globulin (Ig), the specific rates (SR), human immune deficiency viruses (HIV), acquired immune-deficiency syndrome (AIDS), mathematical models (MM), and evolutionary systems (ES).

The general situation about AIDS has been described, for example, in [1, 16]. The chapter is mainly devoted to a formalized description of HIV, HIV population, and IN of an organism separately and also joint MM of a patient with HIV infection, all in all are based on extension and development of MM obtained earlier (see Part I and Ch. 12) as well as the descriptions of HIV given in [1, 16]. The first part of the chapter is devoted to a formalized description of ISO, which is a crucial point of various viral effects on an organism. In this connection, we dwell on a collection of ISO parameters that are necessary for description of IN (Ch. 12). The final part of the chapter is devoted to the preliminary consideration of the applied problem consisting in selection of immune correctors that help the organism to neutralize HIV and to induce ISO to the norm.

2. ON ISO CONCEPT

One can evaluate ISO as a collection of parameters such that the appropriate MM describes IS adequately, i.e. with the assigned accuracy. ISO includes the parameters for IN (see Ch. 12) and for different organs of IS [6], namely: marrow, thymus, spleen, lymphatic glands, and so on.

Instead of the parameters ISO can include various characteristics (grounded with the help of parameters) important instances of which can be given by the indices of efficiency of functioning of α and β-types (Part I) for some subsystems of IS. In the special cases α and β have the meaning of SR for the process to be described by systems of differential equations.

Keeping in mind the great ISO complexity, it is not difficult to understand the presence of the vast variety of IS' behavior and also the fact that a lot of the same IS' behavior can be realized under various combinations of the parameters.

Here, we dwell in detail on the collection of ISO parameters that are necessary for description of IN. According to Ch. 12, the parameters in question are:

1/. The initial numbers $M_{6j+i,0}$ of ICC and Ig for the level j ($j = 0, 1, 2, 3$): T helpers (Th, $i = 1$), T suppressors (Ts, $i = 2$), T effectors (Tef, $i = 3$), B_1 cells ($i = 4$), B_2 cells ($i = 5$) and Ig ($i = 6$). Taking into account that number has exponent 10^6 for various Ag, different levels can amount to 3-5, the main species of ICC and Ig to 6, and essential gradation of the cells by avidity (*affinity*) to Ag about 20-40, the total number of parameters has the exponent

$$N_1 = 10^6 \times 4 \times 6 \times 30 = 0.72 \times 10^9. \tag{1}$$

2/. SR for the proliferating cells (8--10 hours).

3/. SR for decay of Ig (50--100 hours).

4/. SR for decay of the mature cells (2--4 days).

5/. SR (β_i) for realization of the following control functions: 10^7 Ig per hour for the proliferating B cells, 10^8 Ig per hour for the plasma cells, 2--4 target cells (one cell per 3 hours) for the proliferating T cells and 6--10 target cells for the mature T cells.

6/. Phases duration for some kind of cell activation (6--30 hours).

7/. x_i, averages and σ_i, variances of the truncated normal distributions F_i for the cells' output from proliferation and also for various interactions among the cells of different levels and between the cells and Ig, $i = 1,..., 8$ (see Ch. 12 and s.4. below).

Assuming similarly to Ch. 12 that the cells and Ig have the same parameters at all the levels for any molecular-disperse Ag or Ig, we have

$$N_2 = 1 + 1 + 1 + 4 + 2 + 16 = 25 \tag{2}$$

as the total number of the main parameters of ISO in conformity with IN.

In Ch. 12 MM of IN are constructed regarding the above mentioned parameters and several (31) immunology hypotheses based on the known experimental data.

We select here some of those hypotheses to exhibit that the decline of Th-cells results the immune disruption.

1) The first level of IN consists of the idio-type-positive (Id) lymphocytes and Ig. The second level consists of the anti-idio-type (AId) lymphocytes and Ig. The third level consists of the anti-anti-idio-type (AAId) lymphocytes and Ig, etc.

2) For the higher order level the Ag are the cells and Ig of the previous level.
3) For lymphocytes of every level Ag can induce two states according to its concentration: stimulation by the mean doses and suppression by the large doses. The part of the cells stimulated by the mean doses of Ag (10^{13} - 10^{15} molecules) and suppressed by the large doses (10^{15} -10^{16} molecules) can be determined according to the concentration of Ag using the (truncated) normal law for its decimal logarithm.
4) The following three properties of the lymphocytes are employed: a) the start of proliferation of the lymphocytes is impossible without their stimulation; b) the proliferating lymphocytes characterized by refractoriness with respect to the stimulus; c) the lymphocytes that began to proliferate stopped their proliferation spontaneously in 2-5 days after the stimulation, the number of days has a normal truncated distribution. The lymphocytes that stopped to proliferate can be stimulated again.
5) Among the B cells of every level we distinguish the B_2 cells (90% of all B cells) which are lunching into proliferation and differentiation only by Th lymphocytes.
6) Ag and Id-AId bridges cause the interaction of the cells within each level and between levels. Therefore, IdTh help IdB and AIdB; IdTs suppress IdTh, IdTef, IdB, AIdTh, AIdTef, and AIdB; IdTef eliminate IdTh, IdTs, IdTef, IdB, AIdTh, AIdTs, AIdTef, and AIdB; AIdTh help IdB, AIdB, and AAIdB; AIdTs suppress IdTh, IdTef, IdB, AIdTh, AIdTef, AIdB, AAIdTh, AAIdTef, AAIdB, and so on.
7) One molecule Ig eliminates on the average one molecule of Ag.

Regarding these hypotheses, under the condition of the decline of Th cells, Ag-determinants of HIV can increase up to large doses such that induce the suppression of ICC of a certain spectrum. In general, the decline of Th cells, perhaps, results the tolerance to any Ag from the respective spectrum and a cellular immune response to suppressed ICC. The latter can result different diseases including auto-immune ones.

3. SOME PUBLISHED INFORMATION ON HIV

According to [1, 2, 3, 8, 16]:

The size of viral grains is 85-200nm.

The size of thorns extending from the external surface area of HIV membranes and presenting apparently the main medium of infection is 15 nm in diameter and 9 nm in height.

HIV coincide with retrovirus structure by their main morphological marks.

Three classes of virus-specific RNA are found in infected cells: RNA 9.3 kb, 4.3 kb, and 1.8 kb related to the full genome, gene *env,* and genome region EI (3' orf), correspondingly.

Pro-viral DNA consists of 9213 nucleotides and contains at least 9 genes, namely, *gag, pol, env, vif, tat, rev, nef, vpr,* and *vpu.*

As a whole, the genes coding structural proteins are concentrated to the left part

(5') of genome, and the control genes (*tat, rev, nef,* (*vpr, vpu*)) are adjacent to the 3'-end.

The control mechanism of *nef* and *rev* includes negative feedback and that of the *tat* includes a positive one that can enlarge a level of gene expression approximately 1000 times.

Gene *pol* is associated with synthesis of the viral enzyme, mainly OT enzyme that takes up key position in HIV biology.

OT is magnesium-dependent, has molecular mass of 53-100 kD and is highly immune-genic (serum of 98% of AIDS patients contains Ig for OT).

Strong HIV variation connected, in particular, with that of their genes is the most important factor of HIV pathogenesis.

Th cells of an organism, more exactly the so-called CD4+ T cells, are the main HIV targets.

A protein (GP120) on the surface of the virus has a high affinity for the CD4 protein on the surface of the T cell. Binding takes place, and the contents of the HIV is injected into the host T cell. After the host cell has duplicated the DNA of the virus, it is reassembled and new virus particles bud from the surface of the host cell.

This budding can take place slowly sparing the host cells or rapidly bursting and killing the host cells.

Death rate of infected T cell population is 0.5--1.0 per day.

Death rate of uninfected T cell population is 0.02 per day.

Number of the free virus produced by the busting infected cell is 1000.

Source of new T cells from thymus $s(t)$ is 1000.

For the new hypothesis about auto-immune aspects of HIV infection see [5, 15].

4. MM OF HIV IN VIVO

It is natural to consider HIV in vivo as an example of ES [7]. So (see Part I and Ch. 11), *base*, minimal aggregate *MM of HIV* have the form

$$m(t) = \int_{a(t)}^{t} \alpha(t,s)u(s)y(s)m(s)ds,\ M(t) = \int_{0}^{t} m(s)ds,\ 0 \le a(t) \le t,\ c(t) = \int_{b(t)}^{t} \beta(t,s)v(s)[1-y(s)]m(s)ds,\ C(t) = \int_{t^\wedge}^{t} c(s)ds,\ 0 \le b(t) \le t,\ P(t) = \int_{a(t)}^{t} y(s)m(s)ds + \int_{b(t)}^{t} [1-y(s)]m(s)ds,$$

$$0 \le y(t) \le 1,\ G(t) = M(t)-P(t),\ f^\wedge(t) \ge f(t) = m(t) + c(t) = k(f^{\wedge\wedge})m(t),\ t \ge t^\wedge \ge 0, \qquad (3)$$

where $t^\wedge$ is the starting time of modeling; $m(t)$ is the rate of creation of the new HIV bio-mass (another interpretation is the number of active or reactive centers that can be considered as analogy of work places in economics) at the time instant t that provides internal needs or internal functions of HIV as ES; $y(s)m(s)$ is a share of $m(s)$ for the fulfillment of the internal functions in the subsystem A HIV for recreation of $m(t)$; $\alpha(t,s)$ is the efficiency index of functioning of the subsystem A, i.e., the number of units of $m(t)$ to be created for the unit of the time starting from the instant t per one unit of $y(s)m(s)$; $a(t)$ is the temporal threshold for liquidation of the obsolete products in the subsystem A; $c(t)$ is the rate of creation of the new thorns (proteins GP 120 and GP 41, and maybe some other HIV components) which provide their external functions (in particular, the initial and final interaction between HIV and the target Th cells); $[1-y(s)]m(s)$ is a share of $m(s)$ for the fulfillment of the external functions in the subsystem B HIV for creation of $c(t)$; $\beta(t,s)$ is the efficiency index for the functioning of the subsystem B that is similar to $\alpha(t,s)$; $b(t)$ is the similar temporal threshold to $a(t)$, but for the subsystem B; $P(t)$ is the total HIV bio-mass functioning at the instant t; $G(t)$ is the total out-of-date HIV bio-mass at the instant t; $f^{\wedge}(t)$ is the rate of the inflow of the external substrate accessible to HIV ($m(t)$ and $c(t)$ are measured in the same units as $f^{\wedge}(t)$); $k(f^{\wedge\wedge})$ is the coefficient, $0 < k(f^{\wedge\wedge}) \leq 1$, characterizing a share m in f and depending on qualitative composition of $f^{\wedge}$, named $f^{\wedge\wedge}$; $[0,t^{\wedge}]$ is the prehistory of HIV for which all the functions are considered as given.

Thus, the MM (1) is a nonlinear system (with the prehistory) which consists from 8 equalities and 9 inequalities connecting 14 different values. The relations (3) are faithful by definition (and here α and β can depend on all the elements of the model) provided that out-of-date bio-mass in bio-systems A and B is never used, but the new bio-mass created after $a(t)$ and $b(t)$ is used entirely. Assuming the exogenous-type factors α, u, β, v, f, $f^{\wedge}$, $k(f^{\wedge\wedge})$, and t are being assigned and regarding partitioning y which can be usually obtained from some extreme principles is being also assigned, we conclude that the system of the equalities in (3) is the determined system over the desired values m, a, b, c, M, C, P, and G.

On the strength of (3), we easily find

$$m(t) = k(f^{\wedge\wedge})f(t),\ c(t) = [1-k(f^{\wedge\wedge})]f(t) \leq [1-k(f^{\wedge\wedge})f^{\wedge}(t),$$

$$P(t) = \int_{a(t)}^{t} y(s)k(f^{\wedge\wedge})f(s)ds + \int_{b(t)}^{t} [1-y(s)]k(f^{\wedge\wedge})f(s)ds,\ G(t) = M(t)-P(t),\ t \geq t^{\wedge}. \qquad (4)$$

where $a(t)$ and $b(t)$ can be found from

$$\int_{a(t)}^{t} \alpha(t,s)u(s)y(s)k(f^{\wedge\wedge})f(s)ds = k(f^{\wedge\wedge})f(t),\ \int_{b(t)}^{t} \beta(t,s)v(s)[1-y(s)]k(f^{\wedge\wedge})f(s)ds =$$

$$[1- k(f^{\wedge\wedge})]f(t). \qquad (5)$$

It follows from (4) that under $f = f^{\wedge}$ the value $c(t)$, aggressive factor of HIV, depends only on composition and quantity of $f^{\wedge}(t)$. Under "favorable conditions" for HIV when $k(f^{\wedge\wedge}) \approx 1$, the value $c(t) \approx 0$. If $k(f^{\wedge\wedge})$ is decreasing for some reasons or others, $c(t)$ will be able to increase, which can result in killing of the host cells. Under the fixed quantity of $f^{\wedge}(t)$ only $k(f^{\wedge\wedge})$ affects the result. From here, it follows, first, that under the favorable conditions HIV can exist in vivo a long time without noticeable change for an organism and, secondly, for the support of the above mentioned conditions "feeding" $f^\wedge(t)$ must mainly contain the component $m(t)$. Thirdly, when additional needs of proteins of $m(t)$-type in the cells infected with HIV appear, the contents of those proteins into $f^\wedge(t)$ can decrease, which can induce a growth of $c(t)$. And finally, with regard to the properties of functional Ag specificity [6], for HIV's sparing behavior it will be necessary, perhaps, to support the required quantity of Ig against $c(t)$ and the tolerance condition to $m(t)$.

On the contrary, for resolving elimination of HIV we need to destroy $m(t)$. The most weak or sensitive point for destroying any dynamic system is its structure, i.e., the components of α, β, a, b, y -type. If we do not know how to destroy the structure by special means, we may count on "desperate cure". One of these sorts is stressing effect on HIV. To do so, we need to create the conditions such that HIV forces to keep most of $y \approx 1$. The more precise definitions of all the elements of the model allow us to simplify estimation of the desired strategy.

Let $f^\wedge(t)$ permit significant variation of the solutions of (3). Then assuming P, G, and f are to be given and

$$y(t) = 1,\ t^\wedge < t < t^*,\ y(t) = 0,\ t > t^*, \tag{6}$$

where t^* is the instant of switching from internal to external functioning (which also is given or has to be found during the model identification), we can find m due to

$$m(t) = [P(t)+G(t)]'; \tag{7}$$

and a, b from the relations

$$P(t) = \int_{a(t)}^{t} m(s)ds,\ G(t) = \int_{0}^{a(t)} m(s)ds,\ t^\wedge < t < t^*;\ P(t) = \int_{a(t)}^{t^*} m(s)ds + \int_{b(t)}^{t} m(s)ds,$$

$$G(t) = \int_{0}^{a(t)} m(s)ds + \int_{t^*}^{b(t)} m(s)ds,\ t^* < t < t',\ a(t') = t^*,\ a(t) \le t^* \le b(t);\ P(t) = \int_{b(t)}^{t'} m(s)ds,$$

$$G(t) = \int_{0}^{b(t)} m(s)ds,\ t' < t < t'',\ b(t'') = t';\ P(t) = 0,\ G(t) = \int_{0}^{t'} m(s)ds,\ t > t''; \tag{8}$$

and c due to

$$c(t) = f(t) - m(t). \tag{9}$$

After this we can find αu and βv presupposing that our object has the piecewise stationary dynamics, that is, the stationary dynamics on each interval $(t^\wedge+(j-1)dT,\ t^\wedge+jdT)$, $j = 1, 2, \ldots$, and selecting the temporal intervals of the standard behavior of HIV on which u and v can be considered as constants. Indeed, we have

$$m(t) = u\int_{a(t)}^{t} \alpha_j(t-s)m(s)ds = u\int_{0}^{t-a(t)} \alpha_j(s)m(t-s)ds,\ t \in (t^\wedge+(j-1)dT,\ t^\wedge+jdT),$$

$j=1,\ldots,j^*$, $m(t) = u\int_0^{t^*-a(t)} \alpha_f(s)m(t-s)ds$, $t\in(t\hat{}+(j-1)dT, t\hat{}+jdT)$, $j=j^*,\ldots,j'$,

$j'dT = t'$, $j^*dT = t^*$, $c(t) = v\int_{b(t)}^{t} \beta_f(t-s)m(s)ds = v\int_0^{t-b(t)} \beta_f(s)m(t-s)ds$,

$t\in(t\hat{}+(j-1)dT, t\hat{}+jdT)$, $j=j^*+1,\ldots,j'$, $c(t) = v\int_{b(t)}^{t'} \beta_f(t-s)m(s)ds =$

$$v\int_0^{t'-b(t)} \beta_f(s)m(t-s)ds,\ t\in(t\hat{}+(j-1)dT,\ t\hat{}+jdT),\ j=j'+1,\ldots,j'',\ j''dT=t'', \tag{10}$$

and hence

$$u\alpha_f(t-a(t))m(a(t))(1-a'(t)) = m'(t) - u\int_0^{t-a(t)} \alpha_f(s)m'(t-s)ds,$$

$$t\in(t\hat{}+(j-1)dT,\ t\hat{}+jdT),\ j=1,\ldots,j^*,$$

$$-u\alpha_f(t^*-a(t))m(t-t^*+a(t))a'(t) = m'(t)-u\int_0^{t^*-a(t)} \alpha_f(s)m'(t-s)ds,$$

$$t\in(t\hat{}+(j-1)dT,\ t\hat{}+jdT),\ j=j^*,\ldots,j',$$

$$v\beta_f(t-b(t))m(b(t))(1-b'(t)) = c'(t) - v\int_0^{t-b(t)} \beta_f(s)m'(t-s)ds,$$

$$t\in(t\hat{}+(j-1)dT,\ t\hat{}+jdT),\ j=j^*+1,\ldots,j',$$

$$-v\beta_f(t'-b(t))m(t-t'+b(t))b'(t) = c'(t) - v\int_0^{t'-b(t)} \beta_f(s)m'(t-s)ds,$$

$$t\in(t\hat{}+(j-1)dT,\ t\hat{}+jdT),\ j=j'+1,\ldots,j''. \tag{11}$$

Thus, the problem has been reduced to the solution of Volterra-type equations of the second kind. After the solution of the equations in (11) we can find $u(t)$ and $v(t)$ on the temporal intervals with nonstandard behavior of HIV. By the same token, we can detect the control mechanism of *nef* and *rev* including negative feedback, and that of the *tat* including a positive one, probably with the help of $u(t)$, $y(t)$, and $v(t)$ respectively.

It should be noted that in connections with the properties of HIV (see above) we likely have to accept the alternating step $(dT)_j$ instead of dT. In addition, we likely need to accept the system of the relations (6)-(11) for the viruses for which $t\hat{}$ is the instant of injection into the host cells, and to use these relations until new viruses particles bud from the surfaces of the host cells. What is more, the part of HIV which are binding the host cells at the instant $t\hat{}$ can probably be determined with the help of the truncated normal law F_7 (see Ch. 12) for the decimal logarithm of number of all the viruses, or a certain new law parameters of which have to be found on the stage of the model's testing.

On the whole, we can result MM of HIV combining the above mentioned relations for different values $t^\wedge$.

Obtained α and β can be used for their prediction in the future and hence for estimating in the future the more exact values of m and c on the strength of (3) and (7)-(9).

The relations (6)-(11) are the experimental-theoretical methods for achieving the model. In a similar way, we can achieve the model under the more general than above intensity of utilization of m-type bio-mass (see Ch. 2 and Ch. 12).

One of the possible optimization principles (for HIV) consists in

$$C(T) = \int_{t^\wedge}^{T} c(t)dt = \max\,(uy) \tag{12}$$

by means of the factors of strengthening or weakening of efficiency of functioning and the distribution between two subsystems. The problem (12) has been investigated in detail in Ch.4. Repeating arguments of Ch.4, we obtain the following important principle I (under rather wide assumptions relatively to α and β).

For the solution of the problem (12) *on the "small" segment* $[t^\wedge, T]$ *the values of uy is minimally possible, but for "large" T-t the desired uy may differ from the minimally possible on the larger initial part of the segment* $[t^\wedge, T]$ *and only on the smaller final part of* $[t^\wedge, T]$ *the desired uy is minimally possible.*

The notions "small" and "large" depend on the values of functions α and β, namely, the greater are the functions in question, the nearer to $t^\wedge$ is the boundary between "small" and "large" segments. The obtained result was in essence an implication of the fact that for "large" T-$t^\wedge$ a growth of $\alpha(t,s)u(s)y(s)m(s)$ overtakes a diminution of $\beta(t,s)v(s)[1-y(s)]m(s)$ on the average. We would like to remind here that the result has received, in the sequel, the important qualitative very general interpretation:

the record of external function for any ES can be obtained only under the conditions of its sufficiently comfortable guarantee, that is, under the very significant fraction of resources sent to the internal needs of ES.

It follows from the principle I, at least, the next probable strategy for HIV suppression: creation of sufficiently prolonged conditions of minimization of HIV m-type components for fulfillment of HIV active center restoration. As follows from the results of Ch. 4, one of possible realizations of this strategy is creation of prolonged strong external effects (stresses) localized on HIV.

Possible complications and refinements of HIV MM under consideration can be obtained by analogy with Ch. 2. In particular, we believe that new "technologies" α, u and β, v (another interpretation is the number of appropriate *genes*) are also the "products" of HIV of m-type. So, admitting for simplicity that

$$\alpha(t,s)u(s) = \beta(t,s)v(s) = \gamma(s),\ a(t) = b(t), \tag{13}$$

we have instead of (3) the following MM:

$$m(t) = \int_{a(t)}^{t} \gamma(s)y(s)m(s)ds,\ \gamma'(t) = \int_{a(t)}^{t} \gamma(s)x(s)m(s)ds,$$

$$c(t) = \int_{a(t)}^{t} \gamma(s)[1-x(s)-y(s)]m(s)ds,\ P(t) = \int_{a(t)}^{t} m(s)ds,$$

$$G(t) = M(t) - P(t),\ 0 \le a(t) \le t,\ 0 \le x, y,\ 1-x-y \le 1,$$

$$T \ge t \ge t^\wedge \ge 0 = a(t^\wedge), \tag{14}$$

For MM (14), P or G is usually given (as well as all the elements on the prehistory $[0,t^\wedge]$); x and y are found from the solution of some optimization problems and m, a, γ, c, M, and G (or P) are found as the solution of the system (14). Now the problem (12) takes the form

$$C(T) = \int_{t^\wedge}^{T} c(t)dt = \max\ (x,y). \tag{15}$$

Detailed investigation (see Ch. 4) of the problem (15) has led to the following principle II.

For "large" $T-t^\wedge$ the optimal distributions $x^*(t)$ and $y^*(t)$ possess the characteristics

$$x^*(t) = 1,\ t\in[t^\wedge,\theta];\ y^*(t) = 1,\ t\in[\theta,\Theta];\ x^*(t) = y^*(t) = 0,\ t\in[\Theta,T];$$

$$\theta - t^\wedge \asymp T - t^\wedge,\ \Theta - \theta \asymp (T-t^\wedge)^{1/2},\ T - \Theta \asymp 1, \tag{16}$$

where $\asymp$ means the asymptotic equality in order when $T-t^\wedge$ tends to ∞.

The result qualitatively means that

under the optimal distributions of the given resources among different subsystems of HIV the main priority probably belongs to the subsystem of "G"-type creating the new "more effective" technology (genes); on the second place by priority comes the subsystem A' creating the new active centers (products of m-type) and finally comes the subsystem B of production of the external products (c-type).

From the principle II, it follows, at least, the next probable strategy for HIV suppression: creation of sufficiently prolonged conditions of minimization, mainly, of HIV m-type components' share for creation of $\gamma(t)$, i.e., HIV *genes*, there upon of HIV m-type component share for restoration of $m(t)$, and at last of HIV m-type component share for creation of $c(t)$.

In the general case assuming the optimal strategy of HIV as that of maximization, we have the *mini-max problem* (see, for example, [11])

$$C(T) = \int^{T}_{t^{\wedge}} c(t)dt = \min \max (x, y, u, v) \, (w \in W), \qquad (17)$$

where W is the set of various factors, which can stands against HIV. Among them there are the immunology means of AIDS treatment [9, 12].

5. MM OF HIV POPULATION

Now by analogy with Ch. 12, we consider the theoretical course to achieve the model, the continuous one. We suppose that stimulated and infected Th cells are the same and the share of stimulated cells are determined by the truncated normal law F_7 (see Ch. 12) with the parameters $(3, 0.66^2)$ for the decimal logarithm of the number $V = V(t)$ of all alive viruses at the instant t, $t > 0$.

In contrast to Ch. 12, the number of the proliferating infected Th cells

$$M_1{}^P(t) = \int^{t-12}{}_0 M_1{}^P(t,r)dr, \; t>12, \; g(t) = \lg V(t),$$

$$M_1{}^P(t,r) = f(t,r+12)M_1{}^P(r+12,r)\exp\{\int^{t}{}_{r+12} [0.08 - 0.75/24 - 0.02/24 - u(s) -$$

$$c_1{}^P(s)]ds\} [1 - F_8(t-r-12)], \; M_1{}^P(r+12,r) = f(r+12,r)dM_1{}^s(r)\exp[-\int^{r+12}{}_r u(s)]ds],$$

$$dM_6{}^s(r) = (M_{60} + s(r))F_7(g(r))\exp[-\int^{r}{}_0 u(s)ds], \; r>0, \qquad (18)$$

where the notations are the same as in Ch. 12. It follows from (18) that *MM of HIV population*

$$V(t) = 1000 \cdot 0.75/24u(t)\int^{t}{}_0 M_1{}^P(s)ds\exp[-\int^{t}{}_0 v(s)ds], \qquad (19)$$

where $u(t)$ is a share of infected $M_1{}^P$ cells producing free HIV, $v(s)$ is the death rate of free HIV.

Certain information about u, v one can find in M. Nowak's Shanks lecture and the abstract [13]. In particular, the half-life of free HIV is about 6 hours, and 99% of plasma viruses are produced from the cells that have a half-life less than 2 days, but most HIV infected cells contain defective pro-virus and have life longer than 100 days.

To determine $M_1{}^P(t)$ and $V(t)$, we have to add the other equations of IN from above that are similar to (19), but without the number 0.75/24. It seems unreliable that IS cannot eliminate $V(t)$ and infected Th cells. Indeed, rapid turnover of HIV should result enough $M_2^{p,m}(t)$ and $M_3^{p,m}(t)$ (Ts and Teff proliferating and mature cells) which can probably eliminate infected Th cells and $V(t)$. To be certain, we can either add new assumptions to the list in Ch.12, for example, the natural assumption about suppression of Ts and Teff by a large doze of HIV, or go to another interpretation of HIV as Ag.

Choosing the law F_7 above, we have considered HIV as corpuscular Ag similar to cells. However, we can consider HIV as a very large collection of different types of molecular-disperse Ag such that for their individual representatives MM of IN were constructed in Ch. 12. Let us assume the following guess regarding N of miscellaneous Ag (compare to [10], pp. 86-103):

1) If there exists one so-called immune-dominant molecular-disperse *Ag* determinant [6], we can reduce the situation to the considered case.

2) With N "equivalent" *Ag*, instead of (1), Ch. 12 we should take

$$z_i = k_i x_i y_i = 1/3 F_7(\lg Y_i)/(NY_i) x_i y_i\,,\ i = 1,\ldots, N, \tag{20}$$

where $3k_i$ is a probability of contact of two cells x- and y-types that are ICC with respect to Ag with number i, Y_i are all control cells for x_i.

3) If there exist N "nonequivalent" *Ag*, we should take

$$z_i = 1/3 q_i F_7(\lg Y_i)/(NY_i) x_i y_i,\ \sum^{N}_{i=1} q_i = 1,\ q_i \geq 0,\ \ i = 1,\ldots, N, \tag{21}$$

where q_i are appropriate weights.

4) Let $M_i(t)$ be the total maximal number of ICC of all levels to Ag with number i at the instant t. If

$$\sum^{N}_{i=1} M_i(t) \leq M^{+}, \tag{22}$$

where M^{+} is the maximum admissible number of IN cells, then SR for cell proliferation does not vary.

5) Otherwise, SR reduction for cell proliferation takes place in a way that preserves the relation (22).

Instead of (18) we have now

$$M_6^{p}(t,i) = \int^{t-12}_{0} M_6^{p}(t,u,i)du,\ t > 12,$$

$$M_6^{p}(t,u,i) = M_6^{p}(u+12,u,i) f_1(t,u+12,i) \exp\Big(\int^{t}_{u+12} [0.08-0.75/24-$$

$$0.02/24-u_i(s)-c_{1i}^{p}]ds\Big)[1-F_8(t-u-12)],$$

$$M_6{}^P(u+12,u,i) = dM_6{}^s(u,i)f_1(u+12,u,i)\exp[-\int^{u+12}_u u_i(s)]ds],\ u > 0,$$

$$dM_6{}^s(u,i)= [M_{60}+s(u)]\{f^\wedge{}_{1,2}(u,0,i)[1-F_1(y_{0i}(u))]F_2(y_{0i}(u))+$$

$$f_{1,2}(u,0,i)([1- F_1(y_{0i}(u))][1-F_2(y_{0i}(u))]-1)F_7(z_i(u))\}\exp[-\int^u{}_0 u_i(s)ds],$$

$$f_1(t,s,i) = \lim \prod \{[1-F_1(y_{0i}(k\Delta t))][1- F_5(y_{0i}(k\Delta t))][1-$$

$$F_6(y_{5i}(k\Delta t))]\}\ (k\Delta t = s,\dots, t),\ \Delta t \to 0,$$

$$f_{1,2}(t,s,i) = \lim \prod \{[1-F_1(y_{0i}(k\Delta t))][1- F_2(y_{0i}(k\Delta t))][1-F_5(y_{0i}(k\Delta t))][1-$$

$$F_6(y_{5i}(k\Delta t))]\}\ (k\Delta t = s,\dots,t),$$

$$f^\wedge{}_{1,2}(t,s,i) = \lim \prod \{[1- F_1(y_{0i}(k\Delta t))]F_2(y_{0i}(k\Delta t))[1-F_5(y_{0i}(k\Delta t))][1-$$

$$F_6(y_{5i}(k\Delta t))]\},\ (k\Delta t = s,\dots, t), \tag{23}$$

where F_j is one of the functions in (2, Ch. 12); the subscript i means that the respective formula in Ch. 12 has to be applied for ith Ag.

And instead of (19) we have now the following *MM of HIV population:*

$$V(t) = 1000\cdot 0.75/24\ u(t)\sum\int{}^t{}_0 M_1{}^P(s,i)ds\exp[-\int{}^t{}_0\, v(s)ds]\ (i = 1,\dots, N). \tag{24}$$

We can give multiple interpretations of causes resulting in helplessness of IN regarding (23) and (24) combined with MM of Ch. 12 and also the hypotheses 1) - 5). But we will not dwell on that explicitly in view of the great complexity of these joint MM. It should be pointed out that the appropriate theorems of existence, uniqueness, positiveness and boundedness for the solutions of our models could be proved by the standard methods analyzed in Ch. 3. In Ch. 4 - Ch. 7 some properties of the structure of the solutions of the problems (12) and (15) as well as some algorithms for the solution of the problem (12) can be found. It should be also noted that [14] contains MM of HIV infection based on assumptions, which are much more simple particular case, but nevertheless certain results of these MM investigations have verified some important well-known concepts relating to HIV.

6. ON MM OF AIDS

It seems that for MM of AIDS we should add to the relations above MM of a certain organ of IS which suffers most from helplessness of IN. Such organ most likely is the thymus. MM of thymus suffering can be introduced by the methods developed in [10] or the methods based on MM of thymus as ES [4]. It has been shown in [4], theoretically and by the respective experiments with the thymuses of mice, that rather long and intensive output of the external function of the thymus ($s(t)$) results in destruction of this organ. If the latter happened, it would mean that an organism becomes defenseless from attacks of great number of dangerous *Ag*. Apparently, HIV "knows" how to fight with IN and IS as a whole with the help of stress effects on them. But at the same time HIV knows "too much" because it usually dies with the HIV patient.

Of course, combined MM of HIV, IN, HIV population, and some of organs IS are capable to contribute to clinicians to fight with HIV. But as one can see these MM are very complicated. Qualitative and numerical investigations of the models require cross-participation of interdisciplinary scientists, especially in the field of software, numerical methods, dynamic systems, in general, immunology, and medicine.

7. ON AIDS IMMUNE-THERAPY

Probably on the whole, the AIDS immune-therapy should be consisted in overcoming the tolerance induced by HIV to numerous Ag.

First, we should know ISO of the patient infected with HIV including the main indices of the organs of IS, especially the thymus [4]. The most important factor of their restoration probably the restoration of the organs of IS. It has been shown not once, especially in Ch. 11 that for restoration of any subsystem of the patient the conditions for consuming of essential part of all resources to the internal needs of the subsystem should be created. In the case of the thymus it means that in spite of necessity of giving the source of new Th cells the conditions should be created for temporal decreasing of this source to restore the "health" of the thymus.

It is emphasized in [13] that the evolution of HIV in vivo during individual infection can explain the mechanism of disease progression. But s. 4. here is devoted namely to the problem evolution or development of HIV.

From the analysis above, it results that probable temporal AIDS immune-therapy consists in creation of the conditions such that IN will be both tolerant to the HIV Ag determinants of $m(t)$ and aggressive to the HIV Ag determinants of $c(t)$. The main *Ag* determinants of $c(t)$ are those of OT.

We have considered in Ch. 12 a possible application of MM of IN and the *hybridoma* technology [12] for elimination of hypothetical Ag similar to HIV Ag determinants. It should be noted that for realization of this idea we should, in addition, consider the guess 1) - 5) so that the problem is getting much more complicated.

It is desirable to create conditions for the intake of Ig to Ag-determinants of $c(t)$ such that the value $C(T)$ to be minimized for the given ISO. This problem cannot be likely solved without computer-based modeling with the use of MM of HIV and IN.

It follows from the mentioned principles I and II that both the $C(T)$ and $M(T)$ can be minimized under sufficiently prolonged conditions of minimization of the special shares of HIV *m*-type components.

Of course, the further investigation and perfection of MM of HIV, IN, and IS as a whole are necessary to consider very fast possible variation of HIV Ag-determinants and to have sufficiently accurate prediction for these determinants and determination of the necessary actions with immune correctors of IS.

As to the auto-immune aspects of HIV infection [5, 15], IN role can only grow.

In conclusion, we would like to emphasize that this Ch. mainly devoted not to the means, but the methods of AIDS treatment. At present, there have been some immunology means of AIDS treatment (see, for example, [1, 9, 12, 16], where [9] is associated with the IN concept), but the question about the best condition for the intake of the known agents is not trivial with regard to the peculiarities of the given patient and complexity of AIDS. It can be solved by mathematical simulation faster and more qualitatively than without that on the basis of application of the necessary MM and the appropriate software and computers.

REFERENCES

1. AIDS is the Acquired Immune-Deficiency Syndrome, Kiev: Zdorov'ya, 1988, 232 pp. (in Russian).
2. Anderson D.J. et al., Effects of Disease Stage and AZT Therapy on the Detection of HIV-1 in Semen, Journal of the American Medical Association, 267, 1992, 2769-2774.
3. Coffin J.M., HIV Populations Dynamics in Vivo: Implication for Genetic Variations, Pathogenesis and Therapy, Science 267, 1995, 483-489.
4. Glushkov V.M., Ivanov V.V. et al., Modeling of Lymphocytes Adaptive Redistribution and Regenerating Accumulation Process for Thymus, Kiev: Inst.of Cybernetics, Reprint 82-30, 1982, 31 pp. (in Russian).
5. Hoffman G.W. and Gran M.D., Hypothesis: AID is Auto-immune Disease Caused by HIV Plus Allogenic Cells, in International Conference on AIDS. The Scientific and Social Challenge. Montreal, Quebec, Canada June 4-9, 1989. Abstracts, 613.
6. Immunology. Reference Book, Kiev: Naukova dumka, 1981, 480 pp. (in Russian) (translated from IMMUNOLOGIE ein Nachschlagewerk, Akademic Verlay, Berlin, 1978).
7. Ivanov V.V., Korzhova V.N., Mathematical Model of the Immune Network of the Patient Infected with HIV, M.: VINITI, N 133-B90, 1990, 19 pp. (in Russian).
8. Kirschner D., Using Mathematics to Understand HIV Immune Dynamics, Notices of the AMS, 43, N 2, 1996, 191-202.
9. Leturcq D., Winslow B., Johnson P., Thornton G., Neutralization of HIV-1 by Anti-Idio-Typical Antibodies, 554 in the same book as 3.
10. Marchuk G. I., Mathematical Models in Immunology, M.: Nauka, 1985, 264 pp. (translated into English, the same title, Optimization Software Inc., New York, 1983).
11. Mini-max and Application, Ed. by Du D.-Z. and Pardalos P.M., Kluwer Academic Publishers, 1995, 294 pp.
12. Moses V., Moses S., Exploiting Biotechnology, harwood academic publishers, 1995, 332 pp.
13. Nowak M., Mathematical models of virus infection and evolution, Abstract, Conference on Mathematical Models in Medical and Health Sciences, Vanderbilt University May 28-31, 1997.
14. Theoretical Immunology, The Proceedings of the Theoretical Immunology Workshop, June 1987, Santa Fe, New Mexico, Addison-Weley. Publ. Co., 1987, Part two, Vol.3, 407 pp. (see pp. 50-100).
15. Via C.S., Morse III H.C. and Shearer G.M., Altered Immune-regulation and Auto-immune Aspects of HIV Infection: Relevant Murine Models, Immunology Today, 11, N 7, 1990, 250-255.
16. What does Science Know about AIDS, Scientific American, v. 259, 1988, 132 pp.

CHAPTER 14

MORE APPLICATIONS OF MM OF DEVELOPMENT

Besides MM of ES applications considered earlier, we dwell here on the following:

i) Modeling of plant development
ii) Modeling of animal development
iii) Modeling of business development
iv) Space crew as ES
v) Health care as ES
vi) Formalization of civilization conception
vii) A few addenda: 1. New computers as ES. 2. Science as ES. 3. Education as ES. 4. MM of cells as ES. 5. MM of bio-fields as ES 6. Modeling of natural ES origin problems. 7. Modeling of life longevity problems.

i) MODELING OF PLANT DEVELOPMENT (a sugar beet as example)

1. MM OF A PLANT UP TO THE START OF PHOTOSYNTHESIS

This model has the form

$$m(t) = \int_0^t \alpha(t,v)\lambda(t,v)y(v)m(v)dv + m(0),\ M(t) = M(0) + \int_0^t m(v)dv,\ c(t) =$$

$$\int_0^t \beta(t,v)\mu(t,v)[1-y(v)]m(v)dv,\ C(t) = \int_0^t c(v)dv,\ P(t) = M(0) + \int_0^t \{y(v)\lambda(t,v) +$$

$$[1-y(v)]\mu(t,v)\}m(v)dv,\ 0 \le y,\ \lambda,\ \mu \le 1,\ p = P',$$

$$G = M - P,\ g = G',\ f = p + g + c,\ F(t) = \int_0^t f(v)dv,\ \ 0 \le t \le t^\wedge, \tag{1}$$

where m is the factor of the new plant *bio-mass* or *phyto-mass* quantity; ym is the factor of the new plant *bio-mass* quantity in the subsystem A of "means of production" m; $(1-y)m$ is the factor of the new plant *bio-mass* quantity in the subsystem B of "production" of the external product (O_2); c is the factor of the external product (O_2) quantity; α and β are the *indices* of efficiency of functioning along the channels y -- m and $(1-y)m$ -- c respectively; λ and μ are the coefficients of *intensity*; P is the total functioning plant bio-mass; G is the total "heap"; F is the total substratum; $M(0)$ is the bio-mass of semen; $m(0)$ is the starting rate of plant *bio-mass* creation; 0 is the instant of the semen germination; t^ is the instant of appearance of leaves.

Knowing from an experiment P, G, and F, we can find

$$m(t) = [P(t) + G(t)]',\ c(t) = [F(t) - P(t) - G(t)]'. \tag{2}$$

After this we can probably find $\lambda(t,v)y(v)$, $\alpha(t,v)$, $\mu(t,v)[1-y(v)]$, $\beta(t,v)$ from different data and postulates about this period of plant's development. There upon, in turn, we can find prognostic values of $m(t)$, $M(t)$, $P(t)$, $G(t)$, and $c(t)$.

A certain data for the model (1) for a sugar beet see [1] and [7, Ch. 2].

2. MM OF A PLANT IN THE FIRST VEGETATIVE PERIOD

This model has the form

$$m(t) = \int_{t^\wedge}^{t} \alpha(t,v)\lambda(t,v)y(v)m(v)dv + m(t^\wedge),\ M(t) = M(t^\wedge) + \int_{t^\wedge}^{t} m(v)dv;$$

$$c_i(t) = \int_{t^\wedge}^{t} \beta_i(t,v)\mu_i(t,v)z_i(v)m(v)dv + c_i(t^\wedge),\ C_i(t) = C_i(t^\wedge) + \int_{t^\wedge}^{t} c_i(v)dv,\ i=1,\ 2;$$

$$c_1(t^\wedge) = 0,\ c_2(t^\wedge) = c(t^\wedge),\ C_2(t^\wedge) = C(t^\wedge),\ 0 \le y,\ z_1,\ z_2,\ \lambda,\ \mu_1,\ \mu_2 \le 1,\ y+z_1+z_2 = 1;$$

$$P(t) = P(t^\wedge) + \int_{t^\wedge}^{t} [\lambda(t,v)y(v) + \mu_1(t,v)z_1(v) + \mu_2(t,v)z_2(v)]m(v)dv;$$

$$F = M - M(t^\wedge) + C_1 + C_2,\ G = M - P,\ p = P',\ f = F',\ g = G',\ T \ge t \ge t^\wedge > 0, \tag{3}$$

where $m+c_1$ is the growth factor of the new plant bio-mass quantity; c_1 is the growth factor of the accumulated bio-mass quantity; c_2 is the growth factor of the product quantity, leaving a plant; ym, z_1m, and z_2m are the factors of the bio-mass quantity creating m, c_1, and c_2 respectively; α is the capacity (specific rate) along the channel ym -- m: the factor of the new bio-mass quantity m at the instant t per one unit of the bio-mass quantity ym at the instant v; β_1 and β_2 are the specific rates along the channels z_1m--c_1 and z_2m --c_2 respectively (both are determined similarly to α); λ, μ_i, $i = 1,\ 2$, are the coefficients of utilization intensity at the instant t of the products at the instant v; $P(t)$ is the total plant's bio-mass at the instant t; G is the "heap", i.e., the total plant's bio-mass left a plant; f is the factor of substratum inflow; t^ is the instant when plant leaves are to be appeared; T is the end of the first vegetative period; 0 is the instant when the semen is to be germinated. Note that instead of bio-mass in several applications there are reasons to consider active centers of plants (analogy of work places), which are of three main components: ferments, membranes, etc. (tools or means of reproduction); saccharose, glucose, etc. (raw material), adenosine-d(i)-phosphate (ADP), adenosine-t(ri)-phosphate (ATP), etc. (energy).

In addition, we introduce the static relation, *balance equation*

$$f(t) = m(t) + c_1(t) + c_2(t) = kc_2(t), \tag{4}$$

where k is the coefficient ($k > 1$), arising because the invariable shares of substratum f go to $m+c_1$ and c_2. For sugar-beets the main bio-mass is created during *photosynthesis* from carbonic acid CO_2, relatively small fractions of nitrogen N, phosphorus P, potassium K, and hydrogen H and oxygen O_2 that are the parts of incoming water H_2O. During H_2O hydrolysis hydrogen H and oxygen O_2 go to energetic consumption of plants and oxygen O_2 from CO_2 goes to atmosphere. Thus

$$f(t)/c_2(t) = k \approx (\text{mass}(CO_2) + \text{mass}(H_2O))/\text{mass}(O_2) \approx 2. \tag{5}$$

It should be noted that the process of plant's breathing during a day and especially a night results in consumption of O_2 and secretion of CO_2, but the total plant bio-mass is practically invariable; the equation (5) is the static one, not regarding the respective transients; and the ranges of different elements of the model and the values of the main products to the end of the vegetative period T for a sugar beet one can find in [1].

3. MM OF A PLANT REGARDING ITS LEAVES AND ROOTS

This model has the form

$$m = \int_{t^\wedge}^{t} \alpha(t,v)\lambda(t,v)y(v)m(v)dv,\ m = \{m_i,\ i = 1, 2\},\ \alpha\lambda y = \{\alpha_{ij}\lambda_{ij}y_{ij},\ i, j = 1, 2\},$$

$$c = \int_{t^\wedge}^{t} \beta(t,v)\mu(t,v)z(v)m(v)dv,\ c = \{c_i,\ i = 1, 2\},\ \beta\mu z = \{\beta_{ij}\mu_{ij}z_{ij},\ i, j = 1, 2\},$$

$$C_i(t) = \int_{t^\wedge}^{t} c_i(v)dv,\ P(t) = P(t^\wedge) + \int_{t^\wedge}^{t} [\lambda(t,v)y(v) + \mu(t,v)z(v)]m(v)dv,\ M_i(t) =$$

$$\int_{t^\wedge}^{t} m_i(v)dv,\ M = M_1 + M_2 + C_1,\ \ G = M - P,\ \sum y_{ij} + z_{ij} = 1\ (i = 1, 2), j = 1, 2;$$

$$0 \le y, z, \lambda, \mu \le 1,\ m_1 + m_2 + c_1 = kc_2,\ T \ge t \ge t^\wedge > 0, \tag{6}$$

where M_1 is the bio-mass of leaves and roots, $M_2 + C_1$ is the remained part of plant's bio-mass, and the other elements have the similar sense to above.

4. SEVERAL SPECIAL CASES

1^. Let in (1)

$$\lambda = \mu = 1,\ \alpha = \alpha^\wedge m(v)/[M(v)y(v)],\ \beta = \beta^\wedge m(v)/[M(v)(1-y(v))], \tag{7}$$

where $\alpha^{\wedge}$ and $\beta^{\wedge}$ are constants. Then

$$m' = \alpha^{\wedge} m^2/M,\ c' = \beta^{\wedge} m^2/M, \tag{8}$$

from which under $\alpha^{\wedge} \neq 1$

$$m = m(0)\{1+m(0)/M(0)(1-\alpha^{\wedge})t\}^{\alpha^{\wedge}/(1-\alpha^{\wedge})},\ M = M(0)\{1+m(0)/M(0)(1-\alpha^{\wedge})t\}^{1/(1-\alpha^{\wedge})},$$

$$c = \beta^{\wedge}/\alpha^{\wedge}[m-m(0)],\ C = \beta^{\wedge}/\alpha^{\wedge}[M-M(0)-m(0)t]; \tag{9}$$

and under $\alpha^{\wedge} = 1$

$$m = m(0)\exp[m(0)/M(0)t],\ M = M(0)\exp[m(0)/M(0)t]. \tag{10}$$

One can see that bio-mass of a plant in the case (7) can increase fast. Under $\alpha^{\wedge} > 1$ we should put

$$m(0)/M(0)(\alpha^{\wedge}-1)t^{\wedge} < 1. \tag{11}$$

2^. Let in (3)

$$\lambda = 1,\ \mu_i = 1,\ i=1,\ 2;\ \alpha = \alpha^{\wedge} m(v)\exp[h(v-t)]/[M(v)y(v)],$$

$$\beta_2 = \beta^{\wedge}_2 m(v)\exp[h(v-t)]/[M(v)z_2(v)], \tag{12}$$

where $\alpha^{\wedge}$ and $\beta^{\wedge}_2$ are constants, h is the decrement of decay at the instant t for the bio-mass created at the instant v ($v \leq t$). The form of non-linearity assumed is typical for biology [1, 14, 16] and corresponds to the qualitative conclusion that under a simple growth of number elements and their deterioration, the system productivity is decreased. Regarding (12), we have

$$m' + hm = \alpha^{\wedge} m^2/M,\ m(t^{\wedge}) = m^{\wedge},\ \ M(t^{\wedge}) = M^{\wedge},\ c_2' + hc_2 = \beta^{\wedge}_2 m^2/M,\ c_2(t^{\wedge}) = c_2^{\wedge},$$

$$f = m + c_1 + c_2 = kc_2,\ y + z_1 + z_2 = 1,\ T \geq t \geq t > 0, \tag{13}$$

Let set $m' = dm/dt = dm/dMm$. Then under $\alpha^{\wedge} \neq 1$ solving first the linear differential equation and then the *Bernoulli equation*, we have

$$m = M' = [m^{\wedge} + hM^{\wedge}/(1-\alpha^{\wedge})](M/M^{\wedge})^{\alpha^{\wedge}} - h/(1-\alpha^{\wedge})M,\ M = M^{\wedge}\{1 + (1-\alpha^{\wedge})m^{\wedge}/(M^{\wedge}h)(1-$$

$$\exp[-h(t-t^{\wedge})])\}^{1/(1-\alpha^{\wedge})},\ m = m^{\wedge}\exp[-h(t-t^{\wedge})]\{1+ (1-\alpha^{\wedge})m^{\wedge}/(M^{\wedge}h)(1-\exp[-h(t-t^{\wedge})])\}^{\alpha^{\wedge}/(1-\alpha^{\wedge})},$$

$$c_2 = c_2\hat{}\exp[-h(t-t\hat{})] + \beta_2\hat{}\int_{t\hat{}}^{t} m^2(s)/M(s)\exp[-h(t-s)]ds,$$

$$c_1 = kc_2 - m, f = m + c_1 + c_2. \tag{14}$$

Note that in the case of a sugar beet $\alpha\hat{}$ and $\beta\hat{}$ are less than 1 [1]. From (14), it follows that the graphs of M, C_1, C_2, and F are the curves of logistic-type, which is consistent with the experimental data [1].

3^. Let set in (6)

$$\lambda_{ij} = \mu_{ij} = 1,\ i, j = 1, 2;\ \alpha_{ij} = \alpha\hat{}_{ij} m_i/[M_j y_{ij}]\exp[-h(t-v)]. \tag{15}$$

Then

$$m'_1 + hm_1 = \alpha\hat{}_{11} m_1^2/M_1 + \alpha\hat{}_{12} m_1 m_2/M_2,$$

$$m'_2 + hm_2 = \alpha\hat{}_{21} m_1 m_2/M_1 + \alpha\hat{}_{22} m_2^2/M_2,\ m_i(t\hat{}) = m\hat{}_i,\ i = 1, 2, \tag{16}$$

from which denoting $m_i/M_i = u_i$, $i = 1, 2$, we have

$$u_1' = (\alpha\hat{}_{11} - 1)u_1^2 + \alpha\hat{}_{12} u_1 u_2 - hu_1,\ u_2' = \alpha\hat{}_{21} u_1 u_2 + (\alpha\hat{}_{22} - 1)u_2^2 - hu_2,$$

$$u_i(t\hat{}) = m\hat{}_i/M\hat{}_i. \tag{17}$$

Setting the right parts of the system (17) to 0, we have two singular points

$$u_{11} = u_{21} = 0,\ u_{12} = h(1 + \alpha\hat{}_{12} - \alpha\hat{}_{22})/[\alpha\hat{}_{12}\alpha\hat{}_{21} - (1 - \alpha\hat{}_{11})(1 - \alpha\hat{}_{22})],$$

$$u_{22} = h(1 + \alpha\hat{}_{21} - \alpha\hat{}_{11})/[\alpha\hat{}_{12}\alpha\hat{}_{21} - (1 - \alpha\hat{}_{11})(1 - \alpha\hat{}_{22})]. \tag{18}$$

We assume that u_{12}, u_{22} are positive.

The *characteristic equation* of the system (17) in the vicinity of the root u_{12}, u_{22} has the form

$$\kappa^2 - p\kappa + q = 0,\ q = -\alpha\hat{}_{12}\alpha\hat{}_{21} u_{12} u_{22} < 0. \tag{19}$$

So, we have the case $\kappa_1 > 0$, $\kappa_2 < 0$. Thus, only the unstable knot of saddle-type is possible. Note that the point $u_{11} = u_{21} = 0$ is the stable knot under h, u_1, $u_2 > 0$.

From these results, it follows that the growth of u_1, u_2, and hence α_{ij}, β_{ij} is possible. We can interpret this as increasing of the system efficiency due to dividing of the functions between its two subsystems.

It is not difficult to introduce the more detailed model with regard to characteristic of both leaves of a plant and its roots. We only note that it turns out to be necessary

[10, Ch. 2] to introduce more complicated description of the decay process:

$$\alpha = \alpha\text{^}m(v)\exp[h(v-t)+h\text{^}(v-t)^2+h\text{^^}(v-t)^3]/[M(v)y(v)], \tag{20}$$

arising, for example, from the fact that decreasing efficiency of functioning of a sugar beet's leaves in the end of vegetative period takes place about three times faster than in the beginning [1].

5. APPLICATION OF THE MODELS

Prediction of plant development on the segment $[T', T]$ from data of the model identification on the segment $[0, T']$ or $[t\text{^}, T']$ $(0 < t\text{^} < T' < T)$ is one of the important problems of application. True prediction allows us to increase number of plants' selection variants with a view to select the most productive kind [1].

Maximization of *productivity* of a *plant* is the second important problem. More accurately, it is required to find plant's characteristics that give the solution of the problem $C_1(T) = \max$ using the control factors y_{ij}, z_{ij}, and some others.

In the general case, as the consequence from the well-known results of the theory of ES, we have the following principle:

So that one can guarantee the maximal productivity of plants during of a vegetative period it is necessary that almost the whole first period of ontogenesis, a greater share of resources to be redistributed on the active internal metabolism and only in the end of ontogenesis on the process of stock creation.

ii) MODELING OF ANIMAL DEVELOPMENT (a Philodina roseola as example).

1. ON DIFFERENCES BETWEEN MM OF PLANTS AND ANIMALS

First, we can apply the models of the section **i)** to modeling of the animal's development, but, of course, with another interpretation of the initial data and results. The model (1) in **i)** now can be interpreted as the model for a prenatal period of the animal's development. The model (3), (4) in **i)** is as the model for a postnatal period of the animal's development, where $m+c_1$ is the growth factor of the new animal's *bio-mass* quantity; c_1 is the growth factor of the accumulated *bio-mass* quantity; c_2 is the growth factor of the product quantity, which goes to the fulfillment of the external functions of the animal's interactions with environment. Determination of the *balance equation* for an animal, which is similar to (5) in **i)**, is essentially different because an animal usually does much more work than a plant, and extracts a significant quantity of heat and the respective waste matter. The model (6) in **i)** can be interpreted as the more detail models (than (3), (4)) for the animal's development, where M_1 is the *bio-mass* of the stomach-intestinal subsystem of an animal, M_2+C_1 is the functioning *bio-mass* of the remained part of an animal, C_1 is the accumulated *bio-mass*.

The *elementary indices* of the form (15) in **i)** for an animal usually take the more general form

$$\alpha_{ij} = \alpha\hat{}_{ij} m_i / [(M_j + d_{ij}) y_{ij}] \exp[-h(t-v)], \tag{1}$$

where nonnegative constants d_{ij} are associated with the *processes of saturation* in liquid mediums of an animal [16].

2. ON MM OF ANIMAL POPULATIONS

Let $m(t)$ be the growth factor of the new bio-mass of animals in population due to their reproduction. Then accepting the simplest base MM of NES (see Ch. 2), we have

$$m(t) = \int^{t}_{a(t)} \alpha(t,s) y(s) m(s) ds,\ P(t) = \int^{t}_{a(t)} m(s) ds,\ G(t) = \int^{a(t)}_{t\hat{}} m(s) ds,\ 0 \le y \le 1,$$

$$c(t) = \int^{t}_{a(t)} \beta(t,s)[1-y(s)] m(s) ds,\ m(t) + c(t) = f(t) = k(f) c(t),\ t\hat{} \le a(t) \le t, \tag{2}$$

where $P(t)$ is the number of living functioning animals in the instant t, $G(t)$ is the number of dead animals, $c(t)$ is the outlet rate of work of interaction of animals with each other and the other external factors, $\alpha(t,s)$ is the factor of the bio-mass rate of animals ($P'(t)$ is the factor of the birthrate of animals), $\beta(t,s)$ is the factor of the intensity of interaction of animals with the external surrounding, and $t\hat{}$ is the starting time for modeling. With MM (2), it is clear that the number of animals at the instant t of age u is $P'(t-u)$ and the life span of "active life" of animals is $t-a(t)$.

After the birth, animals' growth can be described by the models similar to (1), (3), (4), and (6) in s. **i)**. Accepting the base model of animals similar to (3), s. **i)** we have

$$m(t,u) = m(u) + \int^{t}_{u} \alpha(t,v,u) \lambda(t,v,u) y(v,u) m(v,u) dv,\ c(t,u) = c(u) +$$

$$\int^{t}_{u} \beta(t,v,u) \mu(t,v,u) z(v,u) m(v,u) dv,\ P(t,u) = P(u) + \int^{t}_{u} [\lambda(t,v,u) y(v,u) +$$

$$\mu(t,v,u) z(v,u)] m(v,u) dv,\ y + z = 1,\ 0 \le y,\ z,\ \lambda,\ \mu \le 1,$$

$$M(t,u) = M(u) + \int^{t}_{u} m(v,u) dv,\ m + c = kc,\ t \ge r + u \ge t\hat{} > 0, \tag{3}$$

where $m(t,u)$ is the growth factor at the instant t of the bio-mass of animals in the population being born at the instant u, $c(t,u)$ is the growth factor of the product quantity, which goes to the external functions of animals' interactions with environment, $P(t,u)$ is the functioning bio-mass, and $M(t,u)$ is the total bio-mass.

3. PHILODINA ROSEOLA AS EXAMPLE

The example of Philodina roseola for modeling of animals' development was appeared as the result of the author's contact with L.I. Lebedeva [10] who raised the question about selection of the *species* of the age with maximal *productivity* and determination of the appropriate conditions. In the mathematical sense the questions are

$$m(t\text{-}u) = \max\ (t\text{-}u \,|\, c > c^*) \tag{4}$$

and

$$m(t,u) = \max\ (t,\ u \,|\, c(t,u) > c^*), \tag{5}$$

where c^* is the condition of the norm of animals interaction with the external factors.

Since $P(t)$ and $G(t)$ can be usually determined experimentally and $m(x) = P'(x) + G'(x)$, the solution of the problem (4), that is, max $m(t\text{-}u)$ can be found in the point x where $P'+G'$ is max. The relation's (6) should be found under the different external conditions, for example, under the various temperature conditions. It is shown in [10] that the most favorable temperature for reproduction of Philodina roseola is 20 °C.

In general, we have to solve the system (2) with respect to m and c, after that we can find the desired x.

For the problem (5), we can apply the model (3). In the case (1), $\lambda = 1$, we have

$$m'_t(t,u) = \alpha\hat{}m(t,u)^2/[M(t,u)+d]\ - hm(t,u),\ m(u,u) = m(u). \tag{6}$$

So, $m'_t(t,u) = 0$ in the point t^* for which

$$m(t^*,u)/[M(t^*,u)+d] = h/\alpha\hat{}. \tag{7}$$

Setting in (6) $m/[M+d] = v$, it is easy to find

$$v' = -v[(1\text{-}\ \alpha\hat{})v + h],\ v(u) = m(u)/[M(u)+d], \tag{8}$$

from which

$$v = hv(u)/\{[(1\text{-}\ \alpha\hat{})v(u)+h]\exp[h(t\text{-}u)]\ \text{-}\ (1\text{-}\ \alpha\hat{})v(u)\}. \tag{9}$$

From (7)-(9) it follows that

$$t^* = u + 1/h\ln\ m(u)/[(1\text{-}\ \alpha\hat{})m(u)+h(M(u)+d)]. \tag{10}$$

It is not difficult to prove that t^* is the point of max. Indeed, in the virtue of v decreasing $m'_t(t,u)$ is greater than 0 for $t < t^*$ and less than 0 for $t > t^*$. If $c(t^*,u) > c^*$, then t^* gives the solution of the problem (5). Otherwise, we can find $c(t,u)$ from (3) and after this we can find max $m(t,u)$ with the restriction $c > c^*$. Similar approach allows us to simulate the problems of productivity maximization for other animals.

iii) MODELING OF BUSINESS DEVELOPMENT

Business development, in general, should be concerned to the functioning of all the ES: AES, NES, and JES.

1. BASE MM OF BUSINESS AS ES

The *base*, simplest *MM of business* as ES

$$m(t) = \int_{a(t)}^{t} \alpha(t,s)y(s)m(s)ds + \Delta m(t);\ P_1(t) = \int_{a(t)}^{t} y(s)m(s)ds;\ c(t) = \int_{b(t)}^{t} \beta(t,s)[1-y(s)]m(s)ds + \Delta c(t);\ P_2(t) = \int_{b(t)}^{t} [1-y(s)]m(s)ds;\ P(t) = P_1(t) + P_2(t),\ M(t) = \int_{0}^{t} m(s)ds,$$

$$G(t) = M(t) - P(t);\ f^{\wedge}(t) \geq f(t) = m(t)+c(t),\ 0 \leq y \leq 1;\ 0 \leq a(t),\ b(t) \leq t;\ t \geq t^{\wedge} \geq 0, \quad (1)$$

where m is the rate of creation of the new *resource* quantity in the subsystem A that provides the internal functions of ES; α is the efficiency index for functioning of A along the channel $y(s)m(s)$--$m(t)$; Δm is an increment of m; $P(t)$ is the total quantity of resource functioning at the time t; $c(t)$ is the rate of creation of the new resource quantity in the subsystem B that provides the external functions of ES; β and Δc are similar to α and Δm, but for the subsystem B; $G(t)$ is the total quantity of the out-of-date (or obsolete) resource at the time t; $a(t)$ and $b(t)$ are the temporal thresholds for elimination (or reservation) of the out-of-date resources in the respective subsystems A and B; $f^{\wedge}(t)$ is the rate of the new resource quantity flowing in from the outside ES at the instant t; $t^{\wedge}$ is the starting point for modeling; $[0,t^{\wedge}]$ is the prehistory of ES on which all the functions are given (their value will be noted by the same symbol, but with superscript "^"). It is obvious that all the relations in (1) are faithful representations by definition and all the functions in (1) are nonnegative. The functions α and β can depend on the rest ones, in general.

If one need to consider an interaction between AES and NES, their MM should be considered jointly.

We assume that the resource of the AES $f^{\wedge}$ is consisted from a certain parts q_1M^*, q_2C^*, and q_3S^* of the ecological products of the biosphere and its cosmic resource S^*:

$$f^{\wedge} = q_1M^* + q_2C^* + q_3S^*;\ \ 0 < q_i < 1,\ i = 1, 2, 3. \quad (2)$$

We also assume that the waste of the AES $g = G'$ is the part of the resources incoming into the biosphere:

$$f^*(t) = c^*(t-e(t))+g^*(t-h(t))+g(t-u(t))+(1-q_3)s^*(t),\ s^* = S^{*\prime}, \tag{3}$$

where e, h, and u are delays or lags of transformation into f^* on the strength of the bio-geo-chemical cycle in the biosphere.

The MM of business connected with interaction of HA and the biosphere can be now rewritten in the form

$$m(t) = \int_{a(t)}^{t} \alpha(t,s)y(s)m(s)ds,\ M(t) = \int_{0}^{t} m(s)ds,\ c(t) = \int_{b(t)}^{t} \beta(t,s)[1-y(s)]m(s)ds,$$

$$P_1(t) = \int_{a(t)}^{t} y(s)m(s)ds,\ P_2(t) = \int_{b(t)}^{t} [1-y(s)]m(s)ds,\ P(t) = P_1(t) + P_2(t),$$

$$G(t) = M(t) - P(t), f(t) = m(t) + c(t),\ 0 \le y \le 1,\ 0 \le a(t),\ b(t) \le t,\ t \ge t^\wedge \ge 0;$$

$$m^*(t) = \int_{a^*(t)}^{t} \alpha^*(t,s)y^*(s)m^*(s)ds - q_1M^*(t),\ M^*(t) = \int_{0}^{t} m^*(s)ds,$$

$$c^*(t) = \int_{b^*(t)}^{t} \beta^*(t,s)[1-y^*(s)]m^*(s)ds - q_2C^*(t),\ C^*(t) = \int_{t^\wedge}^{t} c^*(s)ds,$$

$$P^*_1(t) = \int_{a^*(t)}^{t} y^*(s)m^*(s)ds,\ P^*_2(t) = \int_{b^*(t)}^{t} [1-y^*(s)]m^*(s)ds,$$

$$P^*(t) = P^*_1(t) + P^*_2(t),\ G^*(t) = [M^*(t) - P^*(t)],\ f^*(t) = m^*(t) + c^*(t),$$

$$0 \le y^* \le 1,\ 0 \le a^*(t),\ b^*(t) \le t,\ t \ge t^\wedge \ge \min[a^*(t^\wedge), b^*(t^\wedge)] = 0. \tag{4}$$

Here the functions with the sign * have approximately the same sense as the functions without * (see Ch. 9 in detail). Given α, β, P_i, $i = 1, 2$, y, α^*, β^*, y^*, e, h, u, s^*, q_i, $i = 1, 2, 3$, as well as the prehistory, (2)-(4) are the systems of 19 equations relative to 19 indeterminate: m, M, a, b, c, f, P, G, m^*, M^*, a^*, b^*, c^*, C^*, f^*, P^*_1, P^*_2, P^*, and G^*.

Using a general mathematical approach, it is not difficult to introduce different generalizations of MM (1)-(4) and some additional MM (see Ch. 2).

2. DEFINITION OF WP AND MODELS OF α AND β

WP is localized in productive space and time by labor means and its organization learned combination of labor functions whose fulfillment during any given temporal period required a labor activity of one worker during a full-time period. As a result, WP is characterizing by many factors, such as technology level, i.e., quality and quantity of machinery (technological ware); quality and quantity of raw materials (raw ware); energy-equipment (energy ware); a share basic fund forced per one WP (fund-equipment); presence of personnel, their organization and structure (personnel and organized ware); personnel qualification with accordance to equipment level; collection of fulfilled labor functions; actual volume and quality of production produced per one WP in the unit of the time; coefficient of WP loading; presence of technical documentation, instructions and descriptions (information ware), and also several social and psychological factors, such as level of technological comfort (presence of convenient and effective apparatuses); level of psychological comfort (level of salary, extent of well-being as far as housing and other problems are concerned, satisfaction of work, a psychological climate in the community, level of personal interest and responsibility of the community members, and so on).

Variable (essential) of any of indices or any of their collections in the list means origin of the new WP. The problem consists in that the new WP was more effective, i.e., the actual volume and/or quality of the production produced on the new WP was higher than on the old WP.

Apparatus and macro-models of business as ES must be such that one can resolve the problem mentioned. Therefore, the main elements of the models of business must be indices of WP efficiency: α for the subsystem A, and β for the subsystem B. In addition, there must be constructed the models of bounds of α and β for every WP depending on the various above-mentioned factors, and also the models of obtaining of the similar indices for the aggregate or standard WP.

Here one is in a need of the solution of, at least, the following *identification problem*: Let

$$\alpha(t,u) = \alpha(u)\exp(-d_1 t),\ \beta(t,u) = \beta(u)\exp(-d_2 t),\ d_1, d_2 > 0, \tag{5}$$

and let $c(t)$,

$$P_1(t) = \int_{a(t)}^{t} y(u)m(u)du,\ G_1(t) = \int_{a(0)}^{a(t)} y(u)m(u)du,\ M_1(t) = P_1(t) + G_1(t),$$

$$P_2(t) = P(t) - P_1(t),\ G_2(t) = G(t) - G_1(t),\ M_2(t) = P_2(t) + G_2(t), \tag{6}$$

and d_1, d_2 are given on the prehistory (d_1 and d_2 can be obtained at the verification stage of the prediction). It is required to find $\alpha(u)$ and $\beta(u)$.

Knowing P_i, G_i, $i = 1, 2$, we can find $m(t)$, $a(t)$, $b(t)$, and $y(t)$:

$$m(t) = [P(t) + G(t)]', \tag{7}$$

$$a(t) = M_1^{-1}[P_1(t) - M_1(t)],\ b(t) = M_2^{-1}[P_2(t) - M_2(t)],\ y(t) = M_1'(t)/m(t), \tag{8}$$

where M_i^{-1}, $i = 1, 2$, are the reciprocal functions for M_i. Knowing in addition c, we can find $\alpha(u)$ and $\beta(u)$:

$$\alpha(u) = [m\hat{}'(u) + \alpha(a(u))y(a(u))m\hat{}(a(u))a'(u)]/[y(u)m\hat{}(u)],\ m\hat{}(u) = m(u)\exp(d_1 u),$$

$$\beta(u) = [c\hat{}'(u) + \beta(b(u))(1-y(b(u))m\hat{}\hat{}(b(u))b'(u)]/[(1-y(u))m\hat{}\hat{}(u)],$$

$$c^\wedge(u) = c(u)\exp(d_2u),\ m^{\wedge\wedge}(u) = m(u)\exp(d_2u), \tag{9}$$

We have two different cases here:

$$t^\wedge = a(t^\wedge) = b(t^\wedge) = \alpha(0) = \beta(0) = 0 \tag{10}$$

and

$$t^\wedge > 0. \tag{11}$$

In the first case $\alpha(u)$ and $\beta(u)$ can be found with the help of (9). In the second case we are in a need of the restoration of data back to the instant 0 of ES origin.

It should be noted that for realization of the formulae (6)- (11), we have to use the technique of the solution of the *ill-posed problem* of numerical differentiation (Ch. 5).

For some other identification problems see Ch. 2, 8, 9, and 11.

3. STATEMENT OF PROBLEMS

We can solve at least the following modeling problems:

3.1. The solution of problems of the type "If . . ., then . . ." and/or "In order to . . ., it is sufficient . . .";

3.2. The solution of optimization problems.

The examples of modeling problems 3.1.are

i) If the business efficiency indices, α and β, are increasing two percentage per year then how fast will $m(t)$ be increasing for the certain values of the distribution function $y(s)$?

ii) In order for the business functioning resources $P(t)$ to be decreasing, that is $P'(t) < 0$, it is sufficient that $y(s)$ will be given certain values.

For examples of optimization problems see Ch. 2, 4, 8- 11. Usually, we have deal with the problems of the best distribution of resources among several subsystems of business as ES. However, there are also other factors.

Let us dwell on the important factor such as the method of construction of the given collection of labor functions (operations) $F_1, F_2,..., F_k$ with the help of the given basic universal or complete set of labor functions (operations) $f_1, f_2,..., f_n$. Universality of the basic set means that for any F_j there exist integers v_{ij}, the number of operations f_i in the appropriate "program" F_j, such that the total number of basic operations for realization F_j

$$N(F_j) = \sum{}^{n}{}_{i=1} v_{ij},\ j = 1,..., k. \tag{12}$$

Let basic operations f_i have different means of reliability for their fulfillment with the total costs γ_i. Then fulfillment of the given set of labor functions $F_1, F_2, \ldots, F_k$ will have the total cost

$$F = \sum_{j=1}^{k} \sum_{i=1}^{n} \gamma_i v_{ij}. \qquad (13)$$

The problem is to minimize the total cost (13) with the help of $\{v_{ij}\}$, where $\{v_{ij}\}$ means all possible collections of integers satisfying the relation's (12). It should be in reality taken into account the cost for the solution of the optimization problem itself.

Yet, the more interesting problem can be that of minimization of F not only on v_{ij}, but also on the initial set of basic operations f_i and k itself. All of these problems are of great importance, especially, in the condition of the flexible business when we need to variable not infrequently the labor functions F_i, $i = 1, 2, \ldots, k$.

Many results about existence, uniqueness, and structure of the solutions of the equations of MM and some optimization problems one can find in Ch. 2, 4, 8-11 (see also [8]). As to the problem (13), the many respective results can be obtained from the general theory of algorithm optimization (see, for example, [3, Ch. 6]).

In particular, from this theory it follows that in the general case there is no simple program for realization any F_j with the help of the fixed set of $f_1, \ldots, f_n$. It means that the number $N(F_j)$ (12) can be arbitrary large. Regarding the importance for the practical case of the so-called NP-complete problems [1, Ch. 7], the number $N(F_j)$ may be exponentially large with respect to the number of F_j initial data.

iv) SPACE CREW AS ES

1. BASE MM

Regarding the structure and physiological functions of a crew as ES, its simplest, *base MM of a space crew* can be represented in the form

$$m(t) = \int_{a(t)}^{t} v(t,u)y(u)m(u)du,\ c(t) = \int_{b(t)}^{t} w(t,u)[1-y(u)]m(u)du,\ P(t) =$$

$$\int_{a(t)}^{t} y(u)m(u)du + \int_{b(t)}^{t}[1-y(u)]m(u)du,\ G(t) = M(t) - P(t),\ M(t) = \int_{0}^{t} m(u)du,\ m(t) =$$

$$k(f\hat{\ })f(t), f(t) = m(t) + c(t),\ 0 \le y,\ k(f\hat{\ }) \le 1,\ 0 \le a(t\hat{\ }), b(t\hat{\ }) \le t\hat{\ }, t \ge t\hat{\ }, \qquad (1)$$

where m is the new resource (per unit of the time) of a crew providing its internal functions (recreation of m); c is the new resource providing external functions of a crew; y is the distribution function; v and w are the efficiency indices of functioning along the channels ym--m and $[1-y]m$--c respectively; $a(t)$ and $b(t)$ are the temporal bounds: at the segments $[a(t^\wedge),a(t)]$ and $[b(t^\wedge),b(t)]$ the respective resources ym and $[1-y]m$ are considered as obsolete; $P(t)$ is the total amount of functioning internal resource of a crew; $G(t)$ is the total amount of the obsolete resource; $k(f^\wedge)$ is the coefficient depending on the qualitative structure $f^\wedge$ of the external resource f; $t^\wedge$ is the starting time for modeling; $[0,t^\wedge]$ is the prehistory time on which all the elements are given.

It is obvious that all the relations (1) are exact by definition and that all the elements of the model are nonnegative; the functions v and w can depend on the other elements in the general case.

The total number of different elements in (1) is 12 of which three, namely y, $k = k(f^\wedge)$, and f can be considered as the control elements of the model. The total number of the equations in (1) is seven. If the exogenous character functions v, w, and the control element y, k, f are given, we will have a determined system of the equations.

Knowing y, k, and f, we can find m and $a(t)$; knowing $c = f - m$, we can find $b(t)$; knowing a and b, we can find P, G, and M.

2. STATEMENT OF PROBLEMS

1/. Problems of the type "*If . . ., then . . .* ":

1. If the control elements y, f, and k and functions v and w are given in a predetermined manner, then what kind of the solution of the systems (1) will the functions m, a, c, b, P, G, and M be.

2. If v, w, y, P, and G are given in a predetermined manner, then what kind of the solution will a, b, m, c, f, k, and M be.

3. *If* v, w, c, P, and y are given, then what kind of the solution will m, a, b, f, k, G, and M be.

2/. *Optimization problems*:

1. It is required to find y, f, and k such that the functional

$$I1 = \int_{t^\wedge}^{T} c(t)dt = \max \tag{2}$$

(maximum of external activities of a crew) under observing of (1) (T is given).

2. It is required to find y, f, and k such that

$$I2 = \int_{t^\wedge}^{T} P(t)dt = \min \tag{3}$$

(minimum number of members of a crew) if the relations (1) are observed.

3. It is required to find y, f, and k for which

$$I3 = T = \min \tag{4}$$

(optimal control by speed) if the relations (1) and the conditions

$$C(T) = \int^{T}_{t^\wedge} c(t)dt \in C^*,\ P(T) \in P^*, \tag{5}$$

where sets C^* and P^* are given, are observed.

4. It is required to find y, f, and k for which

$$I3 = T = \max \tag{6}$$

(maximum of an active crew life) if the relations (1) and the restriction

$$c(t) \geq c^0(t), \tag{7}$$

where $c^0(t)$ is given, are observed.

3/. Identification problems.

We can repeat all what has been described above.

3. MORE GENERAL MM

It is not difficult to introduce different extensions of the two-subsystem model (1). In particular, introducing the equation

$$c^\wedge(t) = \int^{t}_{b(t)} w^\wedge(t,u)x(u)m(u)du, \tag{8}$$

where w^ is the index of efficiency for functioning of a member of a crew along the channel xm--$c^\wedge$, $c^\wedge$ is the new resource providing the external function of this member of a crew, and the replacing 1-y by 1-x-y and $f = m + c$ by $f = m + c + c^\wedge$, we will have the three-subsystem model. Introducing for a member of a crew the model of (1)-type besides the model of a crew, we will have the fore-subsystem model. Introducing for every member of a crew the model of (1)-type, we will have the 2N-subsystem model, where N is the number of members of a crew. Introducing for every member of a crew, in addition, the model of (8)-type, we will have the 3N-subsystem model, and so on.

4. CONCLUSION

We would like to emphasize in conclusion the following advantages of these models:

1. All the elements of the respective models have profound technological and physiological interpretation.
2. All the relations of the models are exact by definition under which we can apply qualitative methods for investigation of different problems.
3. Examination in detail of a member of a crew based on the models of a crew as a whole.
4. The solution of the problem "If . . ., then . . . " reduced to the solution of Volterra-type equations that can be solved with the pre-assigned accuracy by the optimal numerical methods.
5. The solution of the identification problems for the given crew and its given member can be found from the known prehistory of the crew functioning and the relations of the models themselves.
6. The solution of the optimization problems can be also reduced to the solution of Volterra-type equations (see Ch. 6).
7. The models under consideration can be useful, especially, in the self-regulating operating conditions during prolonged action of a crew.

v) HEALTH CARE AS ES

1. INTRODUCTION

The health services or health care (HC) external product is the cure, i.e., the number of people who have improved their health to such extent that they do not need providers of the care.

In the ES language it means that we need to improve indices of efficiency of functioning for both subsystems A and B in HC as ES. Here A is, as usual, the subsystem creating new, more effective WP for itself and for the subsystem B that, in turn, is the aggregate WP with curing as the main labor function. In addition, the subsystem A comprises not only all doctors, so-called organizers of the health system [7], but any doctor who enlarges his/her own qualification and, in general, any person while helping to perfect HC or medicine as ES.

A special feature of HC as ES is the fact that, on its WP in the subsystem B, "raw" materials that should be cured are living organisms that, in turn, can be considered as NES. So, we have to deal with interaction between AES and NES (see Ch. 1 and Ch. 9). We would like to emphasize here again that according to the theory of NES the most difficult cases of pathology and diseases themselves and, what is primary, the reasons for these diseases are related to an infraction in an organism system namely A-type (see Ch. 11 in more detail).

According to the theory of ES, for successful development of HC (under the same resources inflow), it is important that we distribute its internal resources between its subsystems A and B in a reasonable way, and, in the same way, we determine the temporal boundary for the obsolete WP in both A and B and, of course, that we have a competitive market of interaction between HC and its consumers, and so on.

The respective formalized, mathematical apparatus allows us to consider and investigate such kinds of problems both on the qualitative and quantitative level.

2. MM OF HC ORGANIZATIONAL STRUCTURE

This MM has the form

$$\alpha'(t) = \int_{a(t)}^{t} \alpha(t,s)x(s)m(s)ds,\ O(t) = \int_{a(t)}^{t} x(s)m(s)ds,\ 0 \le a(t) \le t,$$

$$\alpha(t,s) = \alpha(s)\exp[-d(t-s)] \vee \alpha(t,s) = \alpha(t-s)\exp(dt),\ d \ge 0,$$

$$m(t) = \int_{a(t)}^{t} \alpha(t,s)y(s)m(s)ds,\ P(t) = \int_{a(t)}^{t} y(s)m(s)ds,\ \beta_i(t) = \int_{a(t)}^{t} \alpha(t,s)z_i(s)m(s)ds,$$

$$Q_i(t) = \int_{a(t)}^{t} z_i(s)m(s)ds,\ \beta_i(t,s) = \beta_i(s)\exp[-d_i(t-s)] \vee \beta_i(t,s) = \beta_i(t-s)\exp(d_it),\ d_i \ge 0,$$

$$c_i(t) = \int_{a(t)}^{t} \beta_i(t,s)u_i(s)m(s)ds,\ R_i(t) = \int_{a(t)}^{t} u_i(s)m(s)ds,\ x(t)+y(t)+\Sigma^n_{i=1}[z_i(t)+u_i(t)] =$$

$$1,\ 0 \le x(t), y(t), z_i(t), u_i(t) \le 1,\ i = 1,\dots,n,\ P^\wedge(t) = O(t) + P(t) + \Sigma^n_{i=1}[Q_i(t) + R_i(t)],$$

$$f(t) = \alpha'(t) + m(t) + \Sigma^n_{i=1}[\beta_i(t) + c_i(t)],\ t \ge t^\wedge \ge 0, \qquad (1)$$

where $O(t)$ is the number of WP in the HC subsystem of recreation of the new, more effective technology $\alpha(t,s)$ at the instant t, $P(t)$ is the number of WP in the HC subsystem of creation of the new, more effective WP $m(t)$, $Q_i(t)$ is the number of WP in the subsystem of creation of the new technology $\beta_i(t,s)$ for ith disease, $R_i(t)$ is the number of WP in the ith HC branch for curing of ith disease, $c_i(t)$ is the rate of number patients have been cured for ith disease, and $[0,t^\wedge]$ is, as usually, the prehistory for which all the elements are supposed to be given.

Let $O(t)$, $P(t)$, d, $Q_i(t)$, $R_i(t)$, and d_i be given. Then with the help of the relation (1) we can find $P^\wedge$ and after this a, m, α, β_i, c_i, and f (see the theorems in Ch. 3).

3. ON MM OF A DISEASE

We should consider first MM of OS (Ch. 11). All the elements in these MM have the profound sense from the point of view of the structure and functions of OS. So, it is natural to suppose that one of the conceptions of the norm of OS functioning consists in the validity of the relations (see also Ch. 11)

$$r^-_s \leq rs \leq r^+_s,\ s = 1,...,10,$$

$$r1 = m,\ r2 = a,\ r3 = \alpha,\ r4 = y,\ r5 = c,\ r6 = \beta,\ r7 = P,\ r8 = G,\ r9 = M,\ r10 = f, \quad (2)$$

where r^-_s and r^+_s are admissible bounds for variations of the introduced values from point of view of the norm. On the other hand, if at least one of the relations in (2) is violated, then it is natural to speak about the pathology or the pathological state of OS. Of course, there can be other values including the traditional ones: temperature, pressure, etc. that are certain functions from those in (2).

The simplest, *base MM of a disease* has the form those MM in Ch. 11 under the condition that at least one of the relations in (2) does not take place.

The disease or pathology can be detailing by discrimination of two classes of (2) violations: A) those of the resource values, namely *m, c, P, G, M,* and *f*-types and B) those of the structural values, namely *a, α, y, b,* and *β*-types. In turn, the indices of class A) can be decomposed into two groups: A') the indices connected with the internal functions of resource recreation, namely *m, P, G, M,* and A") the indices connected with the external functions of reaction centers, namely *c* and *f*.

The treatment of a disease based on the model (1) and (2) can be considered, in general, under three aspects:

1. prophylaxis, when we hold the relation's (2)-type by using the permissible control factors;

2. diagnosis, when we determine the violations of at least one of the relations (2) by using the direct and/or indirect indices of OS;

3. therapy, when we reduce the improper values of the magnitudes in (3) to the norm by using the permissible factors.

To realize this program, we are in a need of the application of all the techniques considered in this book.

4. ON MM OF A DOCTOR'S BUSINESS

The simplest, *base MM of a doctor's business*, according to the general theory of ES, has the form

$$m(t) = \int_0^t \alpha(t,u)\lambda(t,u)y(u)m(u)du,\ c(t) = \int_0^t \beta(t,u)\mu(t,u)[1-y(u)]m(u)du,$$

$$P(t) = \int_0^t \lambda(t,u)y(u)m(u)du + \int_0^t \mu(t,u)[1-y(u)]m(u)du,\ G(t) = M(t) -$$

$$P(t),\ M(t) = \int_0^t m(u)du,\ f(t) = m(t) + c(t) = kc(t),\ t \geq t\hat{} \geq 0, \qquad (3)$$

where m is the new resource (per unit of the time) of a doctor's organism providing its internal functions (recreation of m); c is some new resource providing external functions of a doctor including the function of curing; λ, μ are the intensities of functioning along the channels ym--m, $[1-y]m$--c respectively; $P(t)$ is the total amount of functioning internal resource of a doctor's organism; y, α, β, $a(t)$, $b(t)$, G, k, f, and $t\hat{}$ have the same sense as in (2).

The function of curing decides a doctor's business. The number of patients c_i (see (1)) who have been cured for ith decease in the unit of the time is decided actually by the same function. Therefore, this function should be well estimated. It should be estimated by the total number of the patients who have been successfully cured in the certain period of the time.

The methods of curing or restoration of the norm depend on many factors. We would like to dwell only on some approaches to curing following from the models.

The most simple and extended restoration is that of when the determined values of the group A") are injected into the organism directly or on the contrary are removed from the organism and/or the respective medicines are injected into the organism so that to achieve validity of the necessary bounds (2) relative to this group. However, under the condition that the indices of violations of the group A") are accompanied by those of the group A') and class B) such a simple method is usually the temporal measure and does not deduce restoration.

For increasing (decreasing) of the values of $m(t)$-type, if the values of the other magnitudes are not broken, it is usually necessary to increase (decrease) the values $y(s)$, $s < t$. In this connection it should be noted that the values $c(t)$ can be even increased (see Ch. 4).

Thus, the crucial condition for restoration (under weakened functions of the class A) is the increase of a share of all the resources of OS to its internal requirements. It is possible to hope that any modes of action, among them nontraditional ones [2], including the various physical loading on a certain part of an organism and contributing to increasing of resource distribution in advantage to the internal sphere of the sick subsystem will usually bring an essential positive effect.

The most serious and profound violations are connected with the structural shift of the subsystem, i.e., with the violation of (2) for the indices of the class B. Here, only interference into the *gene* apparatus (*gene* therapy) can likely restore the norm. Although, in principle, in this case there can be also developed the theory similar to (1)-(3) (see Ch. 2 and Ch. 11).

vi) FORMALIZATION OF CIVILIZATION CONCEPTION

According to the well-known methodology of the systems analysis (see, for example, [5, Ch. 8]), in the general plane we should: 1. give the qualitative verbal determination of Civilization that appropriates to update ideas; 2. do parameterization of Civilization; 3. construct MM of Civilization.

We induce the further description along the above scheme.

1. DETERMINATION OF CIVILIZATION

According to [6], p.553 "Civilization (from Latin "Civilis", i.e., civil) is high level, stage or state of social development of material and spiritual culture. *Conception of Civilization* appeared in XVIII century in the tight connection with conception "culture." French philosophers of the Enlightenment called human society civilized if it was based on the principles of reason and justice.

Regarding this general theory, one of definition of Civilization may be the following: *Civilization is the neo-sphere's state responding to determined civilized goals posed by human society before itself.*

Idea "civilized goals", obviously, requires from us clarification here. It is not hard to formulate "desired goals" for development of human society that would suit majority of people if they could be realized. As example of such goals, there can be achievements of determined level of comfort for every member of society during determined period of the time. In particular, achievement of the goal: "Present generation of Soviet people will live under communism" (Chrushchev). However, "wanted", but unattainable goals or attainable goals at the expense of essential decrease of the possibility of achievement of other desired goals should not be included in the civilized ones. So, the civilized goals are naturally the attainable desired goals regarding observance of several civilized limitations. In turn, idea "civilized limitations" requires clarification here.

The historic trial of human beings and the nature interaction and the interaction among people themselves result in the several postulates of common moral and the culture of interaction with the nature. This postulates and objective laws of the interaction naturally should be taken as the civilized limitations.

Enumerating all the possible postulates is rather difficult. In particular, we must enumerate all the commandments of the Bible and all the clauses of the Declaration of the Human rights. Therefore, we would like to note the most popular: 1). Do to everyone what you want to be done to yourself; 2). From every by ability to every by labor (necessary for society); 3). Work hard and carefully and do not harm, but better help; 4). Know yourself and make self-perfection; 5). Keep family and friendship.

Combining the postulates 1)-3) we come, in essence, to the well-known principle of "coordinated optimum". Observance of all the five postulates may probably be as the guarantee for progressive society and the biosphere golden age and also the productive and efficient labor growth. Of course, we must as well define more exactly the ideas: "necessary for society labor," "do not harm," and "self-perfecting." We will formalize them later.

Note also two main achievements of civilization: democracy and control of economy through the market.

"Democracy background is respect of human personality and a law leadership. Democracy is the best guarantee for liberty of anyone's opinion expression, tolerance related to all the groups in the society and equal opportunities for every man" ("Paris Charter for the new Europe", 1990, February). "Human dignity is sacred. To respect and defend it is duty of any state power" (article 1 of the principal law of FRG).

As to control of economy through the market, the position of dynamic system theory is entirely clear: the market is "governor" with negative "feedback", which decreases production in the case it exceeds consumption and increases production of any commodities in the case they become critical. However, in spite of obvious usefulness of such "governor" with the point of view any specialist in the control theory, in the several countries politicians undertook and are undertaking much effort not to tolerate operation of the market laws.

2. PARAMETERIZATION OF CIVILIZATION

From 1., it follows that for parameterization of Civilization it is necessary to formalize the civilized (desired attainable) goals for society's development and the civilized limitations.

Since in general case civilized goals are formed by means of social opinion inquiry with the help of appropriate procedure (see, for example Ch. 8), we confine ourselves here to probable examples that are not occurring earlier.

Let, as usually, t be the time, $t^\wedge$ be the starting time for modeling, $t^\wedge \geq 0$; $m(t)$ be the outcome of the new (aggregated) work places (WP) in the unit of the time at the instant t; $M(t)$ be the total outcome of WP during the period t-0; $c(t)$ be the outcome of the new (generalized) consumption goods and services per unit of the time at the instance t; $C(t)$ be the total outcome of the consumption product during the period t-$t^\wedge$; $\alpha(t,s)$ be the index of efficiency for functioning of WP in the subsystem A of restoration of WP; $\beta(t,s)$ be the index of efficiency for functioning of WP in the subsystem B of production of $c(t)$; $\lambda(t,s)$ and $\mu(t,s)$ be the intensities of utilization of WP in A and B respectively; $P_1(t)$ be the total number of functioning WP in the subsystem A at the instant t; $P_2(t)$ be the total number of functioning WP in the subsystem B at the instant t; $P(t)$ be the total number of functioning WP at the instant t; $G(t)$ be the total number of the obsolete WP at the instant t; $y(s)m(s)$ be a share of $m(s)$ running for restoration of WP in A, $0 \leq y \leq 1$; $m^*(t)$ be the rate of recreation of the new (generalized) ecological product of the neo-sphere (alive substance or bio-mass [10, Ch1]) in its subsystem A* at the instant t; $M^*(t)$ be the total quantity of the bio-mass created by the neo-sphere during the period t-0; $c^*(t)$ be the rate of formation of the new (generalized) ecological product of "consumption" (bio-boned substance) in the subsystem B* of the neo-sphere at the instant t; $C^*(t)$ be the total quantity of the bio-boned substance created by the neo-sphere during the period t-$t^\wedge$; α^*, β^*, λ^*, μ^* be the indices of efficiency and intensity of the neo-sphere functioning similar to the indices α, β, λ, μ; $P^*(t)$ be the total quantity of the alive bio-mass in the neo-sphere in the instant t; $y^*(s)m^*(s)$ be a share of $m^*(s)$ running for creation of $m^*(t)$, $0 \leq y^* \leq 1$; $f(t)$ be the rate of inflow of the external resources for human activity consisting from $q_1M^*(t) + q_2C^*(t) + q_3S^*(t)$, where $0 \leq q_i \leq 1$, $i = 1, 2, 3$, S^* be the total resource inflow into the neo-sphere from the cosmos; $f^*(t)$ be the rate of inflow into the neo-sphere of the generalized resources at the instant t consisting from $G'(t-e(t)) + G^{*\prime}(t-h(t)) + C^{*\prime}(t-u(t)) + (1 - q_3)S^{*\prime}(t)$, where $e(t)$, $h(t)$, $u(t)$ be delays in the process of the bio-geo-chemical cycle into the neo-sphere; k^* be the coefficient such that $f^*(t) = k^*c^*(t)$; $[0, t^\wedge]$ be the prehistory for which all the functions considered are as given. Then we have, for example, the following variants of formalization of the civilized goals with regard to the civilized limitations:

A. $I1 = C(T) = \max (U) | \; U = (y, y^*, q_1, q_2, q_3), (M^*, P^*, C^*) \in A;$

B. $I2 = P(T) = \min (U) | c \in c^*, (M^*, P^*, C^*) \in A;$

C. $I3 = T - t^\wedge = \min (U) | c(T) \in c^*, (M^*, P^*, C^*) \in A;$

D. $I4 = T^* = \max (U) | (\alpha, \beta \in A, (M^*, P^*, C^*) \in A;$

E. $I5 = \{ U | \nexists V : Ii(V) \geq Ii(U), i = 1, 4, Ii(V) \leq Ii(U), i = 2, 3, V \neq U,$

$$c(t) \in c^*, (M^*, P^*, C^*) \in A\}, \tag{1}$$

where T- t^ is the time for achievement of the goals; T^* is the average time for an active useful human life; c^*, A, and A are permissible sets, $\nexists$ is the sign of no existence, / is the sign of the condition. Almost all these values (quantities) are required to be made more exact. As examples of the permissible set we can consider:

$$c^* = \{c(t)/c(t) \geq c^{\check{}}\}, A = \{M^*, P^*, C^*/M^* \geq M^{\check{}}, P^* \geq P^{\check{}}, C^* \geq C^{\check{}}\}. \tag{2}$$

Here $c^{\check{}}$, $M^{\check{}}$, $P^{\check{}}$, and $C^{\check{}}$ are the permissible levels of the respective quantities. The permissible level of consumption $c^{\check{}}$ is almost intelligible quantity that means that under $c(t) < c^{\check{}}$ the civilized standards in the sense of the scientific-based norm of consumption of different kinds of goods and services, and also the behavior's culture may be broken. Civilized standards, in turn, are obtained from long investigation of the conditions for the normal human activity (first, not resulting in different diseases) concerning the peculiarities of environment.

By analogy, every of the values $M^{\check{}}$, $P^{\check{}}$, and $C^{\check{}}$ means that there will be some possible undesirable, morbid phenomena in the biosphere (or the neo-sphere) if any of the values $M^*(t)$, $P^*(t)$, and $C^*(t)$ is appeared less than the respective permissible level. Of course, the standards $M^{\check{}}$, $P^{\check{}}$, and $C^{\check{}}$ must be also estimated from long investigation of the conditions for normal functioning of the neo-sphere.

The value T^* is connected with the idea of "actively useful (civilized) life" in which "active" may mean, that the values of the functions α and β - types for the set of the specialists of age T will satisfy the condition

$$\alpha(t,s) > \alpha\hat{}, \beta(t,s) > \beta\hat{}, \tag{3}$$

where α^ and β^ are also the standards of functioning efficiency of the respective WP, e.g., the standards such that the personnel of these WP has moral satisfaction.

It is clear that the problem E. determines the so-called *set of Pareto* on which simultaneous improvement of the values $I1$, $I2$, $I3$, and $I4$ is impossible. The notion of self-perfection is the personal effort that likely includes the tendency to hold the condition (2), (3), and all those in (1).

Of course, (1)-(3) are a few examples of the relations that may be needed in practice. For the general mathematical decision making theory see [11, 15].

3. MM OF CIVILIZATION

The values under consideration can be connected among themselves. Determination of the appropriate bonds means construction of necessary MM. Thus, the *base MM of civilization* as ES:

$m(t) = \int_0^t \alpha(t,s)\lambda(t,s)y(s)m(s)ds$, $M(t) = \int_0^t m(s)ds$, $c(t) = \int_{t^\wedge}^t \beta(t,s)\mu(t,s)[1-y(s)]m(s)ds$, $C(t) = \int_{t^\wedge}^t c(s)ds$, $P_1(t) = \int_0^t \lambda(t,s)y(s)m(s)ds$, $P_2(t) = \int_0^t \mu(t,s)[1-y(s)]m(s)ds$, $P(t) = P_1(t) + P_2(t)$, $G(t) = M(t)-P(t)$, $f(t) = m(t)+c(t) = q_1M^*(t)+q_2C^*(t)+q_3S^*(t)$, $0 \le y, \lambda, \mu \le 1$, $t \ge t^\wedge \ge 0$;

$$m^*(t) = \int_0^t \alpha^*(t,s)\lambda^*(t,s)y^*(s)m^*(s)ds - q_1M^*(t),\ M^*(t) = \int_0^t m^*(s)ds,$$

$$c^*(t) = \int_{t^\wedge}^t \beta^*(t,s)\mu^*(t,s)[1-y^*(s)]m^*(s)ds - q_2C^*(t),\ C^*(t) = \int_{t^\wedge}^t c^*(s)ds,$$

$$P^*(t) = \int_0^t \{\lambda^*(t,s)y^*(s)+\mu^*(t,s)[1-y^*(s)]\}m^*(s)ds,\ G^*(t) = [M^*(t) - P^*(t)],$$

$$f^*(t) = m^*(t) + c^*(t) = G'(t-e(t)) + G^{*\prime}(t-h(t)) + C^{*\prime}(t-u(t)) +$$

$$(1-q_3)S^{*\prime}(t) = k^*c^*(t),\ 0 \le y^*, \lambda^*, \mu^* \le 1,\ t \ge t^\wedge \ge 0. \qquad (4)$$

Given α, β, P_i, $i = 1, 2$, y, α^*, β^*, y^*, k^*, e, h, u, S^*, q_i, $i = 1, 2, 3$, as well the prehistory, (8) are the systems of 19 equations relative to 18 indeterminate m, M, λ, μ, c, C, f, P, G, m^*, M^*, λ^*, μ^*, c^*, C^*, f^*, P^*, and G^*. Certain results for the model (8) and the problem (1)-(4) see Part I and Ch. 9.

vii) A FEW ADDENDA

1. NEW COMPUTERS AS ES

The principal structure of new computers as ES is similar to that of in the Fig.1, Ch.1, and their MM are also similar to the considered above, but the interpretation, of course, is different. Let the *base MM of new computers*

$$m(t) = \int_{a(t)}^t \alpha(t,s)y(s)m(s)ds,\ M(t) = \int_0^t m(s)ds,\ c(t) = \int_{a(t)}^t \beta(t,s)[1-y(s)]m(s)ds,$$

$$C(t) = \int_{t^\wedge}^t c(s)ds,\ P(t) = \int_{a(t)}^t m(s)ds,\ G(t) = M(t) - P(t),\ f(t) = m(t) + c(t),$$

$$0 \le y \le 1, 0 \le a(t) \le t, t \ge t^\wedge \ge 0. \quad (1)$$

Then $m(t)$ is the rate of creating (in bits) of the new systematic computer software in the computer subsystem A at the instant t, $\alpha(t,s)$ is the index of efficiency of computer functioning along the channel *ym--m* or the index of the subsystem A efficiency, $y(s)$ is a relative share of distribution of $m(s)$ between the subsystems A and B, $c(t)$ is the rate of creation of the new applied software in the subsystem B, $\beta(t,s)$ is the index of the subsystem B efficiency, $P(t)$ is the functioning software, $G(t)$ is the obsolete software, $f(t)$ is the resource inflow from outside.

The optimization problem

$$C(T) = \max (y) \quad (2)$$

under the restrictions (1) now means maximization of quantity of the applied software. From the earlier obtained results it follows that the desired solution has the property: *an essential relative share y must go to the computer subsystem* A. It means that we can create more applied software by the perfect tools of general software.

For the completion of the model (1), we can introduce the following relations:

$$\alpha'(t) = \int_{a(t)}^{t} \alpha(t,s)x(s)m(s)ds,\ m(t) = \int_{a(t)}^{t} \alpha(t,s)y(s)m(s)ds,\ \beta(t) =$$

$$\int_{a(t)}^{t} \beta(t,s)z(s)m(s)ds,\ c(t) = \int_{a(t)}^{t} \beta(t,s)[1-x(s)-y(s)-z(s)]m(s)ds,\ \alpha(t,s)/\alpha(s) =$$

$$\beta(t,s)/\beta(s) = \exp[-d(t-s)],\ P(t) = \int_{a(t)}^{t} m(s)ds,\ f(t) = m(t) + c(t) + \alpha'(t) + \beta(t),$$

$$0 \le x, y, z, 1-x-y-z \le 1, 0 \le a(t) \le t, t \ge t^\wedge \ge 0, \quad (3)$$

where d is the constant subjected to identification, $\alpha'(t)$ and $\beta(t)$ mean not only the indices of efficiency, but also the respective computer hardware. The sense of the rest of the magnitudes is the same as in (1). Knowing x, y, z, and P, we can express $a(t)$ through m, after which we can solve the system of the first two equations with respect to α and m and find the other magnitudes on the strength of their definitions in (3).

It was shown earlier (see Ch. 4) that for the solution of the problem $C(T) = \max (y)$, with the replacement of y by x, y, z, the highest priority has x, next y, and then z.

2. SCIENCE AS ES

The Science (S) final product is the new, more effective technology. Since the new technology results in higher level values of indices of WP efficiency, we can assume that the S external products are these indices of WP efficiency for different ES and both its subsystems A and B. The internal product of the subsystem A of S as ES is new technology (new WP in B) for creation of the external new technology and new technology (new WP in A) for creation or restoration of itself. So, we can accept that indices of WP efficiency in the subsystem A of S and their products are actually the same. But the determined part of these products goes to the subsystem B and the remainder goes back to the subsystem A.

Any scientific center, any branch of S and S as a whole can be considered as ES. At all events the subsystem A of ES is the set of the WP creating new technology for itself and external ES.

Thus, the simplest, *base MM of S* as ES has the form

$$\alpha'(t) = \int_{a(t)}^{t} \alpha(t,s)x(s)m(s)ds, \quad \alpha(t,s) = \alpha(s)\exp[(-d(t-s)], \quad m(t) = \int_{a(t)}^{t} \alpha(t,s)y(s)m(s)ds,$$

$$M(t) = \int_{0}^{t} m(s)ds, \quad \beta_i(t) = \int_{a(t)}^{t} \alpha(t,s)z_i(s)m(s)ds, \quad \beta_i(t,s) = \beta_i(s)\exp[-d(t-s)], \quad i = 1, \dots, n,$$

$$P(t) = \int_{a(t)}^{t} m(s)ds, \quad G(t) = M(t) - P(t), \quad f(t) = m(t) + c(t) + \alpha'(t) + \sum_{i=1}^{n} \beta_i(t),$$

$$x + y + \sum_{i=1}^{n} z_i = 1, \quad 0 \le x, y, z_i \le 1, \quad 0 \le a(t) \le t, \quad t \ge t\hat{} \ge 0, \tag{4}$$

where $\alpha(t,s)$ is the new technology for recreation of scientists' WP whose rate is $m(t)$, and for creation of the new technology $\beta_i(t,s)$ in the ith branch of knowledge; $P(t)$ is the total number of acting scientists; a, d, x, y, z_i, M, G, f, and t^ have the usual sense mentioned above. Knowing d and the numbers of scientists in the different branches of S, we can find the magnitudes m, M, a, α, β_i, x, y, z_i, G, and f.

3. EDUCATION AS ES

Any educational center: school, college, university, etc., and education as a whole can be considered as ES. All the features of ES: WP as the main elements of ES, existence of the subsystem A for realization of the internal functions of the system's development and the subsystem B for realization of the external functions, allocation of the system's resources between its internal and external functions, the out of date WP, the inflow of resources from the outside and existence of the moment of ES origin or ES prehistory, also can be applied to the case of education. Essential difference consists of their WP and their products.

The educational product is the number of specialists of the determined quality; and the index of efficiency of WP that produces specialists is the number of new specialists per unit of WP in the subsystem B in the unit of the time. The quality of specialists can be determined by means of different rules and tests.

The *subsystem A in education* is the set of WP, the product of which is the new, more effective WP in both A and B subsystems of education as ES. Creation of these WP also means training of the respective, more effective specialists in the field of education. Any science-methodical center in the field of education, public educational office, and administration of any educational center will belong to the subsystem A if their functions are creation of new WP. Actually, any educator belongs to A while he or she enlarges the own qualifications. Thus, the primary sign of belonging to the subsystem A or B is not a place, time or person, but a kind of fulfilled labor function.

The model of education and, especially, interpretations of the model elements depend on the teaching philosophy or policy to be adopted. An example of teaching philosophy can be seen in Ch.1.

Thus, the simplest, *base MM of education* as ES has the form

$$\alpha'(t) = \int^{t}_{a(t)} \alpha(t,s)x(s)m(s)ds,\ m(t) = \int^{t}_{a(t)} \alpha(t,s)y(s)m(s)ds,\ \beta_i(t) =$$

$$\int^{t}_{a(t)} \alpha(t,s)u_i(s)m(s)ds,\ \alpha(t,s)/\alpha(s) = \beta_i(t,s)/\beta_i(s) = \exp[-d(t-s)],\ i = 1, \ldots, r,\ n_i(t) =$$

$$\int^{t}_{a(t)} \beta_i(t,s)v_i(s)m(s)ds,\ M(t) = \int^{t}_{0} m(s)ds,\ P(t) = \int^{t}_{a(t)} m(s)ds,\ G(t) = M(t) - P(t),$$

$$0 \le x, y, u_i, v_i \le 1,\ x+y+\sum^{r}_{i=1} u_i + \sum^{r}_{i=1} v_i = 1.$$

$$f(t) = \alpha'(t) + m(t) + \sum^{r}_{i=1} \beta_i(t) + \sum^{r}_{i=1} n_i(t),\ t \ge t^\wedge \ge 0, \quad (5)$$

where $\alpha(t,s)$ is the new technology for recreation of educators' WP whose rate is $m(t)$, and for creation of new technology $\beta_i(t,s)$ in the ith branch of education; $P(t)$ is the total number of acting educators; $n_i(t)$ is the rate of creation of specialists number for the ith branch of human activities; $a, d, x, y, u_i, v_i, M, G, f$ and $t^\wedge$ have the usual sense mentioned above.

Knowing d and the number of educators in the different branches of education as well as all the elements on the pre-history, we can find from (5) the remained magnitudes $m, M, a, \alpha, \beta_i, x, y, u_i, v_i, G$, and f.

4. MM OF CELLS AS ES

The *MM in question*, regarding the spatial structure, has the form

$m(t,X) = \int^t_{a(t,X)} \int^X_0 \alpha(t,X,s,Y)m(^1)(s,Y)dYds,\ M(t,X) = \int^t_0 m(s,X)ds,\ M(t) =$

$\int_V M(t,X)dX,\ c(t,X) = \int^t_{b(t,X)} \int^X_0 \beta(t,X,s,Y)m(^2)(s,Y)dYds,\ C(t,X) = \int^t_{t^\wedge} c(s,X)ds,$

$C(t) = \int_V C(t,X)dX,\ P(t,X) = \int^t_{a(t,X)} \int^X_0 m(^1)(s,Y)dYds + \int^t_{b(t,X)} \int^X_0 m(^2)(s,Y)dYds,$

$P(t) = \int_V P(t,X)dX,\ m(^1)(t,X) + m(^2)(t,X) = m(t,X),\ G(t,X) = M(t,X) - P(t,X) > 0,$

$G(t) = M(t) - P(t),\ f(t) = \int_V m(t,X)dX + \int_V c(t,X)dX = k(f)\int_V m(t,X)dX,$

$$F(t) = \int^t_0 f(s)ds,\ 0 \le a, b \le t,\ t > t^\wedge > 0,\ \ O, X \in V \subset \mathbf{R}^3, \tag{6}$$

where V is the domain that a cell occupies in $\mathbf{R}^3$: $X = (x1, x2, x3)$; $m(t,X)$ is the new cell's bio-mass quantity at the instant t in its point X per unit of time (this means that X is simultaneously notation for the product $m(t,X)$ of X-type and by the same token the various products of a cell are distributed in the space); $M(t,X)$ is the total cell's bio-mass in its point X (created during the time t); $M(t)$ is the total created cell's bio-mass; $P(t)$ is the total functioning bio-mass of a cell; $G(t)$ is the total obsolete bio-mass of a cell; $c(t,X)$ is the new resource quantity in the unit of the time in the point X that is necessary for fulfillment of the external functions of a cell, $\alpha(t,X,s,Y)$ is the index of efficiency of a cell functioning along the channel $m(^1)(s,Y)$ -- $m(t,X)$, etc.

The sense of the values here is similar to ones mentioned before. In general, we can consider that $m(^1)$, $m(^2)$, m, a, b, c, P, and f are the vector functions of t, X; α and β are appropriate matrices. Then the first equality in (6) means

$$m_i(t,X) = \sum^n_{j=1} \int^t_{a_j(t,X)} \int^{x1}_0 \int^{x2}_0 \int^{x3}_0 \alpha_{ij}(t,X,s,Y)m(^1)_j(s,Y)dYds,\ i = 1,\dots,n; \tag{7}$$

the sense of the other equalities in (6) is similar to (7); inequalities for vectors mean, as usually, the appropriate inequalities for the respective components of these vectors.

Setting, in particular, $a_j(t,X)=0$,

$$\alpha_{ij}(t,X,s,Y)m(^1)_j(s,Y) = [F_{ij}(M_1,\dots,M_n)+D_{ij}(M_1,\dots,M_n)\partial^2 M_j/\partial X^2]m_j(s,X)/M_j(t,X), \tag{8}$$

where $M_1,\dots, M_n$ depend on t, X, we result in the system of the nonlinear equations with partial derivatives:

$$m_i(t,X) = \partial M_i/\partial t = F_i(M_1,\dots,M_n)+\sum^n_{j=1} D_{ij}(M_1,\dots,M_n)\partial^2 M_j/\partial X^2],\ i = 1,\dots, n, \tag{9}$$

which describes many well-known self-oscillating processes [9, 12, 14, 16], various dissipative structures [12], and also very complicated processes of cells' proliferation and differentiation [12, 13].

Experience of investigation and application of the more general models of the form (1), (2) shows that for explanation of cells' proliferation and differentiation, there are new possibilities that, in particular, do not connect with the mechanism of stepwise formation of the model elements [13].

It has been shown in Ch. 3 and Ch. 4 that in a certain case of the oscillating functions α and β-types, the equations (6)-type itself and different optimization problems for the class of the models under considreration have non-unique solutions. Assuming, that cells' development pursues the solution of determined optimization problems, e.g., cells' duplication maximizes reliability of their functioning, and cells' differentiation maximizes efficiency of fulfillment of given functions, we result in the possibility of explanation of cells' duplication and differentiation by the non-uniqueness of solutions of the appropriate optimization problems. Starting from the prehistory, for the first cell we can obtain (under decrease of functions of α and β-types stimulated by the internal and/or external factors through the membrane of the cell [12]) two different solutions of model (6), which can be interpreted as the models of development and growth for two new cells (under the increase of α and β values), whose prehistory (and new decrease of α and β) results in creation of the models for four new cells, and so on.

Thus created cells can be probably used, with the help of the program laid in DNA of the cells, for the construction different tissues, organs, subsystems of an organism, and an organism as a whole.

5. MM OF BIO-FIELD AS ES

Let V be the domain in $\mathbb{R}^3$ for isolated bio-object and $\mathbb{R}^3 \setminus V$ be the complement of V to $\mathbb{R}^3$. We accept that the external bio-field with respect to a bio-object will be decided if the vector function

$$c(t,x) = \{c1(t,X), c2(t,X), c3(t,X)\}, X \in \mathbb{R}^3 \setminus V, \tag{10}$$

is given. Here t is the time, $(x1, x2, x3)$ are Cartesian coordinates of the point X; $c1$ is the rate of creation along t,X (i.e., along t, $x1$, $x2$, and $x3$) of the (generalized) new material product of t,X-type in the point X at the instant t; $c2$ is the rate of creation along t,X of the (generalized) new energy product of t,X-type in the point X at the instant t; $c3$ is the rate of creation along t,X of the information product of t,X-type in the point X at the instant t.

As the *base MM of a bio-field*, we can naturally assume the following system:

$$m(t,X) = \int^{t}_{a(t,X)} \int^{X}_{O} \alpha(t,X,s,Y)m(^{1})(s,Y)dYds,\ O,\ X \in V,\ c(t,X) = \int^{t}_{b(t,X)} \int^{X}_{O}$$

$$\beta(t,X,s,Y)m(^{2})(s,Y)dYds,\ X \in \mathbb{R}^{3},\ P(t,X) = \int^{t}_{a(t,X)} \int^{X}_{O} m(s,Y)dYds,\ f(t,X) = m(t,X) +$$

$$c(t,X),\ t \geq t\hat{} > 0,\ m(^{1}) + m(^{2}) = m,\ m(t,X) = \{m_{1}(t,X), m_{2}(t,X), m_{3}(t,X)\},\ X \in V, \qquad (11)$$

where $m_i(t,X)$ are the rates of creation along t,X of the (generalized) internal products of t,X-type respectively bio-mass ($i = 1$), energy ($i = 2$), and information ($i = 3$) in the X at the instant t; α and β are matrices of the order three whose components are indices of efficiencies of functioning of a bio-object along the corresponding channels; the senses of all the other values are similar to the previous ones.

It is natural to call V with (11) as a bio-field of the internal sphere of a bio-object or the internal bio-field and $\mathbb{R}^{3}\backslash V$ with (10), (11) as the external bio-field.

The conditions under which the significant values of $c_2(t,X)$ can arise for the values X outside V and far removed from the boundary of V one can see in [5].

6. MODELING OF NATURAL ES ORIGIN PROBLEMS

Let us consider the MM of NES in the form

$$\alpha'(t) = \int^{t}_{a(t)} \alpha(t,s)x(s)m(s)ds,\ \alpha(t,s) = \alpha(s)\exp[-d(t-s)],\ m(t) = \int^{t}_{a(t)} \alpha(t,s)y(s)m(s)ds$$

$$+ \Delta m(t),\ P(t) = \int^{t}_{a(t)} m(s)ds,\ c(t) = \int^{t}_{a(t)} \alpha(t,s)[1-x(s)-y(s)]m(s)ds,\ 0 < x,\ y,\ 1-x-y < 1,$$

$$0 \leq a(t) < t,\ t \geq t\hat{} \geq 0,\ a(t\hat{}) = 0, \qquad (12)$$

where x, y, and $\Delta m(t)$ are given.

If $t\hat{} = \Delta m(t) = 0$, then the system equation in (12) has only the zero solution. Otherwise, this system under the general condition (see Ch. 3) has non-trivial solutions. Thus, in the presence of the NES typical structure and the external effect Δm-type, we have all necessary conditions for natural ES origin and its subsequent development. The origin of the necessary structure follows, in particular, from the well-known fluctuating systems analysis (see, for example, [4]). The origin DNA code follows naturally from the necessity of economical development.

7. MODELING OF LIFE LONGEVITY PROBLEMS

Let us accept the *base MM of an organism* as a whole in the form

$$m(t) = \int_{a(t)}^{t} \alpha(t,s)u(s)v(s)/v(t)y(s)m(s)ds,\ P(t) = \int_{a(t)}^{t} m(s)ds,\ G(t) = \int_{0}^{a(t)} m(s)ds,$$

$$c(t) = \int_{a(t)}^{t} \beta(t,s)[1-y(s)]m(s)ds,\ f(t) = m(t)+c(t) = k(t)c(t),\ 0 \le y \le 1,\ t \ge 0, \qquad (13)$$

where m is the rate of recreation of the new bio-mass or generalized resource of an organism, $P(t)$ is the total acting resource, $G(t)$ is the total obsolete recourse, $c(t)$ is the rate of creating of the external product, u and v are the general control functions (see Ch. 2, s. 6.2.), y is the distribution function, α and β are the indices of efficiency of functioning, f is the resource from outside, k is a share of c with respect to f.

There are the following independent possibilities for $m(t)$ tendency to zero:

$$i)\ \alpha(t,s)\to 0,\ ii)\ u(s)\to 0,\ iii)\ v(t)\to\infty,\ iv)\ y(s)\to 0,\ v)\ a(t)\to t,\ vi)\ k(t)\to 1,\ vii)\ f(t)\to 0. \qquad (14)$$

We remind that $u(s)$ is usually a non-stationary factor connected with the interaction of OS, $v(s)$ is a factor connected with the interaction of an organism and external surroundings, $y(s)$ is connected with the interaction of OS and external surroundings, $a(t)$ decides the rate of formation of the obsolete recourse, $k(t)$ decides qualitative composition of the external resource $f(t)$. There is extensive literature about impact of these type factors on aging. In particular, [2, and 3] contains nontraditional interesting data about impact of the factors of $k(t)$ and $f(t)$-types. As to the factors u, v, y, and a, they not once have been investigated above.

We dwell on the case *i*) as the strongest one, connected probably with the genome's main program of aging. Let in the case *i*) $\alpha(t,s) \le \alpha^*/t,\ t > t^*$, where α^* and t^* are certain constants. Then from the first equation in (13), we have

$$m(t) \le \alpha^*/t \max\,[u(s)v(s)/v(t)y(s)]\ (s \in [a(t),t])P(t). \qquad (15)$$

Thus, the natural condition

$$u(s)v(s)/v(t)y(s) \le c^*,\ P(t) \le P^*, \qquad (16)$$

where c^* and P^* are positive constants, results the $m(t) \to 0$. But this does not mean that $P(t)$ and $c(t)$ also tend to 0. It has been shown in Ch. 3 (see p. 54) that $P(t)$ can be almost a positive constant under the condition $m(t)$, $c(t) \to 0$. It is true, in this connection, $t - a(t)$ cannot be bounded.

One can see that for the longevity maximization we are in need of holding

$$R^* = \int_{a(t)}^{t} u(s)v(s)/v(t)y(s)m(s)ds,\ c(t) > c^- > 0, \tag{17}$$

where c^- is the desired minimal level of external activities, as large as possible.

REFERENCES

1. Biology and Selection of Sugar Beet, M.: Kolos, 1968, 775 pp. (in Russian).
2. Bolotov B. V., I Teach You Not to Be Ill and Not to Grow Old, Kiev: Informatsionno-izdatel'skoe agenstvo "Ukraina", 1992, 48 pp. (in Russian).
3. Clark, H. R., The Cure for All Diseases; With Many Case Histories, New Century Press, 1995, 604 pp.
4. Glushkov V.M., Fluctuation Systematization, Kibernetika, N 2, 1979, 114-115 (in Russian).
5. Glushkov V.M., On Possible Peculiarities of Physical Fields of Bio Systems, Kibernetika, N 3, 1981, 105-106 (in Russian).
6. Great Soviet Encyclopedia, Vol. 28, 19, 553 (in Russian)
7. Ivanov V.V., Patsalyuk S.S., Yanenko V.M., Efficiency of Work Places Functioning Estimation in the Health Care, Kibernetika, N 6, 1986, 110-113 (in Russian).
8. Ivanov V.V., Kunovskiy Yu.B., Protsenko N.M., ASSI SAD (Automated Systems of Scientific Investigations for Systems of Automated Design) as Evolutionary Systems, Kiev: Glushkov Inst. of Cybernetics, Reprint 91-34, 1991, 25 pp. (in Russian).
9. Ivanitskiy G.R., Krinskiy V.I., Sel'kov E.E., Mathematical Biophysics of Cell, M.: Nauka, 1978, 312 pp. (in Russian).
10. Lebedeva L.I. and Gerasimova T.N., Survival and Reproduction Potential of Philodina roseola (Ehrenberg) (Rotatoria, Bdelloida) under Various Temperature Conditions, Int. Revue ges. Hydrobiol., 72, 6, 1987, 695-707.
11. Lootsma F.A., Fuzzy Logic for Planning and Decision Making, Kluwer Academic Publishers, 1997, 197 pp.
12. Romanovskiy Yu.M., Stepanova N.V., Chernavsky D.S., Mathematical Biophysics, M.: Nauka, 1984, 304 pp. (in Russian).
13. Sendov B.Kh., Mathematical Models of Processes of Cells Proliferation and Differentiation, M.: Izdatel'stvo MGU, 1976, 58 pp. (in Russian).
14. Svirezhev Yu.M., Pasekov V.P., Fundamentals of Mathematical Evolutionary Genetics, Kluwer Academic Publishers, 1990, 384 pp.
15. Yudin D.B., Computational Methods of the Theory of Decision Making, M.: Nauka, 1989, 320 pp. (in Russian).
16. Zhabotinskiy A.M., Concentrating Auto-oscillations, M.: Nauka, 1974, 178 pp. (in Russian).

SUMMARY

For the first time, we have new MM allowing us to model and simulate the problems analysis and synthesis from the unique position for a wide class of ES. One can relate the ecology-economic system as a whole, to different branches of economy, separate enterprises and businesses, scientific and education centers, health care, and so on; plants and animals, human organisms and their subsystems, populations of animals and plants, etc.

There are potentially unlimited utilization of these MM because of unlimited quantity of different ES that can be integrated on any desired level of consideration. Of course, these MM do not belittle the other numerous useful models, but complement them.

We actually already have new general methodology that can allow everyone to integrate multiple component systems. This methodology consists, briefly, in that everyone needed should consider a given system as one of ES, decide its subsystems A and B, the prehistory of their functioning, construct the indices of efficiency α, β-types, make decisions about the initial data and desired results, and use MM of ES in question. Also, everyone has at his/her disposal the multiple components: the constructive magnitudes α, β, $a(t)$, $y(t)$-types and the resources or product magnitudes $f(t)$, $m(t)$, $c(t)$, $P(t)$, $G(t)$-types integrated in the unique entire which we name MM of development or development modeling.

In the case of interaction of two or more ES one can use a certain combination essentially of the same MM.

Under development modeling of different ES, everyone can use, evidently, the new control factors, such as the resources distributions between the subsystems A and B-types of ES and/or temporal bounds for liquidation of the obsolete technology in ES, to optimize ES functioning. In addition, we have the very important qualitative results about regularities of ES optimal functioning.

Science itself can be considered as an example of ES product which is the new, more effective technology. One of the results (see Ch. 4) shows that for well being to be maximized (for a long-term period of the time), almost all resources have to go to Science. Popularization of this result, we think, is very important for Science's development.

Modeling of the health care system as ES allows us, in particular, to describe deeper the main structure and functions of this system as well as the main structure and functions of a human organism both in the norm and in the pathology. In addition, the description is essentially different from the traditional ones, which require us to revise the whole statement and technology of necessary biomedical experiments.

According to MM of ES, we have to learn to measure the new bio-mass quantities of different kinds that are always non-negative. This does not mean that their behavior is always monotone. Increasing the function of $a(t)$-type can decrease other values.

The increase of all the resources of an organism toward its internal needs of restoration is likely playing the crucial role in the case of pathology, in serious cases, of the structural values of α, β, a, and y-types (see Ch. 2 and 9).

All these factors demand reconstruction of the infrastructure of science.

The same demand is related to the infrastructure of engineering. In order to sharpen this statement, it is sufficient to consider the possibility of creation of the new class of computers as ES. They should have structure that is similar to ES. It will cause unlimited development of computers that will be similar to natural ES.

It should be noted that the new technique under consideration is not developed so completely as the traditional, well-known one exploiting, mainly, differential equations. There are many open problems in both general theory and applications of the new techniques. Among them, we would like to name the problems of modeling of phenomena concerning wave, in particular, the so-called solitons, consideration of problems of bifurcation and non-uniqueness of the solutions in great detail, application to modeling of a cancer in great detail, etc.

The problems of design or synthesis of ES at the expense of not only the control functions of x, y, z and u, v-types but also of the functions of α and β-types, which were mentioned in connection with the opportunity of the biotech revolution in Ch. 2, are also still ahead of us.

Everyone should be aware of great difficulties on the way of receiving any new theoretical result in this field and any progress to employ it in practice. It is more so that MM in question are nonlinear systems of integral-functional equations in the presence of some natural restrictions. Experience of work with these MM shows that each qualitative or quantitative result about the new property of ES can be obtained only by hard efforts. But in return, this new property as a rule appears very important and, we may say, beautiful in the theoretical sense.

LIST OF ABBREVIATIONS

A subsystem of ES for fulfillment of internal functions
AES artificial ES
Ag antigen (s)
AIDS acquired immune deficiency syndrome
B subsystem of ES for fulfillment of external functions
C cancer (s)
Ch. chapter (s)
DNA deoxyribonucleic acid (basic constituent of the gene)
E extraction subsystem of ES
ES evolutionary system (s)
FCCP foreign currency conversion problem (s)
G genome subsystem of ES
HA human activity (ies)
HC health care
HIV human immune-deficiency virus (es)
HPO healthy part of an organism
ICC immune competent cells
Id-Aid idio-type anti-idio-typical interaction
Ig immune-globulin
iff if and only if
IN immune network
IS immune system
ISO immune status of an organism
JES joined ES
Lc lymphocyte (s)
MM mathematical model (s)
NES natural ES
OS organism subsystem (s)
R recreation subsystem of ES
RC reaction center (s)
RPO remained part of an organism
S science subsystem of ES
s. section (s)
SP specific rate
STP science-technical progress
Tef T effectors of hyper-sensitivity and T killers Lc
Th T helpers Lc Ts T supressors Lc
V. i. e. Volterra integral equation (s)
WP work place (s)

LIST OF NOTATIONS

$\approx$ - approximately equal
$\asymp$ - asymptotically equal in order
$\lesssim, \gtrsim$ - asymptotically less, greater than $\ll, \gg$ - by far less, greater than
$|$ - condition sign
$[x]$ - dimension of x
$\times, \oplus$ - direct product, sum
$\varnothing$ - empty set
$\square$ - end of the proof
$\succ, \prec$ - follows (succeeds), precedes
∞ - infinity sign
$\mathbb{I}$ - integers
$\cap, \cup$ - intersection, union
$\vee, \wedge$ - logical and, or
$M(f,R), m(f,R)$ - max, min $|f|$ $(t \in R)$
$x = o(y), O(y)$ - mean $x/y \to 0$, $x/y < C$, where C is a positive constant when $y \to 0$ or $y \to \infty$
$x \sim y$ - means $|x-y| < C|y|/t$, $t \to \infty$, where C is a positive constant
$\|x\|$ - measure or norm of x
$\in, \notin$ - member, not a member
$\mathbb{R}^n$ - n-dimensional Euclidean space
$\nexists$ - no existence sign
∂ - partial derivative
$\prod, \sum$ - product, summation
$\subset, \supset$ - proper subset, superset
f^{-1} - reciprocal function or operator for any function or operator f
(x, y) - scalar product of elements x and y
$\to$ - tend to
$seq\ cl(A)$ - the collection of all limits of convergent sequences in A
$C(R)$ - the metric space of continuous functions on R to $\mathbb{R}$
$C(S,X)$ - the metric space of continuous functions on S to X, $S \to X$
$C^l(R)$ - the metric space of continuously differentiable functions on R to $\mathbb{R}$
$C(R)^*$ - the space of functionals l_r on $C(R)$ such that $l_r(c) = c(r)$ for all $c \in C(R)$, $r \in R$.
$L_p(R)$ - the space of functions on R to $\mathbb{R}$ absolute integrable to the power p, $p > 0$
(T, Σ, μ) - the positive finite measure space
$\mathscr{F}(T, \Sigma, \mu, X)$ - the set of all μ-measurable functions on T to X
$frm^+(T)$ - the set of all positive measure in T
$\mathcal{B}(T, \Sigma, \mu, S; X)$ [or $\mathcal{B}(T, S; X)$ or $\mathcal{B}$] - the vector space of functions $\varphi: T \times S \to X$ such that, for all $(t, s) \in T \times S$, $\varphi(t, .) \in C(S, X)$, $\varphi(., s)$ is μ-measurable and there exists a μ-integrable $\psi_\varphi : T \to \mathbb{R}$ with $|\varphi(t,.)|_{\sup} \leq \psi_\varphi(t)$

For simplicity, we denote sometimes $\sum (i \in [a,b])$ and $\int ds$ $(s \in C)$ instead of $\sum^b_{i=a}$ and $\int_C ds$ respectively.

SUBJECT INDEX

ABOUT THE AUTHOR

According to Marquis WHO'S WHO IN THE WORLD 1993-1994, 11th Edition,

IVANOV VIKTOR VLADIMIROVICH, mathematics educator; b. Samara, USSR, June 8, 1929; s. Vladimir Aleksandrovich and Natal'ya Ivanovna (Rudneva) I.; m. A. S. Zhivoglyadova, Mar. 1, 1958 (div. Dec. 1977); children: Yuriy, Vladislav; m. Valentina Nikolaevna Chernogub, Dec. 9 1982; 1 child, Natalya. Degree Rostov U., 1952; Ph D Moscow U., 1956; DSc, Siberian Acad. Sci., 1966. Asst. Polytech. Inst., Novocherkassk, 1952-55; sr. tchr. Agr. Rostov Inst., 1956-57; sr. scis Computing Ctr. Ukrainian Acad. Sci., 1958-64; head of dept. Glushkov Inst. Cybernetics, Kiev, 1965-87; Prof. Kiev U., 1969-78; lead of scis. Glushkov Inst. Cybernetics, Kiev, 1978-92. Author: The Theory of Approximate Methods ..., 1968; Guide Methods of Computations, 1986; author, editor: Methods for Algorithmization of Continuous Industrial Processes, 1975; co-author: (with V. M. Glushkov and V. M. Yanenko) Developing System Modeling, 1983; mem. editorial bd. Ency. of Cybernetics; contbr. over 300 articles to profl. jours. Recipient Valour Labour medal, Presidium of Supreme Soviet of USSR, 1970, 1500th Anniv. of Kiev medal,1982, Vet. of Labor medal, 1985. Mem. Am. Math. Soc., 1972- . Avocations: traveling, problem solving.

He arrived in the USA as a political refugee in August 1992. Since then he was busy improving his English; looking for a job; trying to solve the well-known problem "NP ≠? P"; working as a volunteer in Tomlinson Adult Learning Center, St. Petersburg, and in University of South Florida, College of Engineering, Tampa; writing this book.

GPSR Compliance

The European Union's (EU) General Product Safety Regulation (GPSR) is a set of rules that requires consumer products to be safe and our obligations to ensure this.

If you have any concerns about our products, you can contact us on ProductSafety@springernature.com

In case Publisher is established outside the EU, the EU authorized representative is:

Springer Nature Customer Service Center GmbH
Europaplatz 3
69115 Heidelberg, Germany

Kolibri 360 GmbH
Ferdinand-Jühlke-Straße 7
99095 Erfurt, Deutschland
produktsicherheit@kolibri360.de